SEMICONDUCTORS AND SEMIMETALS

VOLUME 21

Hydrogenated Amorphous Silicon

Part D

Device Applications

Semiconductors and Semimetals

A Treatise

Edited by R. K. WILLARDSON
WILLARDSON CONSULTING
SPOKANE, WASHINGTON

ALBERT C. BEER
BATTELLE COLUMBUS LABORATORIES
COLUMBUS, OHIO

SEMICONDUCTORS AND SEMIMETALS

VOLUME 21

Hydrogenated Amorphous Silicon

Part D

Device Applications

Volume Editor

JACQUES I. PANKOVE

RCA/DAVID SARNOFF RESEARCH CENTER
PRINCETON, NEW JERSEY

1984

ACADEMIC PRESS, INC.
(Harcourt Brace Jovanovich, Publishers)

Orlando San Diego New York London
Toronto Montreal Sydney Tokyo

ACADEMIC PRESS, INC.
Orlando, Florida 32887

United Kingdom Edition published by
ACADEMIC PRESS, INC. (LONDON) LTD.
24/28 Oval Road, London NW1 7DX

Library of Congress Cataloging in Publication Data
(Revised for volume 21B-21D)
Main entry under title:

Semiconductors and semimetals

Includes bibliographical references and indexes.
Contents: v. 1-2. Physics of III-V compounds --v. 3. Optical properties of III-V compounds-- --v. 21, pt. B, C, D. Hydrogenated amorphous silicon.
1. Semiconductors--Collected works. 2. Semimetals--Collected works. I. Willardson, Robert K. II. Beer, Albert C.
QC610.9.S47 537.6'22 65-26048
ISBN 0-12-752150-X (21D)

PRINTED IN THE UNITED STATES OF AMERICA

84 85 86 87 9 8 7 6 5 4 3 2 1

Contents

Chapter 4 Electrophotography

Isamu Shimizu

Chapter 5 Image Pickup Tubes

Sachio Ishioka

Chapter 6 The Development of the a-Si:H Field-Effect Transistor and Its Possible Applications

P. G. LeComber and W. E. Spear

Chapter 7 a-Si:H FET-Addressed LCD Panel

D. G. Ast

Chapter 8 **Solid-State Image Sensor**

S. Kaneko

Chapter 9 **Charge-Coupled Devices**

Masakiyo Matsumura

Chapter 10 **Optical Recording**

M. A. Bosch

Chapter 11 **Ambient Sensors**

A. D'Amico and G. Fortunato

Chapter 12 **Amorphous Light-Emitting Devices**

Hiroshi Kukimoto

Chapter 13 **Fast Detectors and Modulators**

Robert J. Phelan, Jr.

Chapter 14 **Hybrid Structures**

Jacques I. Pankove

Chapter 15 **Electronic Switching in Amorphous Silicon Junction Devices**

P. G. LeComber, A. E. Owen, W. E. Spear, J. Hajto, and W. K. Choi

List of Contributors

Numbers in parentheses indicate the pages on which the authors' contributions begin.

D. G. AST, *Department of Materials Science and Engineering, Cornell University, Ithaca, New York 14853* (115)

M. A. BOSCH, *AT&T Bell Laboratories, Holmdel, New Jersey 07733* (173)

D. E. CARLSON,* *RCA/David Sarnoff Research Center, Princeton, New Jersey 08540* (7)

W. K. CHOI, *Department of Electrical Engineering, University of Edinburgh, Edinburgh EH9 3JL, Scotland* (275)

A. D'AMICO, *Istituto di Elettronica dello Stato Solido, Consiglio Nazionale delle Ricerche, 00156 Rome, Italy* (209)

G. FORTUNATO, *Istituto di Elettronica dello Stato Solido, Consiglio Nazionale delle Ricerche, 00156 Rome, Italy* (209)

J. HAJTO, *Department of Electrical Engineering, University of Edinburgh, Edinburgh EH9 3JL, Scotland* (275)

SACHIO ISHIOKA, *Central Research Laboratory, Hitachi, Ltd., Kokubunji, Tokyo 185, Japan* (75)

S. KANEKO, *Microelectronic Research Laboratories, NEC Corporation, Kawasaki 213, Japan* (139)

HIROSHI KUKIMOTO, *Imaging Science and Engineering Laboratory, Tokyo Institute of Technology, Nagatsuta, Yokohama 227, Japan* (239)

P. G. LECOMBER, *Carnegie Laboratory of Physics, The University of Dundee, Dundee DD1 4HN, Scotland* (89, 275)

MASAKIYO MATSUMURA, *Department of Physical Electronics, Tokyo Institute of Technology, O-Okayama, Tokyo 152, Japan* (161)

A. E. OWEN, *Department of Electrical Engineering, University of Edinburgh, Edinburgh EH9 3JL, Scotland* (275)

JACQUES I. PANKOVE, *RCA/David Sarnoff Research Center, Princeton, New Jersey 08540* (1, 261)

* Present address: Thin Film Division, Solarex Corporation, Newtown, Pennsylvania 18940.

ROBERT J. PHELAN, JR., *National Bureau of Standards, Boulder, Colorado 80303* (249)

ISAMU SHIMIZU, *Imaging Science and Engineering Laboratory, Tokyo Institute of Technology, Nagatsuta, Yokohama 227, Japan* (55)

W. E. SPEAR, *Carnegie Laboratory of Physics, The University of Dundee, Dundee DD1 4HN, Scotland* (89, 275)

G. A. SWARTZ, *RCA/David Sarnoff Research Center, Princeton, New Jersey 08540* (39)

Foreword

This book represents a departure from the usual format of "Semiconductors and Semimetals" because it is a part of a four-volume miniseries devoted entirely to hydrogenated amorphous silicon (a-Si:H). In addition, this group of books—Parts A–D of Volume 21—has been organized by a guest editor, Dr. J. I. Pankove, an internationally recognized authority on this subject. He has assembled most of the who's who in this field as authors of the many chapters. It is especially fortunate that Dr. Pankove, who has made important original contributions to our understanding of a-Si:H, has been able to devote the time and effort necessary to produce this valuable addition to our series. In the past decade, a-Si:H has developed into an important family of semiconductors. In hydrogenated amorphous silicon alloys with germanium, the energy gap decreases with increasing germanium content, while in alloys with increasing carbon content the energy gap increases. Although many applications are still under development, efficient solar cells for calculators have been commercial for some time.

In Volume 21, Part A, the preparation of a-Si:H by rf and dc glow discharges, sputtering, ion-cluster beam, CVD, and homo-CVD techniques is discussed along with the characteristics of the silane plasma and the resultant atomic and electronic structure and characteristics.

The optical properties of this new family of semiconductors are the subject of Volume 21, Part B. Phenomena discussed include the absorption edge, defect states, vibrational spectra, electroreflectance and electroabsorption, Raman scattering, luminescence, photoconductivity, photoemission, relaxation processes, and metastable effects.

Volume 21, Part C, is concerned with electronic and transport properties, including investigative techniques employing field effect, capacitance and deep level transient spectroscopy, nuclear and optically detected magnetic resonance, and electron spin resonance. Parameters and phenomena considered include electron densities, carrier mobilities and diffusion lengths, densities of states, surface effects, and the Staebler–Wronski effect.

The last volume of this miniseries, 21, Part D, covers device applications, including solar cells, electrophotography, image pickup tubes, field effect transistors (FETs) and FET-addressed liquid crystal display panels, solid state image sensors, charge-coupled devices, optical recording, visible light emitting diodes, fast modulators and detectors, hybrid structures, and memory switching.

R. K. WILLARDSON
ALBERT C. BEER

Preface

Hydrogenated amorphous silicon, a new form of a common element, is a semiconductor that has come of age. Its scientific attractions include a continuously adjustable band gap, a usable carrier lifetime and diffusion length, efficient optical transitions, and the capability of employing either *n*- or *p*-type dopants.

Furthermore, it can be fabricated very easily as a thin film by a technology that not only inherently escapes the expense of crystal perfection but also requires significantly smaller amounts of raw materials.

The discovery of a new material endowed with wondrous possibilities for very economical practical applications naturally attracts many researchers who invariably provide new insights and further vision. Their meditation and experimentation build up rapidly and lead to a prolific information flow in journals and conference proceedings.

The initial cross-fertilization generates an overload of data; books are written that attempt to digest specialized aspects of the field with state-of-the-art knowledge that often becomes obsolete by the time the books are published a year or two later.

We have attempted to provide this book with a lasting quality by emphasizing tutorial aspects. The newcomer to this field will not only learn about the properties of hydrogenated amorphous silicon but also how and why they are measured, and the variety of practical applications possible with this method.

In most chapters, a brief historical review depicts the evolution of relevant concepts. The state of the art emerges, and a bridge to future developments guides the reader toward what still needs to be done. The abundant references should be a valuable resource for the future specialist.

We hope that this tutorial approach by seasoned experts satisfies the needs of at least one generation of new researchers.

SEMICONDUCTORS AND SEMIMETALS, VOL. 21, PART D

CHAPTER 1

Introduction

Jacques I. Pankove

RCA/DAVID SARNOFF RESEARCH CENTER
PRINCETON, NEW JERSEY

This first chapter summarizes the contents of this volume. Where relevant, we have added several other new device applications of a-Si:H that have come to our attention since the authors have written their contributions. The authors of this volume have been chosen for their eminent contributions in the various approaches that broaden the usefulness of this material.

In Chapter 2, Carlson reviews and compares various a-Si:H deposition methods that have been used to make solar cells. He points out the importance of various material properties to the efficient performance of solar cells, for example, the absorption edge that depends on hydrogen concentration. Also, alloying Si with other elements such as carbon and germanium affects the absorption edge, thus allowing a stacking of cells with successively narrower energy band gaps to utilize more efficiently the solar spectrum. Carlson discusses the relationship between the collection length and the diffusion length. He stresses the detrimental influence of gap states, those due to dangling bonds and those introduced by various impurities that are often present in a deposition system. There is also concern for defects that may enhance light-induced effects and concern over the doping efficiency that determines the maximum obtainable photovoltage. Several cell structures are described and the performance of the better cells is reported. Light-induced degradation is negligible in thin cells because the resulting large electric field separates the photogenerated electron–hole pairs before they can recombine. Carlson describes techniques for making solar cells on a commercial scale. He considers the economics of solar-cell fabrication for power generation (e.g., most of the cost is in the supporting structure) and projects that a 20% conversion efficiency using stacked cells could be achieved by the end of this decade.

In Chapter 3, Swartz presents a model for the $p-i-n$ solar cell consisting of two transitions (like two portions of a $p-n$ junction) separated by a photoconductive i layer. Various physical properties can be incorporated

ISBN 0-12-752150-X

into the lumped-circuit constants to generate the corresponding $I-V$ characteristics. Conversely, the observed characteristic can be used to evaluate the appropriate physical property. Thus, the light-induced degradation can be attributed to a reduced barrier at the p^+-i layer interface and to a reduction of the $\mu\tau$ product.

In Chapter 4, Shimizu describes the use of a-Si : H as a photoreceptor for transferring an image by electrophotography. Hydrogenated amorphous silicon is endowed with a very low electrical conductivity, a long charge decay time, an excellent spectral sensitivity in the visible, and a good chemical inertness. The retention of a surface charge is helped by blocking layers at both surfaces of the a-Si : H films. An excellent image reproducibility with high resolution and good contrast was demonstrated in a commercial machine whose Se-coated drum was replaced by one coated with a-Si : H.

Hydrogenated amorphous silicon photoreceptors for electrophotography have been coated with a thin layer (< 100 nm) of a-SiO_2 : H by Nakayama *et al.* (1983). This overcoat improves the stability of the photoreceptor against temperature and humidity cycling. A photoreceptor consisting of a 1-μm-thick layer of a-SiGe : H sandwiched between two a-Si : H layers was developed by Nishikawa *et al.* (1983) to match a GaAlAs LED. This device, which has a sensitivity of 0.5 μJ cm^{-2} at 600 nm, was passivated with a thin layer of a-SiC.

In Chapter 5, Ishioka describes the use of a-Si : H in imaging tubes. The a-Si : H is a photoconductor having the vacuum side charged to cathode potential. Light causes charge leakage during a frame time; the signal is the electron-beam current needed to recharge the surface. Blocking contacts are used to reduce the leakage current. Hydrogenated amorphous silicon has a broad spectral response that is especially sensitive in the visible range. The signal current is proportional to the light intensity at all wavelengths, which simplifies color balancing. The spatial resolution is comparable to that of commercial vidicons and exhibits neither blooming nor image burning under strong illumination. When used to intensify x-ray images, the a-Si : H imaging tube has a better resolution than conventional image intensifiers. A single-tube color camera has been demonstrated using striped color filters on the face plate.

Although the development of field-effect transistors (FETs) has already reached a high level of perfection in crystalline silicon (c-Si), a-Si : H offers the advantage that very large arrays of a-Si : H FETs are feasible. Hence, even if their performance level is much lower than that of c-Si FETs, there are still many applications where their characteristics may be adequate. In Chapter 6, LeComber and Spear review the design, fabrication, and performance of a-Si : H FETs. They point out the need for good ohmic connections to source and drain and the problem of achieving a good dielectric under the gate. Sheet conductance of several 10^{-8} mho/□ and switching times on the order of a microsecond have been obtained. These devices appear very resistant to gamma radiation. Among applications briefly discussed are inverter logic and addressable image sensors. The figure of merit given by the gain – bandwidth product is 5 MHz, a limitation that still allows many practical uses.

In Chapter 7, Ast describes the application of a-Si:H FETs to a liquid crystal display (LCD). There is a brief introductory review of the principles involved in changing the orientation of light-polarizing liquid crystals leading to the possibility for high-contrast imaging using either transmitted or scattered light. Ast considers the problem of addressing the numerous elements of an LCD, especially when one needs to avoid a dc bias that might degrade the liquid crystals by electrolytic action. Hence the driving bias must be periodically reversed. Thus, a symmetrical back-to-back pair of Schottky diodes could be used to switch liquid crystals allowing several hundred lines of 1-mm-square pixels to be addressed. If a high-mobility semiconductor were used, electrical considerations would require that the drivers be made at the limits of lithographic resolution. In contrast, the low-mobility a-Si:H is ideally suited for making large FETs having a high OFF-resistance. Furthermore, a-Si:H has the advantage of low-temperature processing so that it can be deposited onto inexpensive soda-lime glass that serves as a transparent substrate. An experimental structure is described that uses a square FET in which the gate surrounds a square drain electrode. This drain serves as a large reflecting electrode for the liquid crystal. The source electrode that surrounds the gate is an extension of the gate. Hence, there is no need to etch the a-Si:H between source and drain. The early model consisting of a 26×26 array has been operated for more than one year without degradation.

In Chapter 8, Kaneko reviews various image sensors suitable for facsimile transmission or for optical character recognition. He then describes a new image sensor consisting of a long array of a-Si:H photocells that sense the light reflected along a line illuminated by a linear array of LEDs. An integrated circuit scans the information stored in the photosensors while the array scans a page one line at a time. Hydrogenated amorphous silicon is more suitable as a photoconductor for this application than other materials because of its flatter response in the visible spectrum and because the fabrication of a long array is readily feasible. Various structures have been explored. One that is particularly successful in reducing the dark current employs blocking layers of Si_3N_4 and *p*-type a-Si:H. Several methods of switching from element to element are considered, such as sequential switching with FETs of various types and even using charge-coupled devices

(CCD). It is expected that a superposition of the a-Si : H sensor and a CCD should produce a performance comparable to that of a vidicon. With the addition of a laminated color filter comprising lines of red-, green-, and blue-pass filters, a color image sensor was demonstrated.

Another potential image sensor is the CCD discussed in Chapter 9 by Matsumura. The major problem in using a-Si : H for CCD applications is the high density of traps or localized states in the band gap. A theoretical analysis assuming an exponential distribution of gap states permits us to predict the transfer characteristics (residual electron density as a function of time). With state-of-the-art material it should be possible to make usable image sensors. Early experimental results with novel test structures have yielded transfer inefficiencies of less than 1% at clock frequencies between 1 and 200 kHz.

In Chapter 10, Bosch reviews optical recording methods, especially those involving a-Si : H. There is the possibility of generating a microstructure that has a high absorption coefficient (low reflectance) but becomes highly reflective upon melting by a laser beam, the surface tension helping the formation of a smooth imprint. A localized amorphous-to-crystalline phase change also provides absorption and reflectance contrast. The most useful result is derived from the thermal dehydrogenation of a-Si : H. Dehydrogenation provides a decrease in energy gap and hydrogen evolution. The latter can form a microscopic blister or a crater that can be easily read optically against a smooth background. Bosch describes a system that greatly reduces the laser power required to record data. In this approach the a-Si : H is used as a photoconductor supporting an externally applied voltage. The writing energy comes from the electrical power dissipated as heat when the photoconductor is triggered into conduction by the laser pulse. This heat can blister or crater either the photoconductor or an outer layer of a more thermally sensitive material. Thus, with an applied voltage, 1.8-μm dots were recorded with 0.1 mW of laser power (two orders of magnitude lower than without applied voltage). The presence of hydrogen that passivates the dangling bonds renders the material corrosion resistant, a quality that is valued for archival information storage.

In Chapter 11, D'Amico and Fortunato report on the use of a-Si : H for measuring the hydrogen concentration in the ambient and also for measuring the temperature of the ambient. In the latter application, the a-Si : H is used as a thermistor. Hydrogen detection utilizes a field-effect transistor in the form of a MISFET in which a palladium gate catalyzes the H_2 dissociation and allows atomic hydrogen to modulate the conductivity of the a-Si : H. They also describe MIS sensors in which the hydrogen concentration is obtained from the $C-V$ characteristics. For practical applications, there is concern over possible long-term changes in barrier height and a need for faster response time and for complete reversibility of characteristics under reset treatment such as O_2 exposure.

In contrast to all the previously described devices, which are based on the light-absorbing properties of a-Si:H, those examined by Kukimoto in Chapter 12 are of the light-emitting variety. Although there are applications for which infrared emission is useful, the most interesting applications for a light emitter of large area—the distinctive advantage of amorphous semiconductors—would be for visible displays. Wide-band-gap emitters that are currently explored are a-Si_xC_{1-x}:H and a-Si:H made by homoCVD. Kukimoto also mentions the potential advantage of quantum well structures if impact excitation were a suitable means for exciting electroluminescence (as it is in phosphors). However, our own feeling is that impact excitation may create dangling bonds that would reduce the luminescence efficiency. This fear is based on our observation that a-Si:H never exhibited cathodoluminescence and on our observation of electron-beam-induced defects (see Volume 21B, Chapter 11).

In Chapter 13 Phelan shows that a-Si:H can be used for fast detectors that take advantage of the high absorption coefficient of this material and of the short transit time of photogenerated carriers in very small structures in the presence of an electric field. Response times in the tens of picoseconds have been obtained. Another application for a-Si:H is as a fast electro-optic modulator. There, the electric field changes the refractive index and shifts the absorption edge (Franz–Keldysh effect). The change in refractive index in the film causes a large spectral shift of the interference pattern. This change results in a large and rapid modulation of a transmitted monochromatic light.

In Chapter 14 Pankove describes several uses for a-Si:H deposited on crystalline Si. The most studied of these applications is as a passivating encapsulant that takes advantage of the abundant hydrogen to tie the dangling bonds at the surface of c-Si and provides a reservoir of hydrogen that ensures the stability of the interface. Another potential application is as a heterojunction to make efficient emitters in bipolar c-Si transistors. Although several attempts have been made at realizing this structure, no beneficial result has been demonstrated thus far. Still another possible hybrid combination of amorphous and crystalline materials is the use of a-Si:H as a waveguide to transfer optical signals over a crystalline integrated circuit without cross talk between optical and electronic interconnections. This application, which capitalizes on the high refractive index of a-Si:H, also has not been demonstrated.

In Chapter 15, LeComber *et al.* describe an intriguing switching phenomenon that may have extensive practical applications. The device is a diode that can be switched to the conducting state by biasing beyond a threshold with one polarity and switched back to the resistive state by exceeding a threshold with opposite polarity. The switching is fast (~ 10 nsec), weakly dependent on temperature, and independent of illumination. The diode remains indefinitely in the ON or OFF state until the appropriate threshold is exceeded.

References

Nakayama, Y., Wakita, K., Nakano, M., and Kawamura, T. (1983). *J. Non-Cryst. Solids* **59/60,** 1231.

Nishikawa, S., Hakinuma, H., Watanabe, T., and Kaminishi, K. (1983). *J. Non-Cryst. Solids* **59/60,** 1235.

CHAPTER 2

Solar Cells

D. E. Carlson†

RCA/DAVID SARNOFF RESEARCH CENTER
PRINCETON, NEW JERSEY

I. Introduction

The photovoltaic effect was first observed in amorphous silicon in 1974 at RCA Laboratories (Carlson, 1977a). The amorphous silicon films were grown by the dc glow-discharge decomposition of silane (SiH_4). Both Schottky-barrier and $p-i-n$ devices were fabricated that year, but the conversion efficiencies were limited to less than 1% mainly because of poor contacts. By 1976 the efficiency of the $p-i-n$ structure had been improved to 2.4% (Carlson and Wronski, 1976), and in 1977 an efficiency of 5.5% was obtained in a small (2-mm^2) Schottky-barrier device (Carlson, 1977b). At about this time, it was becoming clear that hydrogen was playing an important role in assuring the good semiconducting properties of amor-

† Present address: Thin Film Division, Solarex Corporation, Newtown, Pennsylvania.

ISBN 0-12-752150-X

phous silicon and that the films were actually silicon–hydrogen alloys (a-Si : H) (Triska *et al.,* 1975; Connell and Pawlik, 1976).

Continued optimization of the deposition conditions led to efficiencies as high as 6.1% in $p-i-n$ cells (1.19 cm^2) by 1980 (Carlson, 1980a). An efficiency of 6.3% was reported later that year for a small (4.2-mm^2) metal–insulator–semiconductor (MIS) device fabricated from a glow discharge in SiF_4 and H_2 (Madan *et al.*, 1980); the film used in this MIS device was a silicon–hydrogen–fluorine alloy (a-Si : H : F).

In 1981 an efficiency of 7.5% was obtained for a $p-i-n$ structure (3.3 mm^2) in which the p layer was a boron-doped silicon–carbon–hydrogen alloy (a-Si : C : H) (Tawada *et al.,* 1981). A further improvement in conversion efficiency to 8.5% was obtained in 1982 with a stacked junction structure (9 mm^2) that utilized an amorphous silicon–germanium–hydrogen alloy (a-Si : Ge : H) in the back junction of three stacked $p-i-n$ junctions (Nakamura *et al.,* 1982). More recently, an efficiency of 10.1% has been achieved in a $p-i-n$ structure (1.2 cm^2) utilizing p-type a-Si : C : H as a window layer (Catalano *et al.,* 1982).

Commercialization of amorphous silicon solar cells started in 1980 when Sanyo introduced calculators powered only by small solar-cell panels (total area ~ 5 cm^2). Shortly thereafter, Fuji Electric also started producing a-Si : H solar cells for calculators. As of 1983, a-Si : H photovoltaic devices are produced for several other applications such as photodetectors, power supplies for watches, and NiCd battery chargers. Before the end of 1984 one may see a-Si : H solar panels used in larger-scale applications such as irrigation and remote electrification.

In this article, we first review the methods of growing amorphous silicon for solar cells. The next part covers the material properties that are relevant to the development of efficient, stable a-Si : H solar cells. In Part IV, we discuss the fabrication, performance, and scale-up of a-Si : H solar cells, and in Part V, we consider the economics of these cells for various applications. We conclude with some projections for the future of a-Si : H photovoltaics.

II. Methods of Growing Amorphous Silicon for Solar Cells

Several techniques have been used to grow a-Si : H films for solar cells. As shown in Table I, the highest-efficiency cells have been made from films grown in glow discharges in silane (SiH_4). Conversion efficiencies greater than 9% have been reported for devices made from silane glow discharges using rf external capacitive coupling (Hamakawa *et al.,* 1983), rf internal capacitive coupling (Kuwano *et al.,* 1983), and the dc proximity mode (Catalano *et al.,* 1982). Comparable performance has also been reported for devices that were apparently made from glow discharges in silicon tetrafluoride (SiF_4) and hydrogen (Hack and Shur, 1982). Glow discharges in disilane

TABLE I

SOLAR-CELL EFFICIENCIES FOR VARIOUS GROWTH TECHNIQUES

Deposition technique	Solar-cell structure	η (%)	Reference
Glow discharge in SiH_4	Glass/SnO_2/p^{*a}–i–n/Ag	10.1	Catalano *et al.*, 1982
Sputtering in Ar and H_2	ITO/p–i–n/steel	4.0	Moustakas and Friedman, 1982
CVD–higher silanes	ITO/n–i–p/n–i–p/n–i–p/steel	4.0	Dalal, 1982
Photo-CVD with Hg	glass/ITO/SnO_2/p^*–i–n/metal	4.39	Inone *et al.*, 1983
CVD-SiH_4 + posthydrogenation	Pt/i–n/$n^+$$Si_x$[b]	2.7	Hirose, 1981

[a] p^* is p-type a-Si : C : H.
[b] Si_x is crystalline silicon.

(Si_2H_6) have also been used to make a-Si:H solar cells, but the best efficiency to date is ~8.1% (A. W. Catalano, private communication, 1983).

Detailed descriptions of the rf and dc glow-discharge deposition techniques are presented by Hirose and Uchida in Volume 21A, Chapters 2 and 3.

There are some general considerations that pertain to all glow-discharge systems that are used to make solar cells. The substrate temperature during film growth is usually in the range of 200–300°C. Polymer formation $(SiH_2)_n$ generates defects at lower substrate temperatures (Knights *et al.*, 1979), and hydrogen out-diffusion creates dangling bonds at higher substrate temperatures (Pankove and Carlson, 1977). The power density during deposition cannot be too high since energetic ions and neutrals can damage the growing film (Carlson and Magee, 1979). Power densities are typically less than 0.1 W cm^{-2} for most glow-discharge deposition systems. Also, all discharge systems must be relatively clean and free of air leaks since impurities can reduce the conversion efficiency of solar cells and lead to light-induced effects (see Sections 6 and 11).

As shown in Table I, several other techniques have been used to make a-Si : H solar cells, but the efficiencies are all rather low ($\lesssim$4.0%). However, some of these approaches are relatively new, and further optimization will undoubtedly lead to higher efficiencies. The sputter deposition of a-Si : H is described by Moustakas in Volume 21A, Chapter 4. Sputtered a-Si : H films appear to contain more defects than glow-discharge-produced films (Viktorovitch *et al.*, 1981), possibly as a result of more bombardment damage or the inclusion of Ar in the growing film.

The formation of a-Si : H films by the chemical vapor deposition (CVD) of higher silanes is covered by Hirose and Scott in Volume 21A, Chapters 6 and 7. In general, CVD techniques involve slow deposition rates or long posthydrogenation treatments in order to make a-Si:H films of reasonable quality.

III. Relevant Material Properties

1. Optical Absorption

Efficient solar energy conversion requires that the photovoltaic material be capable of absorbing a significant fraction of the energy in sunlight. As shown in Fig. 1, the absorption coefficient of undoped a-Si:H is greater than 10^4 cm^{-1} over most of the visible light region (~ 1.9 eV $< h\nu < \sim 4.0$ eV). Thus, an a-Si:H film need be only ~ 1 μm thick to absorb most of the solar energy.

As described in Volume 21B, "Hydrogenated Amorphous Silicon: Optical Properties," the absorption coefficient of undoped a-Si:H is strongly influenced by the deposition conditions. For example, the optical gap usually increases as the substrate temperature decreases, and this effect has been attributed to an increase in the hydrogen content (Zanzucchi *et al.*,

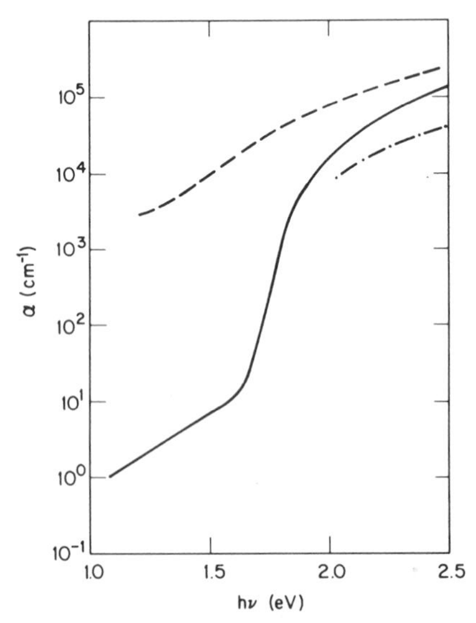

Fig. 1. The optical absorption coefficient as a function of photon energy for undoped a-Si:H (solid curve), *p*-type a-Si:H (dashed curve), and *p*-type a-Si:C:H (dashed–dotted curve). Both *p*-type films contains a few atomic percent of boron, and the a-Si:C:H film also contains ~ 20 at. % of carbon.

1977). The optical properties of a-Si:H appear to be directly influenced by the alloying effect of hydrogen and the nature of the hydrogen bonding. Cody *et al.* (1981) have presented evidence that in some cases the optical properties are determined by structural disorder, which is influenced indirectly by the presence of hydrogen through its ability to relieve strain. In any event, changing the optical properties of a-Si:H by varying the deposition conditions also changes the electronic properties. The data shown in Fig. 1 for undoped a-Si:H represent device quality material made under optimum deposition conditions (diffusion length $\geq 0.5\ \mu m$).

As described in Part IV, high performance a-Si:H solar cells are usually made in a $p-i-n$ configuration so that some light is also absorbed in the doped layers. Since the carrier lifetime in doped layers is usually very short, most of the light absorbed in these layers is lost to recombination. For boron-doped a-Si:H, the absorption coefficient is generally much larger than that for intrinsic or undoped a-Si:H (see Fig. 1), so that ~20% of the useful light may be lost by absorption in the p layer (~10 nm thick) of a $p-i-n$ cell (Carlson, 1980b). This absorption loss can be reduced significantly by alloying the p layer with carbon (Pankove, 1978) as shown in Fig. 1. Another approach is to alloy the doped layers with nitrogen since it also increases the optical gap (Kurata *et al.*, 1981), but a-Si:N:H layers have not yet led to high-performance devices. The optical absorption in the doped layers can also be reduced by using microcrystalline Si:H layers (Carlson and Smith, 1982).

For stacked junction solar cells, one would like to tailor the optical properties so that the active material in the first junction has a wide band gap, and the active layers in other junctions have progressively narrower band gaps (see Section 8). As mentioned earlier, either carbon or nitrogen alloying can be used to open the optical gap, while alloying with tin or germanium can reduce the optical gap. Representative optical data for undoped a-Si:C:H and a-Si:Ge:H films are shown in Fig. 2.

2. Transport

A high-performance solar cell must efficiently collect the carriers photogenerated in the active region of the device. This condition can only be met if the diffusion length (l_d) is greater than the thickness of the active region. For a-Si:H cells with a thickness of ~$0.5\ \mu m$, this translates into a mobility lifetime product ($\mu\tau$) greater than $10^{-7}\ cm^2\ V^{-1}$ using the expression $l_d = (kT\mu\tau/q)^{1/2}$.

In most a-Si:H solar cells, a built-in electric field (F) assists in the collection of photogenerated carriers, and efficient collection occurs as long as the drift length ($\mu\tau F$) is significantly larger than the film thickness. Crandall (1982) has shown that the transport in $p-i-n$ cells can be charac-

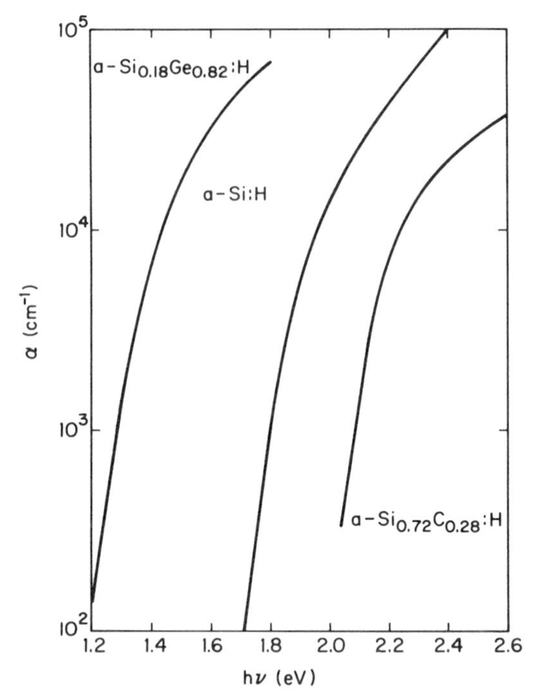

FIG. 2. The optical absorption coefficient as a function of photon energy for a-Si:H ($E_{opt} \simeq 1.73$ eV), a-$Si_{0.18}Ge_{0.82}$:H ($E_{opt} \simeq 1.3$ eV) (von Roedern *et al.*, 1982), and a-$Si_{0.72}C_{0.28}$:H ($E_{opt} \simeq 2.0$ eV) (Morimoto *et al.*, 1982).

terized by a collection length $l_c = (\mu_n\tau_n + \mu_p\tau_p)F$ that is the sum of the electron and hole drift lengths. In high performance cells, the collection length is typically $\gtrsim 5\ \mu$m in the short-circuit mode. The collection length decreases as the cell goes into forward bias, and transport via diffusion dominates as the cell approaches the open-circuit condition.

Diffusion lengths on the order of 1 μm have been measured in undoped a-Si:H by using the constant surface photovoltage technique (see Chapter 7 by Moore in Volume 21C) and similar values have been inferred by Faughnan *et al.* (1983) from measurements of the drift length. Even larger diffusion lengths can be inferred from the measurements of Komuro *et al.* (1983), who used two interfering lasers to create a transient grating of photogenerated carriers and then measured the decay rate with a probe laser. They were able to measure carrier lifetimes in the range of 3–5 μsec. Lifetimes in the range of 10–30 μsec have been measured by the junction recovery technique for injection current densities of ~10 mA cm^{-2} (Snell *et al.*, 1981). These lifetimes imply mobilities in the range of 10^{-1}–10^{-2} cm^2 V^{-1} sec^{-1} for material with a diffusion length of ~1 μm.

Hall effect measurements indicate mobilities of $\sim 10^{-1}$ cm^2 V^{-1} sec^{-1} for both electrons (Dresner, 1980) and holes (Dresner, 1983). Tiedje *et al.* (1981) have measured drift mobilities of ~ 1 cm^2 V^{-1} for electrons and $\sim 10^{-3}$ cm^2 V^{-1} sec^{-1} for holes. However, Silver *et al.* (1982) have estimated that the electron mobility is $\gtrsim 100$ cm^2 V^{-1} sec^{-1} by using the reverse recovery technique.

Although there is some question about the values of μ and τ, it is clear that the product of $\mu\tau$ is on the order of 10^{-7} cm^2 V^{-1} in good-quality a-Si : H, and thus photogenerated carriers can be efficiently collected.

3. Density of Gap States

Large diffusion lengths can only be obtained in a-Si : H material that possesses a low density of gap states. (See Chapter 2 by Cohen of Volume 21C for detailed information on the measurements of the density of states in a-Si : H.) The principal recombination center in a-Si : H is a silicon dangling bond that in the neutral state can trap either electrons or holes (Street, 1982). Some dangling bonds are created during the growth of the films due to hydrogen out-diffusion (Fritzsche *et al.*, 1978). Others appear to be associated with microstructural imperfections such as polymer chains (Knights *et al.*, 1979) or with impurities such as oxygen (Pontuschka *et al.*, 1982) and carbon (Morimoto *et al.*, 1982).

The defect level associated with the neutral dangling bond appears to be located $\sim 1.0-1.25$ eV below the conduction band edge with the singly charged negative state, $\sim 0.25-0.45$ eV higher (Morigaki *et al.*, 1982; Jackson, 1982). In crystalline silicon, the neutral state of the divacancy is 0.81 eV below the conduction-band edge, and the correlation energy for adding an electron is $+0.25$ eV. These values are in rough agreement considering the wider band gap of a-Si : H ($\sim 1.6-1.8$ eV) as compared to crystalline silicon (1.12 eV).

Other defect levels may arise from interactions between nearby dangling bonds that may form weak or stretched bonds, from dangling bonds on impurity atoms, and from weak bonds between silicon atoms and impurity atoms. The net result is that there is a distribution of defect levels throughout the band gap of a-Si : H, and this distribution depends strongly on deposition conditions. Generally, good quality a-Si : H has only been obtained after a comprehensive empirical optimization of the deposition conditions for a particular type of system.

4. Impurities

As mentioned earlier, impurities can create defect levels in the gap of a-Si : H. The most common impurities are oxygen, carbon, and nitrogen, with concentrations typically in the range of $10^{18}-10^{20}$ cm^{-3} (Magee and

Carlson, 1980). The source of oxygen may be either an air leak, outgassing of H_2O, CO, or CO_2 from walls of the vacuum system, or contaminants such as $(SiH_3)_2O$ (disiloxane) in the SiH_4 cylinder. Nitrogen may also come from either an air leak, outgassing from walls, or N_2 in the SiH_4 cylinder. The source of carbon may be either outgassing of CO and/or CO_2, or hydrocarbons that originate primarily from pump oils.

The effects of impurities on solar cells have been investigated by deliberately adding gases such as H_2O, N_2, and CH_4 to the SiH_4 discharge during the deposition of the a-Si:H (Carlson, 1977c, 1982a). Cell efficiency was reduced by ~15–30% by adding either ~0.2% H_2O or ~1% N_2 or ~10% CH_4 to the SiH_4 discharge. Delahoy and Griffith (1981) found that the presence of both oxygen and nitrogen in the SiH_4 causes a greater reduction in solar-cell efficiency than that caused by the presence of either gas by itself. This synergistic effect suggests that some of the recombination centers are nitrogen–oxygen complexes, and defects such as NO_2 have been observed by electron spin resonance in x-irradiated a-Si:H films (Pontuschka *et al.*, 1982).

In addition to siloxanes and nitrogen, other contaminants that have been detected in SiH_4 cylinders are monochlorosilane (SiH_3Cl) (Delahoy and Griffith, 1981) and tetrahydrofuran (C_4H_8O) (A. Gallagher and J. Scott, unpublished results, 1982). Delahoy and Griffith (1981) showed that the solar-cell performance was seriously degraded by the presence of ~500 ppm of Cl in the a-Si:H material. They also showed that small quantities of phosphine (~10 ppm) in SiH_4 could drastically reduce device performance. However, small quantities of diborane (a few parts per million) in SiH_4 often improve the conversion efficiency of a-Si:H solar cells due to compensation of donorlike impurities (Carlson, 1980c; Haruki *et al.*, 1983).

Work at RCA has shown that impurities such as oxygen, carbon, nitrogen, and chlorine can reduce the diffusion length (l_d) in a-Si:H (Carlson *et al.*, 1983a). As shown in Table II, significant reductions are caused by ~0.2% disiloxane or ~2% nitrogen in SiH_4. Both contaminants also cause a large increase in the space-charge density in the dark leading to small depletion widths (W_0). Compositional analyses of these films showed that the first one contained ~0.4 at. % oxygen and the second, ~0.5 at. % nitrogen. The addition of ~2% methane to the SiH_4 discharge (~0.6 at. % carbon in the film) causes both l_d and W_0 to decrease by about 40%. Adding 0.07% dichlorosilane (~40-ppm Cl in the film) causes a similar reduction in l_d but does not significantly affect W_0.

5. Microstructure

The disorder inherent in hydrogenated amorphous silicon gives rise to tails in the density of state distribution near the band edges. Calculations by

TABLE II

EFFECTS OF IMPURITIES ON DIFFUSION LENGTHS

Discharge atmosphere	l_d (μm)	W_0 (μm)	Impurity concentrations (cm^{-3})			
			O($\times 10^{19}$)	C($\times 10^{19}$)	N($\times 10^{19}$)	Cl($\times 10^{17}$)
SiH_4 (control)	0.52	1.47	1.7	2.0	0.03	1.2
~0.2% $(SiH_3)_2O$ in SiH_4	0.18	0.33	15.0	0.8	1.0	1.6
2% N_2 in SiH_4	0.12	0.23	3.8	0.2	24.0	0.4
2% CH_4 in SiH_4	0.32	0.83	5.2	25.0	—	0.7
0.07% SiH_2Cl_2 in SiH_4	0.30	1.56	0.3	0.3	0.15	17.0

Singh (1981) show that variations in the dihedral angle are responsible for features near the top of the valence band, whereas topological disorder affects the edge of the conduction band. Variations in bond lengths and bond angles contribute also to tail states, but these contributions are relatively small. Both the optical gap and the extent of the tail states appear to be influenced by the hydrogen content. Under illumination tail states can become recombination centers as quasi-Fermi levels approach the band edges. This loss mechanism is not present in good-quality crystalline semiconductors, and thus a-Si : H has a somewhat lower theoretical efficiency for photovoltaic energy conversion than an ideal semiconductor with the same band gap (Tiedje, 1982).

Microstructural imperfections such as polymer chains, clustered hydrogen, and microvoids can adversely affect the electronic properties of a-Si : H. Dangling bonds are often associated with these microstructural defects probably as a result of weaker Si–H bonds than are found in high-density, bulk a-Si : H, where distributed monohydride bonding dominates (Reimer *et al.*, 1981). Under certain deposition conditions (e.g., low substrate temperatures, Ar dilution, high power), the a-Si : H films may grow with a columnar morphology in which $(SiH_2)_n$ groups are associated with the connective material between the columns (Knights *et al.*, 1979).

The morphology and density of a-Si : H vary with the deposition conditions, and generally the films become more dense as the substrate temperature is increased (D'Antonio and Konnert, 1981). Under certain deposition conditions, even films grown at relatively high substrate temperatures (~330°C) can be porous and exhibit properties that change with time (Carlson *et al.*, 1979/1980).

The transport in a-Si : H may be strongly influenced by the presence of inhomogenities that give rise to potential fluctuations (Overhof and Beyer, 1981). The model predicts a thermally activated mobility and roughly accounts for the Meyer–Neldel behavior ($\sigma_D = \sigma_0 \exp(-E_\sigma/kT)$ where $\ln \sigma_0 \propto E_\sigma$). The inhomogenities may be compositional variations due to microvoids or hydrogen clustering.

6. Metastable States

Staebler and Wronski (1977) were the first to observe light-induced changes in the properties of a-Si : H. The Staebler–Wronski effect is now known to affect the performance of a-Si : H solar cells through the creation of recombination centers and charged traps (Carlson *et al.*, 1983b). These light-induced centers are metastable and can be annealed out at temperatures of ~150–200°C.

The light-induced creation of recombination centers causes the diffusion length to decrease with exposure time as shown in Fig. 3 (Carlson *et al.*, 1984b). The diffusion length was measured by the surface photovoltage method (Dresner *et al.*, 1980), and similar results were obtained from an analysis of device characteristics (Faughnan *et al.*, 1983).

Some of the metastable centers have been associated with impurities such as oxygen (Crandall, 1981) and carbon (Crandall *et al.*, 1983). Deep-level transient spectroscopy (DLTS) has been used to identify a center associated with oxygen that has a characteristic activation energy of ~1.0 eV, whereas

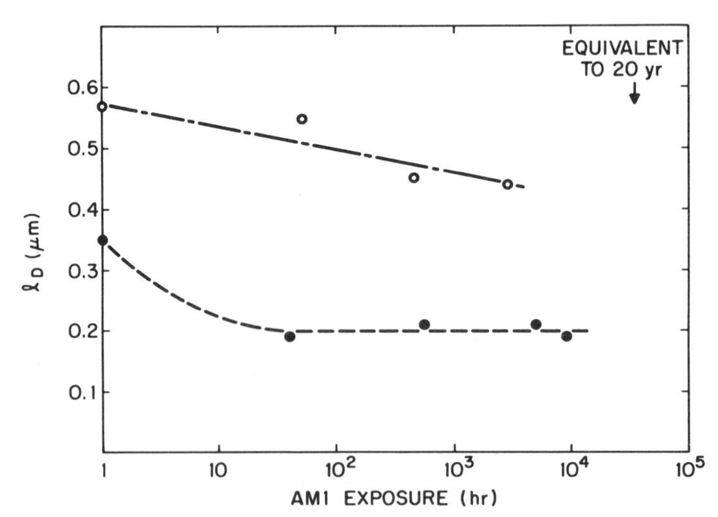

FIG. 3. Diffusion length as a function of exposure time to simulated 1-sun illumination (~100 mW cm^{-2}). Most device-quality films exhibit behavior intermediate to the curves shown here.

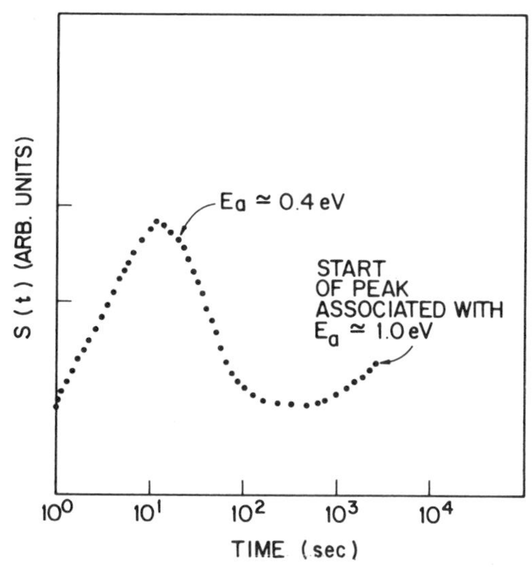

FIG. 4. Capacitance transient DLTS signal for a $p-n$ junction at $T = 437°K$ containing 10 at. % carbon in the p layer. [From Crandall *et al.* (1983).]

that associated with carbon has an activation energy of ~0.4 eV. Figure 4 shows that the transient voltage signals from these two centers are clearly distinguishable because of their different response times.

Other metastable centers appear to be associated with microstructural imperfections since light-induced effects are influenced by annealing treatments (Staebler and Pankove, 1980) and by variations in deposition conditions (Carlson, 1982b; Hirabayashi *et al.*, 1982). For a more thorough discussion of metastable effects see Chapter 11 by Schade of Volume 21B.

7. DOPING EFFICIENCY

Another requirement for efficient photovoltaic energy conversion is that the solar cells have a semiconductor junction with a large built-in potential (V_{bi}). The doped layers in a $p-i-n$ cell are mainly responsible for determining V_{bi}. Ideally, one would want wide-band-gap doped layers that are degenerate or highly conductive. In such a case, the Fermi levels would lie very close to the band edges, and the built-in potential would approach the band gap.

There are two mechanisms that limit the built-in potential in a-Si:H solar cells. One is the existence of band-tail states as mentioned earlier, and the other is the low doping efficiency of a-Si:H. Spear and LeComber (1976) used conductivity and field-effect measurements to estimate that ~30% of the phosphorus incorporated in a-Si:H was electronically active as shallow donors. Using photoemission, von Roedern *et al.* (1979) determined that ~10% of the phosphorus and boron in a-Si:H is active as donors and acceptors, respectively. LeComber *et al.* (1980) estimated doping efficiencies of ~0.5% for a-Si:H that had been ion implanted with boron and ~0.2% for phosphorus-implanted a-Si:H.

More recently, Faughnan and Hanak (1983) have used spectral response data to determine that the concentration of acceptors is $\sim 10^{19}$ cm^{-3} for *p*-type a-Si:H layers containing $\sim 10^{21}$ boron atoms cm^{-3} (as determined by SIMS) for a doping efficiency of ~1%. Dresner (1983) has estimated that the doping efficiency of boron in a-Si:H is ~0.1% for films containing between 10^{19} and 10^{21} boron atoms cm^{-3}. Thus, more recent estimates of the doping efficiency are in the range 0.1 – 1.0%. Apparently, many of the dopant atoms do not go into electronically active substitutional sites.

Street *et al.* (1981) have shown that doping a-Si:H with boron or phosphorus creates new defect states that are in some cases associated with dangling bonds. Other defects produced by doping appear to be associated with dopant–hydrogen complexes (Carlson *et al.*, 1982). Nuclear magnetic resonance studies have shown that nearly all the boron in *p*-type a-Si:H is threefold coordinated in either Si_3B or Si_2BH configurations (Greenbaum *et al.*, 1982).

The net result is that doped a-Si:H films are not very conductive (typically 10^{-3}–10^{-2} Ω^{-1} cm^{-1}). The Fermi level is ~0.2 eV below the conduction band in phosphorus-doped a-Si:H (Spear, 1977) and is ~0.5 eV above the valance band in boron-doped a-Si:H (Jan *et al.*, 1980). Since the optical gap of undoped a-Si:H is typically about 1.7 eV, the built-in potential of a-Si:H *p–i–n* solar cells is about 1.0 eV (Williams *et al.*, 1979). Improving the conductivity of the doped layers should lead to larger built-in potentials and consequently higher conversion efficiencies. The conductivity can be increased significantly by forming microcrystalline-doped Si:H films (Matsuda *et al.*, 1980), but since these films contain both amorphous and crystalline phases, there is no significant increase in the built-in potential (Carlson and Smith, 1982).

The built-in potential of a-Si:H *p–i–n* cells has been increased by alloying the *p* layer with carbon (Tawada *et al.*, 1982). However, as shown in Fig. 5, the resistivity of the *p* layer increases as the optical gap (or carbon content) increases. Thus, the carbon alloying is decreasing the doping efficiency in this case. The increase in V_{bi} with increasing carbon content of the *p* layer is apparently associated with a suppression of the dark current by the wide-band-gap *p* layer.

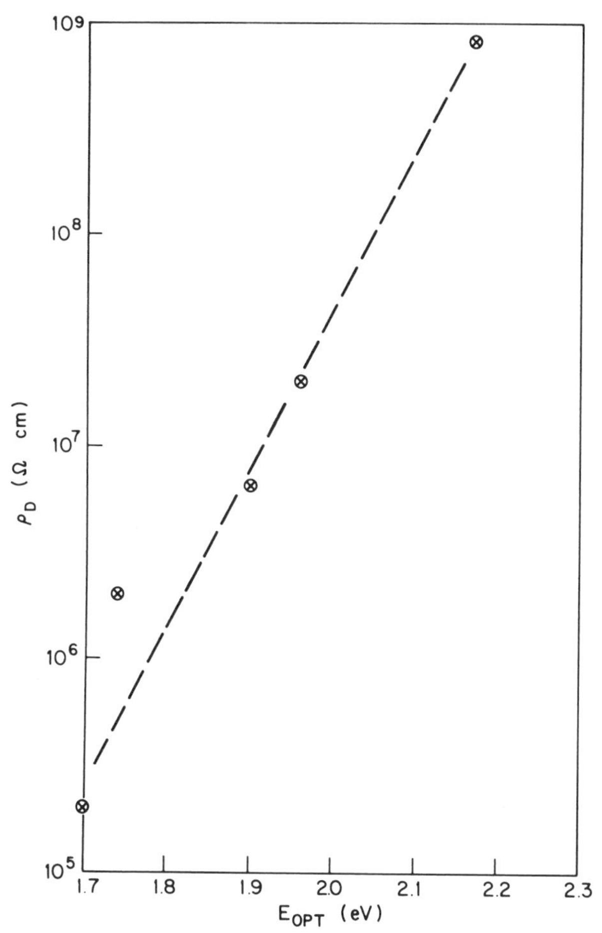

FIG. 5. Resistivity as a function of optical gap for boron-doped, amorphous silicon–carbon–hydrogen alloys.

IV. Solar-Cell Fabrication and Performance

8. STRUCTURES

High performance ($\eta > 7\%$) a-Si:H solar cells have been fabricated in three different structures as shown in Figs. 6 and 7. The highest-performance devices ($\eta = 10.1\%$) have been fabricated in the structure glass/SnO_2/ p–i–n/Ag, in which the p layer is an alloy of a-Si:C:H (Catalano *et al.*, 1982). The p layer typically contains about 20–30 at. % carbon (Morimoto

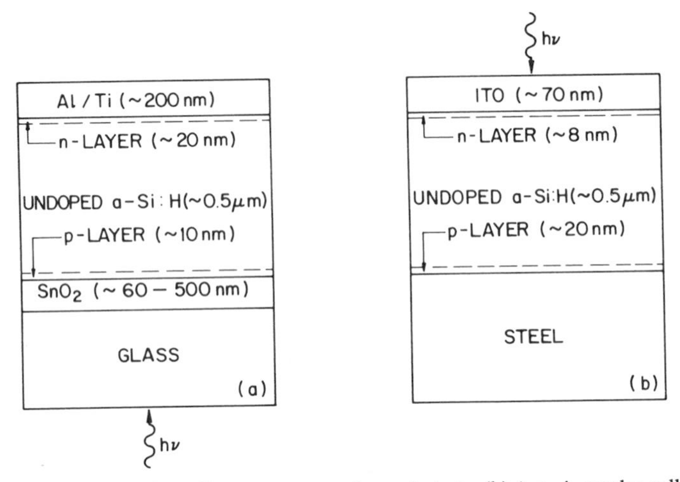

FIG. 6. (a) A $p-i-n$ solar-cell structure on a glass substrate. (b) A $p-i-n$ solar-cell structure on a steel substrate.

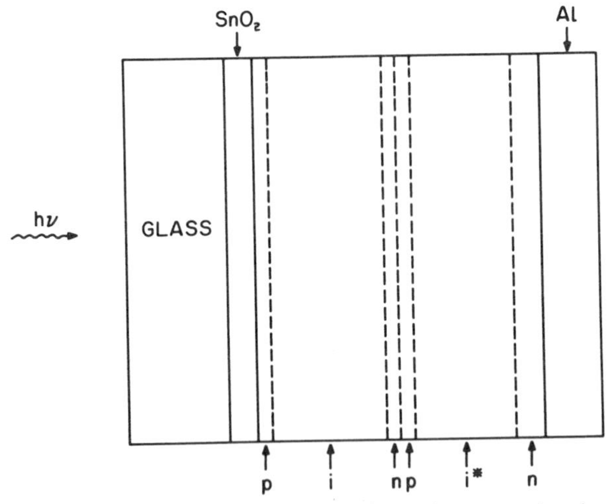

FIG. 7. A stacked or multijunction structure. The i^* layer is a narrow-band-gap alloy such as a-Si:Ge:H.

et al., 1982) and has an optical gap of ~2.0 eV (Tawada *et al.*, 1982). Generally, the *p* and *n* layers contain on the order of 1 at. % boron or phosphorus, respectively. The SnO_2 layer may be chosen to be ~60 nm thick so that it acts as an antireflection layer, or it may be much thicker so that the sheet resistance is low (~ 1 – 10 Ω/□). Silver can be used as a back metal contact so that most of the unabsorbed light reaching the back contact is reflected back into the *i* layer. However, either Al or Ti (~5 – 10 nm)/Al are more practical metal contacts that are also relatively reflective.

Another structure that has been used to make relatively efficient a-Si : H solar cells is shown in Fig. 6b. In this case, the *p* layer is deposited on a steel substrate, and indium tin oxide (ITO) is electron-beam evaporated onto the *n* layer. The ITO serves as both a top contact layer and an antireflection coating. The steel substrate may be coated with Cr to improve the back surface reflection. Generally, in all *p – i – n* cells the top doped layer is thin (~ 10 nm) in order to minimize losses due to absorption and recombination in that layer.

The stacked junction structure is fabricated by growing one *p – i – n* junction directly on top of another (see Fig. 7). Ideally, one wants to adjust the band gaps of the *i* layers so that each layer produces the same photocurrent and the device efficiency is maximized. Thus, the first *p – i – n* junction would employ a wide-band-gap layer (e.g., 1.9 eV) and the second junction would have a much smaller band gap (e.g., 1.1 eV). The stacked junction structure can be extended to three or more junctions where the band gap progressively decreases toward the back of the cell. The band gaps can be tailored by using alloys such as a-Si : C : H and a-Si : Ge : H (see Section 1). Since the junctions are in series, the voltages are additive. The doped layers must be relatively thin (~ 10 nm) to avoid significant absorption and recombination losses in those layers, and they must also be relatively conducting so that the *n – p* junctions are effectively shorted because of tunneling.

There are other structures that have been used to make a-Si : H solar cells such as Schottky-barrier and MIS configurations, but since the conversion efficiencies are generally less than 6%, we will not discuss these structures here (for more information on these structures, see Carlson, 1982b).

9. Solar-Cell Performance

The principal method of characterizing solar-cell performance is the measurement of conversion efficiency while the cell is exposed to 1 sun illumination (~ 100 mW cm^{-2}). The conversion efficiency is determined by measuring the current – voltage characteristic (see Fig. 8), locating the maximum power point ($P_m = J_m V_m$), and also measuring the solar insolation

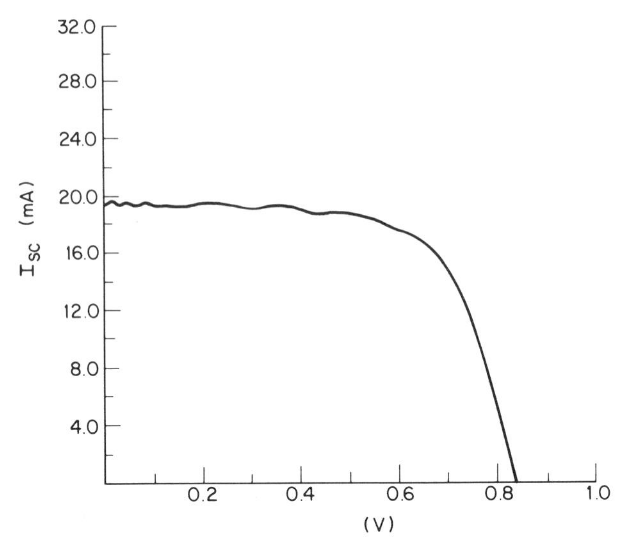

FIG. 8. Current–voltage characteristic of a $p-i-n$ cell fabricated on a glass substrate; $\eta = 10.1\%$, area $= 1.09\ \text{cm}^2$. $V_{oc} = 0.84$ V, $J_{sc} = 1.78\ \text{mA cm}^{-2}$, FF $= 0.676$, illumination $= 98.62\ \text{mW cm}^{-2}$, $T = 25.7°\text{C}$. [From Catalano *et al.*, Attainment of 10% conversion efficiency in amorphous silicon solar cells. *Conf. Rec. IEEE Photovoltaic Spec. Conf.*, Vol. 16, ©1982 IEEE.]

(P_i). The conversion efficiency is then given by

$$\eta = \frac{J_m V_m}{P_i} = \frac{(\text{FF}) J_{sc} V_{oc}}{P_i}, \tag{1}$$

where J_{sc} is the short-circuit current density, V_{oc} the open-circuit voltage, and FF the fill factor as defined by Eq. (1).

The $I-V$ characteristic shown in Fig. 8 is for a cell with the structure glass/SnO_2/$p-i-n$/Ag, where the p layer is alloyed with carbon (Catalano *et al.,* 1982). The conversion efficiency was 10.1% with $V_{oc} = 0.84$ V, $J_{sc} = 17.8\ \text{mA cm}^{-2}$, and FF $= 0.676$. The spectral response (or external quantum efficiency) of the same 10.1% cell is shown in Fig. 9. An integration of the quantum efficiency with the AM1 solar spectrum gave a current density of $17.6\ \text{mA cm}^{-2}$, in good agreement with the measured value of J_{sc} (Catalano *et al.,* 1982).

The performance characteristics of a variety of different types of a-Si : H solar cells are listed in Table III. The last structure listed is actually a hybrid configuration where the a-Si : H cell is stacked on top of a crystalline silicon cell (Hamakawa *et al.,* 1983).

Since the current density of a solar cell in the dark obeys the relation

$$J = J_0[\exp(qV/nkT) - 1], \tag{2}$$

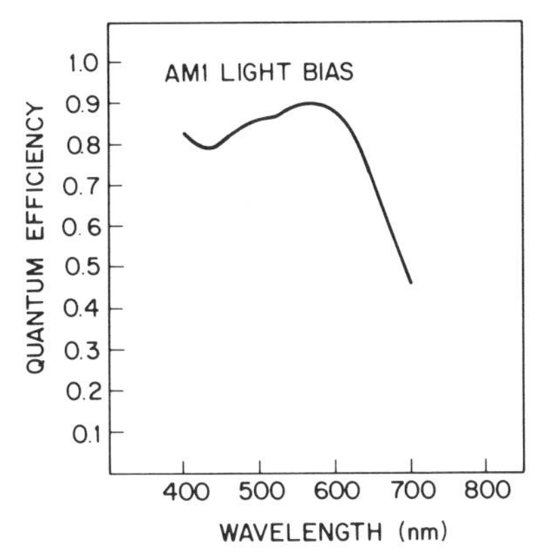

FIG. 9. Quantum efficiency as a function of wavelength for a $p-i-n$ cell with $\eta = 10.1\%$ [From Catalano *et al.*, Attainment of 10% conversion efficiency in amorphous silicon solar cells. *Conf. Rec. IEEE Photovoltaic Spec. Conf.*, Vol. 16, ©1982 IEEE.]

one can also characterize a cell by its reverse saturation current density (J_0) and its diode quality factor (n). Relatively efficient $p-i-n$ cells ($\eta > 7\%$) typically exhibit values of J_0 in the range 10^{-12}–10^{-11} A cm^{-2} and diode quality factors of $\sim 1.5-1.8$. A diode quality factor under illumination (n^1) can be determined from the expression

$$V_{oc} = (n^1kT/q)\ln(J_{sc}/J_0) + 1 \quad (3)$$

and for good $p-i-n$ cells, $n^1 \approx 1.0$.

The performance of a $p-i-n$ cell as a function of temperature is shown in Fig. 10 (G. Swartz, private communication, 1982). The solar-cell structure is glass/SnO_2/$p-i-n$/Ti/Al, where the p layer is alloyed with carbon. From room temperature to 65°C, $1/\eta\ (\partial\eta/\partial T) \approx -0.0032$ °C^{-1} as V_{oc} fell from 0.764 V to 0.630 V; J_{sc} increased from 11.75 mA cm^{-2} to 12.15 mA cm^{-2}, and FF increased from 0.662 to 0.678. As the temperature increased beyond 65°C, $1/\eta\ (\partial\eta/\partial T) \approx -0.0072$ °C^{-1} while J_{sc} leveled off at ~ 12.2 mA cm^{-2} and the fill factor started decreasing.

Han *et al.* (1982) studied the temperature dependence of $p-i-n$ cells deposited on steel using the structure shown in Fig. 6b. For temperatures ranging from 25 to 100°C, they found the conversion efficiency to be relatively constant. Although V_{oc} decreased with increasing temperature ($\partial V_{oc}/\partial T \approx -2.67$ mV °C^{-1}), it was compensated by the increase in J_{sc}. These cells exhibited values of $J_0 \approx 3 \times 10^{11}$ A cm^{-2}, $n = 2$, and $n^1 = 1.2$.

10. Modeling of Solar-Cell Operation

There have been a number of theoretical papers that model the operation of a-Si:H solar cells by starting with Poisson's equation, the continuity

TABLE III

PERFORMANCE OF VARIOUS TYPES OF a-Si:H SOLAR CELLS

Structure	V_{oc} (V)	J_{sc} (mA cm^{-2})	FF	η (%)	Area (cm^2)	Reference
Glass/SnO_2/p^*–i–n/Ag[a]	0.84	17.8	0.676	10.1	1.09	Catalano *et al.*, 1982
Glass/SnO_2/p^*–i–n/metal	0.83	15.8	0.70	9.28	0.04	Kuwano *et al.*, 1983
Glass/SnO_2/p^*–i–n/metal (nine cells in series)	7.31	14.1	0.662	7.58	100	Kuwano *et al.*, 1983
Glass/TCO/p^*–i–n'/Ag[b,c]	0.88	15.9	0.669	9.39	0.003	Hamakawa *et al.*, 1983
ITO/n–i–p/metal	0.836	16.7	0.66	9.2	0.05	Hack and Shur, 1982
ITO/n–i–p/n–i–p/n–i^{**}–p/metal[d]	2.2	6.74	0.57	8.5	0.09	Nakamura *et al.*, 1982
ITO/n'–i–p/n'/poly Si/Al	1.42	13.4	0.651	12.37	0.033	Hamakawa *et al.*, 1983

[a] p^* is p-type a-Si:C:H.
[b] TCO is a transparent conductive oxide such as SnO_2.
[c] n' is n-type microcrystalline Si:H.
[d] i^{**} is undoped a-Si:Ge:H.

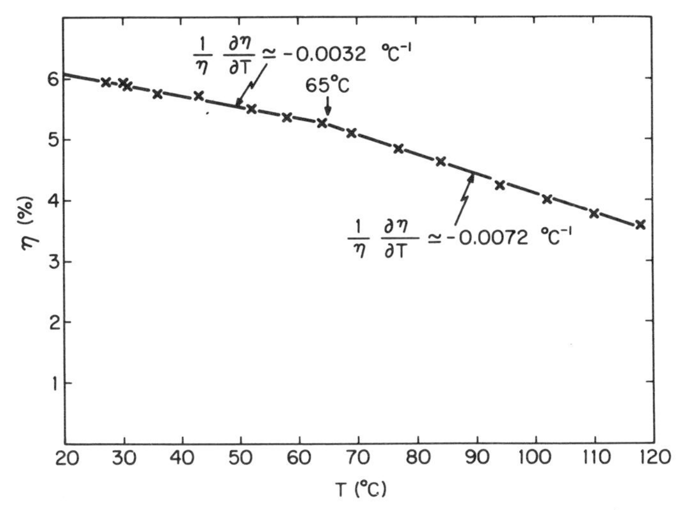

FIG. 10. Conversion efficiency as a function of operating temperature for a $p-i-n$ cell deposited on glass (see Fig. 6a).

equations for electrons and holes, and appropriate boundary conditions (Swartz, 1982; Crandall, 1982; Kuwano *et al.*, 1982; Okamoto *et al.*, 1983). These theoretical efforts have all produced results that are in reasonable agreement with the $I-V$ characteristics of a-Si : H solar cells under illumination, and in some cases the theory has been developed to account for the spectral response (Swartz, 1982; Okamoto *et al.*, 1983) and the $I-V$ characteristics in the dark (Okamoto *et al.*, 1983).

To clarify the physics relevant to the operation of a-Si : H solar cells, we shall restrict our discussion to the structure shown in Fig. 6a. An energy band diagram is shown in Fig. 11 for a $p-i-n$ cell in the short-circuit mode. In many high-performance a-Si : H solar cells, the electric field in the i layer is almost uniform since both the trapped charge and the free-carrier space charge are negligible. In this case, Crandall (1982) has shown that the photocurrent can be written as

$$J = qGl_c[1 - \exp(-d/l_c)], \tag{4}$$

where G is the generation rate of electron–hole pairs per unit volume per second, d is the thickness of the i layer, and $l_c = \mu_n\tau_n\varepsilon + \mu_p\tau_p\varepsilon$ (the sum of the electron and hole drift lengths).

Faughnan and Crandall (1984) have developed a simple model that calculates the fill factor of $p-i-n$ cells by using Eq. (4). They found that the collection length (at short circuit) must be 25 times greater than the thickness of the i layer to obtain a fill factor of 0.72. Figure 12 shows how the fill

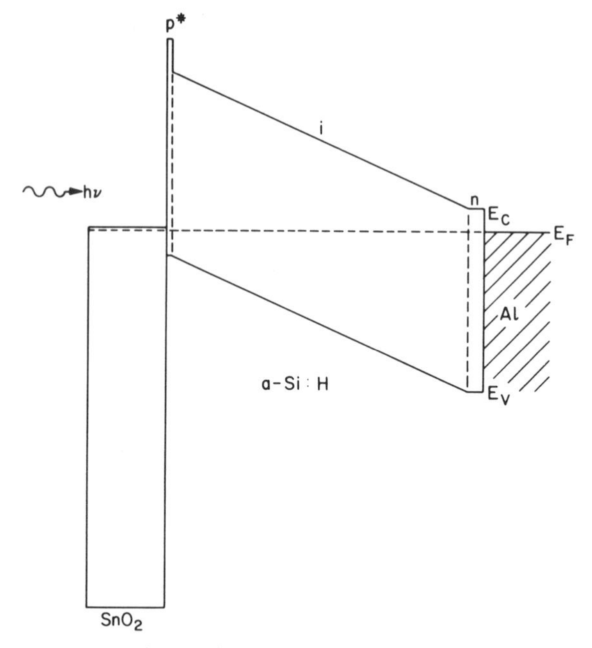

FIG. 11. An energy band diagram for a $p-i-n$ cell in the short-circuit mode.

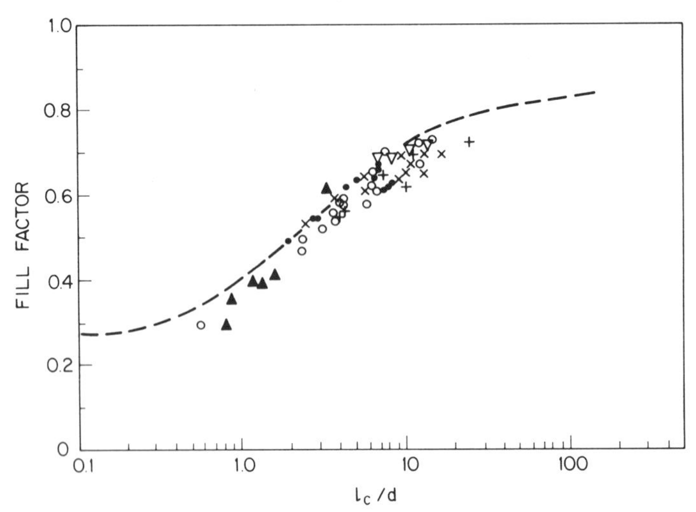

FIG. 12. Fill factor as a function of collection length for a-Si:H $p-i-n$ cells. [From B. W. Faughnan and R. S. Crandall (1984).]

factor varies with the collection length for both the theory and a large number of $p-i-n$ cells.

In many low-performance a-Si : H solar cells, the net space charge in the i layer is not negligible, and a quasi-neutral region is present in the central portion of the i layer. In this case, depletion regions or space-charge fields exist near the $p-i$ and $i-n$ interfaces. If the diffusion length is smaller than the width of the quasi-neutral region, then many of the photogenerated carriers are lost to recombination. Since the quasi-neutral region will grow as the cell goes into forward bias, both the fill factor and J_{sc} will be reduced compared to the uniform field case.

Early $p-i-n$ cells exhibited poor quantum efficiencies at short wavelengths because of recombination in the p layer (Carlson and Wronski, 1976). As discussed in Section 1, absorption losses associated with the front p layer can be reduced by alloying the p layer with carbon. Increasing the band gap of the p layer may also reduce recombination associated with the back-diffusion of local minority carriers. As shown in Fig. 11, the p layer may present a "bump" in the conduction band that reflects both thermal and hot electrons back into the i layer. Thus, the good short-wavelength response shown in Fig. 9 can be attributed to the reflection of minority carriers (or a small surface recombination velocity).

Deckman *et al.* (1983) have shown that the long-wavelength response of a-Si : H solar cells can be increased by using textured surfaces to promote a light-trapping effect. They demonstrated that texturing can enhance the short-circuit current density of $p-i-n$ cells by ~25%.

11. Stability of a-Si : H Solar Cells

The only significant source of degradation present in a-Si : H $p-i-n$ cells is associated with the metastable states discussed in Section 6. As shown in Fig. 3, the diffusion length (l_d) can decrease with prolonged exposure to illumination due to the creation of new recombination centers. This decrease in the recombination lifetime can cause degradation even when the internal electric field remains relatively uniform since the photocurrent decreases as $\mu\tau$ (or l_c) decreases [see Eq. (4)]; the effect of decreasing collection length on fill factor is clearly seen in Fig. 12. In a-Si : H solar cells that possess a quasi-neutral region, the decrease in l_d with light soaking can cause significant decreases in J_{sc} and FF.

Another complexity is that the net space charge often increases with light soaking in a-Si : H cells (Carlson *et al.*, 1983). This effect is shown in the surface photovoltage profiling data in Fig. 13. This profiling technique allows one to determine the potential distribution in portions of an a-Si : H solar cell by selectively sputter etching and measuring the change in surface photovoltage with vibrating Kelvin probe (Goldstein *et al.*, 1981). As shown

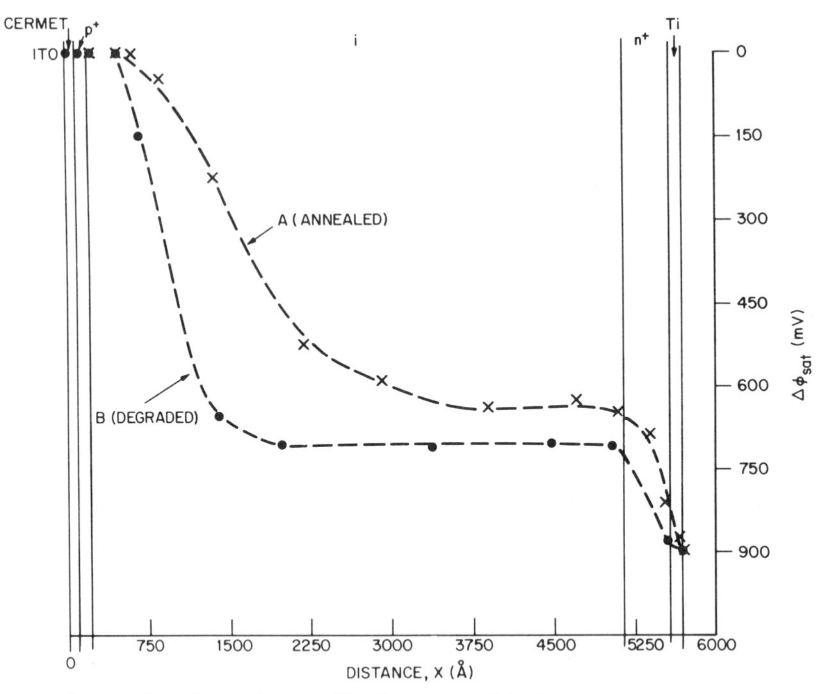

FIG. 13. A surface photovoltage profile of a $p-i-n$ cell in the annealed state ($V = 640$ mV, $w = 3200$ Å, $N_b = 8.1 \times 10^{15}$ cm^{-3}) and in the degraded or light-soaked state ($V = 700$ mV, $w = 1500$ Å, $N_6 = 4.2 \times 10^{16}$ cm^{-3}).

in Fig. 13, the net space charge in the vicinity of the $p-i$ interface increased with light soaking leading to a wider quasi-neutral region. If light soaking causes both an increase in the net space charge and a decrease in l_d, then the solar-cell degradation may be quite severe.

Some metastable centers may be associated with doped layers or with interface states. The light-induced generation of these centers appears to increase the surface recombination velocity in some cells, causing a decrease in the spectral response at short wavelengths. This effect and the others mentioned above are reversible; annealing the a-Si:H cells at ~200°C for several minutes restores the cells to their initial conditions.

Some representative light-soaking data are shown in Fig. 14 for a-Si:H cells grown on both glass and steel substrates. Most a-Si:H solar cells reach a steady-state condition after ~ 10^2– 10^3 hr of continuous AM1 illumination (~ 100 mW cm^{-2}). The steady-state condition results from a balance between the rate of creation of new metastable centers and the rate of removal of old metastable centers by annealing. The efficiency of the cell in the

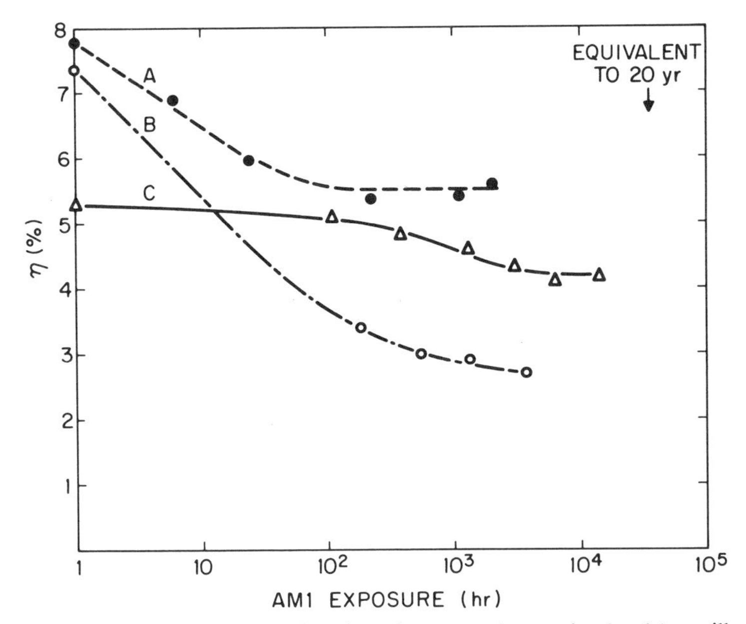

FIG. 14. Conversion efficiency as a function of exposure time to simulated 1-sun illumination with no encapsulation (~ 100 mW cm^{-2}). For curves A and B, the cell structure was glass/SnO_2/p^*–i–n/Ti/Al, where p^* is a-Si:C:H. For curve C, the cell structure was ITO/n–i–p/steel.

steady-state mode depends on the activation energy associated with the metastable centers and the temperature of the cell during operation.

In Fig. 14, both curves A and B represent data for cells grown on glass substrates (see Fig. 6a) by using identical deposition conditions except that 1.5% CH_4 was added to the SiH_4 discharge in the case of curve B (Crandall *et al.*, 1984). The carbon content of this device ($\sim 1.7 \times 10^{22}$ cm^{-3}) was $\sim 10^2$ times larger than that associated with curve A. Metastable defects associated with the carbon caused both a decrease in the diffusion length and an increase in the net space-charge density. Similar effects were observed for cells deliberately contaminated with oxygen (Carlson, 1982a).

Even when the a-Si:H contains a relatively large concentration of metastable centers, stable solar-cell operation can be achieved by making the i layers thin ($\lesssim 0.2$ μm) (Hanak and Korsun, 1982). The data in Fig. 15 show that no degradation is observed in p–i–n cells with thin i layers, but as the i layer thickness exceeds ~ 0.2 μm, all the photovoltaic parameters (V_{oc}, J_{sc}, FF) exhibit some degradation, with the largest changes occurring in the fill factor. The improved stability of thin cells is related to the observation that the metastable centers are activated by recombination

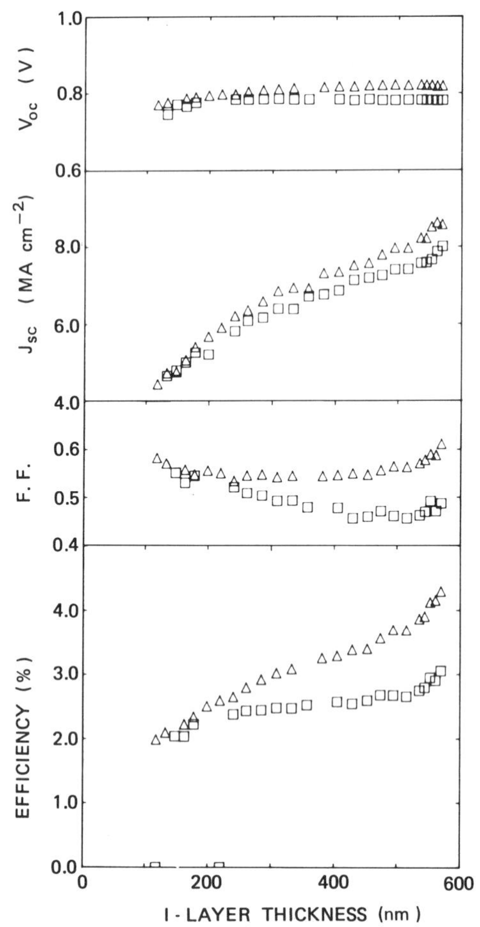

FIG. 15. Photovoltaic parameters (V_{oc}, J_{sc}, FF, and η) as a function of the thickness of the *i* layer in *p–i–n* cells (with deposition temperature 238°C) before (△) and after (□) a 67-hr light soak (~AM1). [From J. J. Hanak and V. Korsun, Optical stability studies of a-Si:H solar cells. *Conf. Rec. IEEE Photovoltaic Spec. Conf.*, Vol. 16, ©1982 IEEE.]

(Guha *et al.*, 1983), and thus strong internal fields can inhibit degradation by efficiently extracting photogenerated carriers from the *i* layer (Hanak and Korsun, 1982; Staebler *et al.*, 1981).

12. PROCESSING OF a-Si:H PANELS

The fabrication of low cost a-Si:H solar cells requires a large-scale, automated process. Currently, two approaches are being pursued. One involves the fabrication of a large (~ 1-ft^2) single cell on sheets of steel using the structure shown in Fig. 6b. This approach can utilize a multichamber deposition system with a continuous feed of steel sheet running from one reel to another (Ovshinsky, 1983). Each cell must have a current collection grid deposited on top of the ITO, and a number of cells must then be wired together and placed in a package with a protective cover (e.g., a sheet of glass) to make a solar panel.

The other approach involves the fabrication of a series-connected monolithic panel on sheets of glass (Hanak, 1979) using the structure shown in Fig. 6a. The monolithic panels are fabricated by first patterning the SnO_2-coated glass into electrically isolated strips, performing another patterning step after the a-Si : H deposition and a final patterning step after the deposition of the metal. The completed panel consists of a multitude of narrow strip cells connected in series (see Fig. 16) to produce a high-voltage, low-current output.

The processing sequence that is used to make a monolithic panel is shown schematically in Fig. 17 (Firester and Cralson, 1983). In this process, the glass is first cleaned (a) and then coated with SnO_2 (b). The SnO_2 is then patterned by laser scribing (c), followed by the deposition of the three a-Si : H layers (d) to form a $p-i-n$ junction. After another laser scribing step (e), a

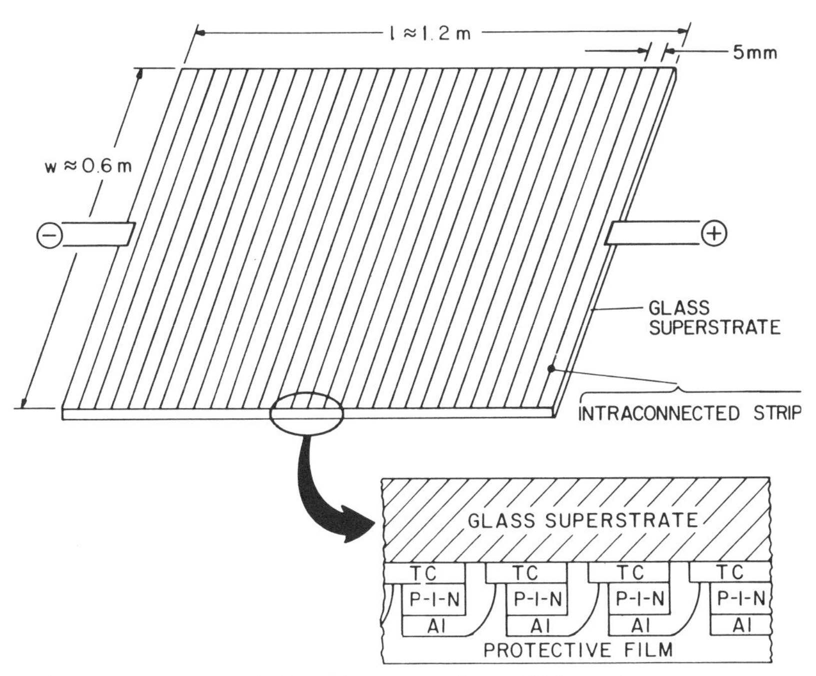

FIG. 16. The series-connected monolithic panel.

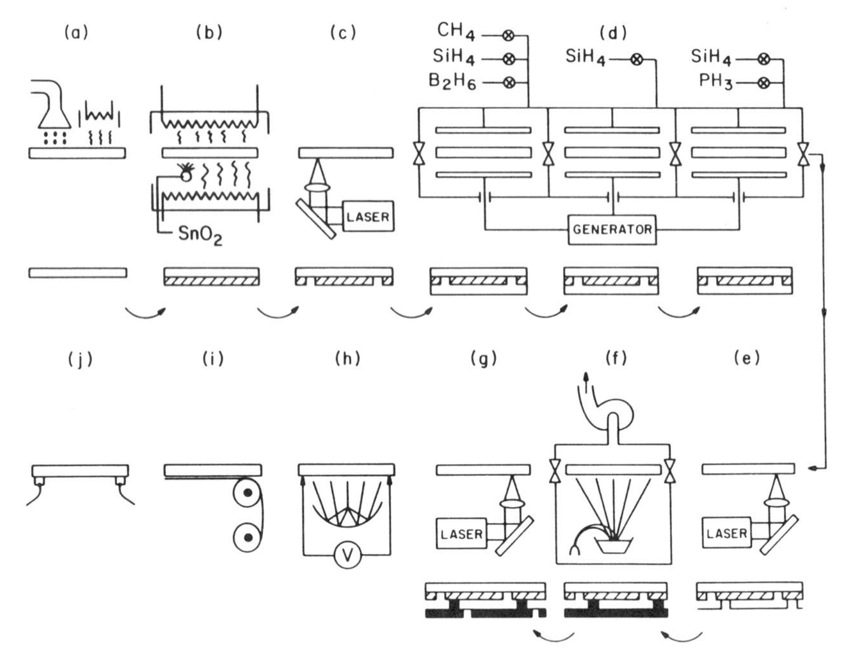

FIG. 17. The processing sequence for fabricating a-Si : H monolithic panels. [From Firester and Carlson (1983).]

thin metal film is deposited to form the back contact (f). The final laser scribing step (g) separates the back metallization so that all the strip cells are series connected. The panel is then tested (h), laminated (i), and connected to power outlets (j).

The monolithic panel can be scaled up in size without increasing the thickness of any of the deposited layers. As the width w is increased, each strip cell is lengthened so that the panel current is increased (see Fig. 16). As the length of the panel l is increased, the number of strip cells increases and so does the panel voltage. Since the interconnections are made by the laser scribing process, the inactive area of the panel can be limited to ~1–2% (Hanak, 1981).

V. Cost Projections

Single-crystal silicon, solar-cell modules presently cost ~\$5–10 per peak watt (W_p), and the world market for 1983 is estimated to be ~18–20 MW_p. By 1990, the annual sales for photovoltaic modules may be greater than \$1 billion (Brandhorst *et al.*, 1982).

Amorphous silicon photovoltaic modules are likely to capture a growing percentage of future markets mainly on the basis of cost. Amorphous silicon modules should be relatively inexpensive to manufacture for several reasons. First, the material costs are low since the substrate is the single most expensive item (glass costs ~\$4 m^{-2}). Moreover, the processing energy requirements are modest since the substrate temperatures are typically 250°C. In addition, production costs are expected to be low since a-Si:H films can be deposited over large areas in reasonably short times.

As shown in Fig. 18, the production costs of a-Si:H modules are projected to decrease significantly as production volume increases (Kuwano and Ohnishi, 1981). At the 50-MW_p yr^{-1} production level, half of the total cost is in the materials. The cost of the capital equipment can be less than 10% of the total manufacturing costs (Firester and Carlson, 1983).

VI. Future Directions

As mentioned in Section 9, the highest conversion efficiency observed to date for an amorphous silicon solar cell is 10.1% (Catalano *et al.*, 1982). The theoretical limit for the conversion efficiency of a single-junction a-Si:H cell can be estimated to be ~20%. This follows from an upper limit of ~22 mA cm^{-2} for J_{sc} as determined from optical absorption data (optical path length ~2 μm), and from upper limits of ~1.0–1.05 V for V_{oc} and ~0.86 for the fill factor (Tiedje, 1982).

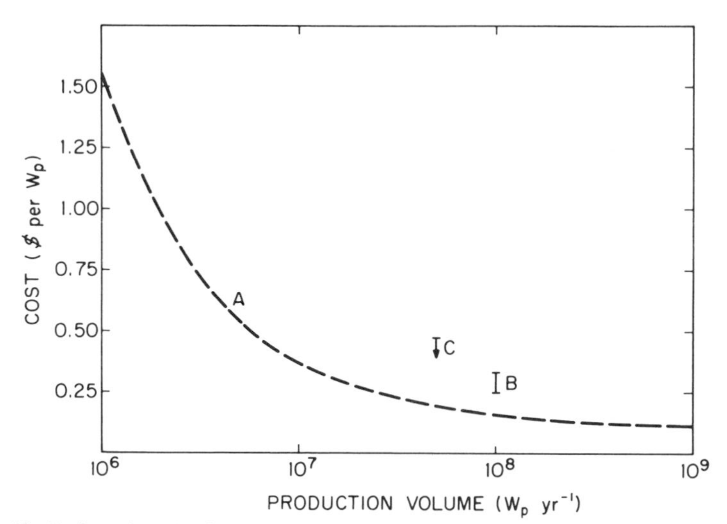

FIG. 18. Projected production costs as a function of production volume for a-Si:H modules. (A, Kuwano and Ohnishi, 1981; B, Hamakawa, 1982; C, Firester and Carlson, 1983.) All projections assume a conversion efficiency of 10% and are adjusted to 1981 dollars.

As shown in Fig. 19, the conversion efficiency of a-Si:H solar cells has improved dramatically in recent years, and we project that efficiencies of ~12–14% will be obtained in the next few years for single-junction cells. Most of the future improvement will probably come from the development of better alloys for doped layers.

Conversion efficiencies as high as 20% might be achieved by the end of the decade if good-quality intrinsic alloys can be developed for stacked junction cells (see Section 8). Wu and Williams (1983) have estimated a theoretical limit of 38% for a structure employing two stacked junctions, but their calculations did not include the effect of tail states. Kuwano *et al.* (1983) have estimated that the maximum practical efficiency is ~24% for a structure with three stacked junctions.

A conversion efficiency greater than 10% would not only lead to lower manufacturing costs (see Part V) but would also reduce the cost of the balance of systems. For a utility solar-cell power plant, the balance of system includes the land, fencing, wiring, inverter, support structure, and so on. Many of these costs are area related so that increasing the array efficiency reduces costs by reducing the area of the power station. Utility applications also require long-term reliability so that a-Si:H cells must be developed with high efficiencies maintained over more than 20 years of operation.

If the projections for production costs (see Fig. 18) and conversion efficiencies (see Fig. 19) are realized, then a-Si:H solar cells may dominate a

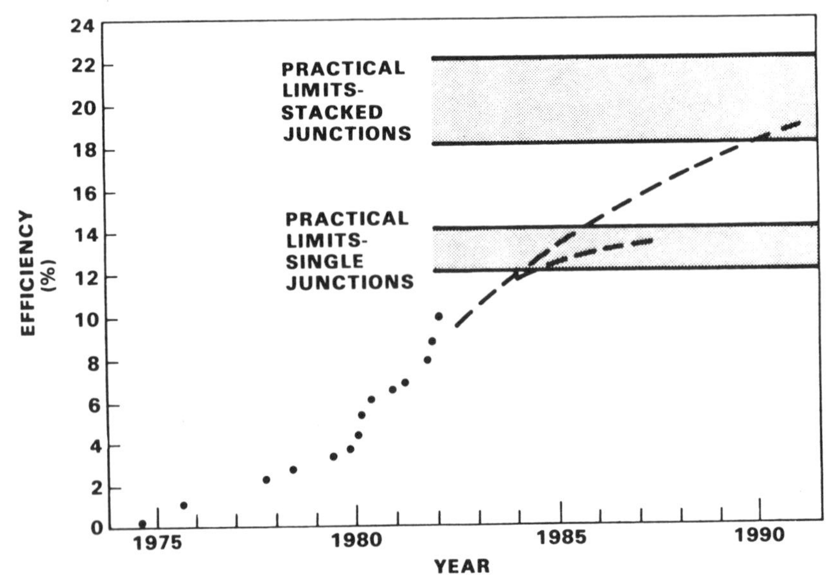

FIG. 19. Past and projected conversion efficiencies for 1-cm² a-Si:H solar cells.

business that could have sales of around $10 billion per year by the year 2000 (Brandhorst *et al.,* 1982).

REFERENCES

Brandhorst, H. W., Barnett, A. M., Evans, J. C., Feucht, D. L., Little, R., McGinnis, R., Maycock, P. D., and Murray, W. (1982). *Conf. Rec. IEEE Photovoltaic Spec. Conf.* **16,** 1478.

Carlson, D. E. (1977a). U.S. Patent 4,064,521.

Carlson, D. E. (1977b). *IEEE Trans. Electron Devices* **ED-24,** 449.

Carlson, D. E. (1977c). *Tech. Dig.—Int. Electron Devices Meet.,* p. 214.

Carlson, D. E. (1980a). *Sol. Energy Mater.* **3,** 503.

Carlson, D. E. (1980b). *Conf. Rec. IEEE Photovoltaic Spec. Conf.* **14,** 291.

Carlson, D. E. (1980c). U.S. Patent 4,217,148.

Carlson, D. E. (1982a). *J. Vac. Sci. Technol.* **20,** 290.

Carlson, D. E. (1982b). *Sol. Energy Mater.* **8,** 129.

Carlson, D. E., and Magee, C. W. (1979). *Eur. Community Photovoltaic Sol. Energy Conf., 2nd, Berlin, 1979,* p. 312.

Carlson, D. E., and Smith, R. W. (1982). *J. Electron. Mater.* **11,** 740.

Carlson, D. E., and Wronski, C. R. (1976). *Appl. Phys. Lett.* **28,** 671.

Carlson, D. E., Magee, C. W., and Thomas, J. H., III (1979/1980). *Sol. Cells* **1,** 271.

Carlson, D. E., Smith, R. W., Magee, C. W., and Zanzucchi, P. J. (1982). *Philos. Mag. B* **45,** 51.

Carlson, D. E., Catalano, A., D'Aiello, R. V., Dickson, C. R., and Oswald, R. S. (1984a). *Conf. Rec. IEEE Photovoltaic Spec. Conf.* To be published.

Carlson, D. E., Moore, A. R., and Catalano, A. (1984b). *J. Non-Cryst. Solids.* To be published.

Carlson, D. E., Moore, A. R., Szostak, D. J., Goldstein, B., Smith, R. W., Zanzucchi, P. J., and Frenchu, W. R. (1983b). *Solar Cells* **9,** 19.

Catalano, A., D'Aiello, R. V., Dresner, J., Faughnan, B., Firester, A., Kane, J., Schade, H., Smith, Z. E., Swartz, G., and Triano, A. (1982). *Conf. Rec. IEEE Photovoltaic Spec. Conf.* **16,** 1421.

Cody, G. D., Tiedje, T., Abeles, B., Moustakes, T. D., Brooks, B., and Goldstein, Y. (1981). *Phys. Rev. Lett.* **47,** 1480.

Connell, G. A. N., and Pawlik, J. R. (1976). *Phys. Rev. B* **13,** 787.

Crandall, R. S. (1981). *Phys. Rev. B* **24,** 7457.

Crandall, R. S. (1982). *J. Appl. Phys.* **53,** 3350.

Crandall, R. S., Carlson, D. E., Catalano, A. W., and Weakliem, H. A. (1983). *Appl. Phys. Lett.* **44,** 200.

Dalal, V. (1982). *Conf. Rec. IEEE Photovoltaic Spec. Conf.* **16,** 1384.

D'Antonio, P., and Konnert, J. H. (1981). *AIP Conf. Proc.* **73,** 117.

Deckman, H. W., Wronski, C. R., Witzke, H., and Yablonovich, E. (1983). *Appl. Phys. Lett.* **42,** 968.

Delahoy, A. E., and Griffith, R. W. (1981). *Conf. Rec. IEEE Photovoltaic Spec. Conf.* **15,** 704.

Dresner, J. (1980). *Appl. Phys. Lett.* **37,** 742.

Dresner, J. (1983) *J. Non-Cryst. Solids* **58,** 353.

Dresner, J., Goldstein, B., and Szostak, D. (1980). *Appl. Phys. Lett.* **38,** 998.

Faughnan, B. W., and Crandall, R. S. (1984). *Appl. Phys. Lett.* **44,** 537.

Faughnan, B. W., and Hanak, J. J. (1983). *Appl. Phys. Lett.* **42,** 722.

Faughnan, B. W., Moore, A. R., and Crandall, R. S. (1984). *Appl. Phys. Lett.* **44,** 613.

Firester, A. H., and Carlson, D. E. (1983). *RCA Eng.* **28,** 40.

Fritzsche, H., Tsai, C. C., and Persans, P. (1978). *Solid State Technol.* **21,** 55.
Goldstein, B., Redfield, D., Szostak, D. J., and Carr, L. A. (1981). *Appl. Phys. Lett.* **39,** 258.
Greenbaum, S. G., Carlos, W. E., and Taylor, P. C. (1982). *Solid State Commun.* **43,** 633.
Guha, S., Yang, J., Czubatyj, W., Hudgens, S. J., and Hack, M. (1983). *Appl. Phys. Lett.* **42,** 588.
Hack, M., and Shur, M. (1982). *Conf. Rec. IEEE Photovoltaic Spec. Conf.* **16,** 1429.
Hamakawa, Y. (1982). *Sol. Energy Mater.* **8,** 101.
Hamakawa, Y., Fujimoto, K., Okuda, K., Kashima, Y., Nonomura, S., and Okamoto, H. (1983). *Appl. Phys. Lett.* **43,** 644.
Han, M.-K., Anderson, W. A., and Wiesmann, H. (1982). *Conf. Rec. IEEE Photovoltaic Spec. Conf.* **16,** 1102.
Hanak, J. J. (1979). *Solar. Energy* **23,** 145.
Hanak, J. J. (1981). U.S. Patent 4,292,092.
Hanak, J. J., and Korsun, V. (1982). *Conf. IEEE Photovoltaic Spec. Conf.* **16,** 1381.
Haruki, H., Sakai, H., Kamiyama, M., and Uchida, Y. (1983). *Sol. Energy Mater.* **8,** 441.
Hirabayashi, I., Morigaki, K., and Yoshida, M. (1982). *Sol. Energy Mater.* **8,** 153.
Hirose, M. (1981). *J. Phys. Colloq. Orsay, Fr.* **42,** C4-705.
Inoue, T., Konagai, M., and Takahashi, K. (1983). *Appl. Phys. Lett.* **43,** 775.
Jackson, W. (1982). *Solid State Commun.* **44,** 477.
Jan, Z. I., Bube, R. H., and Knights, J. C. (1980). *J. Appl. Phys.* **51,** 3278.
Knights, J. C., Lucovsky, G., and Nemanich, R. J. (1979). *J. Non-Cryst. Solids* **32,** 393.
Komuro, S., Aoyagi, Y., Segawa, Y., Namba, S., Masuyami, A., Okamoto, H., and Hamakawa, Y. (1983). *Appl. Phys. Lett.* **42,** 79.
Kurata, H., Hirose, M., and Osaka, Y. (1981). *Jpn. J. Appl. Phys.* **20,** L811.
Kuwano, Y., and Ohnishi, M. (1981). *J. Phys. Colloq. Orsay, Fr.* **42,** C4-1155.
Kuwano, Y., Tsuada, S., and Ohnishi, M. (1982). *Jpn. J. Appl. Phys.* **21,** 235.
Kuwano, Y., Nakano, S., Fukatsu, T., Ohnishi, M., Nishiwaki, H., and Tsuda, S. (1983). *Photovoltaics Sol. Energy* **407** (*SPIE Tech. Symp. East, 1983 Arlington, Virginia.* To be published).
LeComber, P. G., Spear, W. E., Müller, G., and Kalbitzer, S. (1980). *J. Non-Cryst. Solids* **35/36,** 327.
Madan, A., McGill, J., Czubatyj, W., Yang, J., and Ovshinsky, S. R. (1980). *Appl. Phys. Lett.* **37,** 826.
Magee, C., and Carlson, D. E. (1980). *Sol. Cells* **2,** 365.
Matsuda, A., Yamasaki, S., Nakagawa, K., Okushi, H., Tanaka, K., Iizima, S., Matsumura, M., and Yamanioto, H. (1980). *Jpn. J. Appl. Phys.* **19,** L305.
Morigaki, K., Sano, Y., Hirabayashi, I., Konagai, M., and Suzuki, M. (1982). *Solid State Commun.* **43,** 751.
Morimoto, A., Miura, T., Kumeda, M., and Shimizu, T. (1982). *J. Appl. Phys.* **53,** 7299.
Moustakas, T. D., and Friedman, R. (1982). *Appl. Phys. Lett.* **40,** 515.
Nakamura, G., Sato, K., Kondo, H., Yukimoto, Y., and Shirahata, K. (1982). *Eur. Community Photovoltaic Sol. Energy Conf., 4th, Stressa, Italy,* p. 616. D. Reidel, Doredrecht, Holland.
Okamoto, H., Kida, H., Nomura, S., and Hamakawa, Y. (1983). *Sol. Cells* **8,** 317.
Overhof, H., and Beyer, W. (1981). *Philos. Mag. B* **43,** 433.
Ovshinsky, S. R. (1983). *Photovoltaics Energy* **407** (*SPIE Tech. Symp. East, 1983, Arlington, Virginia.* To be published).
Pankove, J. I. (1978). U.S. Patent 4,109,271.
Pankove, J. I., and Carlson, D. E. (1977). *Appl. Phys. Lett.* **31,** 450.

Pontuschka, W. M., Carlos, W. E., Taylor, P. C., and Griffith, R. W. (1982). *Phys. Rev. B* **25,** 4362.
Reimer, J. A., Vaughnan, R. W., and Knights, J. C. (1981). *Phys. Rev. B* **24,** 3360.
Silver, M., Giles, N. C., Snow, E., Shaw, M. P., Cannella, V., and Adler, D. (1982). *Appl. Phys. Lett.* **41,** 935.
Singh, J. (1981). *Phys. Rev. B* **23,** 4156.
Snell, A. J., Spear, W. E., and LeComber, P. G. (1981). *Philos. Mag.* **43,** 407.
Spear, W. E. (1977). *Adv. Phys.* **26,** 312.
Spear, W. E., and LeComber, P. G. (1976). *Philos. Mag.* **33,** 935.
Staebler, D. L., and Pankove, J. I. (1980). *Appl. Phys. Lett.* **37,** 609.
Staebler, D. L., and Wronski, C. R. (1977). *Appl. Phys. Lett.* **31,** 292.
Staebler, D. L., Crandall, R. S., and Williams, R. (1981). *Appl. Phys. Lett.* **39,** 733.
Street, R. A. (1982). *Appl. Phys. Lett.* **41,** 1060.
Street, R. A., Bieglesen, D. K., and Knights, J. C. (1981). *Phys. Rev. B* **24,** 969.
Swartz, G. (1982). *J. Appl. Phys.* **53,** 712.
Tawada, Y., Kondo, M., Okamoto, H., and Hamakawa, Y. (1981). *Conf. Rec. IEEE Photovoltaic Spec. Conf.* **15,** 245.
Tawada, Y., Tsuge, K., Kondo, M., Okamoto, H., and Hamakawa, Y. (1982). *J. Appl. Phys.* **53,** 5273.
Tiedje, T. (1982). *Appl. Phys. Lett.* **40,** 627.
Tiedje, T., Cebulka, J. M., Morel, D. L., and Abeles, B. (1981). *Phys. Rev. Lett.* **46,** 1425.
Triska, A., Dennison, D., and Fritzsche, H. (1975). *Bull Am. Phys. Soc.* **20,** 392.
Viktorovitch, P., Moddell, G., Blake, J., and Paul, W. (1981). *J. Appl. Phys.* **52,** 6203.
von Roedern, B., Ley, L., Cardona, M., and Smith, F. W. (1979). *Philos. Mag. B* **40,** 433.
von Roedern, B., Paul, D. K., Blake, J., Collins, R. W., Moddell, G., and Paul, W. (1982). *Phys. Rev. B* **25,** 7678.
Williams, R. H., Varma, R. R., Spear, W. E., and LeComber, P. G. (1979). *J. Phys. C* **12,** L209.
Wu, C. H., and Williams, R. (1983). *J. Appl. Phys.* **54,** 6721.
Zanzucchi, P. J., Wronski, C. R., and Carlson, D. E. (1977). *J. Appl. Phys.* **48,** 5227.

CHAPTER 3

Closed-Form Solution of *I* – *V* Characteristic for a-Si : H Solar Cells

G. A. Swartz

RCA/DAVID SARNOFF RESEARCH CENTER
PRINCETON, NEW JERSEY

List of Symbols

J_{sc}	Short-circuit current density
J_d	Diffusion current density through junctions
J_p	Photocurrent density
J_T	Total current density
J_0	Reverse saturation current density of junctions
V_{oc}	Open-circuit voltage
V	Voltage at cell terminal
FF	Fill factor of cell
FF(0)	Fill factor of cell with short-circuit current density extrapolated to zero
k	Boltzmann constant
T	Cell temperature
q	Electronic charge
β	Junction quality factor
R_c	Current-independent series resistance
R_s	Photoconductive resistance of *i* layer
V_s	Voltage drop across photoconductive resistance of *i* layer
ΔV	Slope of dV/dJ_T at V_{oc} versus $1/J_{sc}$
l	Thickness of *i* layer
μ	Charge carrier mobility
τ	Charge carrier lifetime
σ	*i*-layer conductivity
n	Electron carrier concentration

ISBN 0-12-752150-X

I. Introduction

The recent development of hydrogenated amorphous silicon solar cells with energy conversion efficiencies as high as 10% makes this photovoltaic device an important factor in the future large-scale commercialization of photovoltaics (Catalano *et al.,* 1982). To achieve improvements in efficiency and understand factors that may cause performance degradation, a mathematical model of the device, which can describe cell performance, is extremely useful. Several computer simulations (Swartz, 1982; Hack and Shur, 1982) and analytical models (Debney, 1978; Miyamoto *et al.,* 1981; Gutkowicz-Krusin, 1981; Gutkowicz-Krusin *et al.,* 1981; Dalal, 1980; Dalal and Alvares, 1981; Crandall, 1981, 1982; Reichman, 1981; Rothwarf, 1981; Faughnan and Crandall, 1984) have been discussed in the literature. The cell performance predicted by the computer simulations are in reasonable agreement with data taken on normal p^+-i-n^+ or n^+-i-p^+ cells. Two of the analytical models (Dalal, 1980; Dalal and Alvares, 1981; Rothwarf, 1981) utilize front and back junctions with an intervening i layer. The p^+-i-n^+ cell with an amorphous silicon carbide p^+ layer that forms a heterojunction at the p^+-i interface presents a structure whose performance is not adequately described by any of these models or computer simulations. This paper also describes a two-junction model with an intervening i layer that gives a closed-form solution for the $I-V$ characteristics, and is applicable to the normal p^+-i-n^+ structure and the corresponding heterojunction structure with a wide-band-gap p^+ layer. The model incorporates the mechanisms responsible for the high cell performance of the heterojunction structure and the performance degradation after periods of cell illumination.

II. Proposed Model

The proposed model for the amorphous silicon solar cell as shown in Fig. 1 consists of two junctions (one at the p^+-i interface and one at the n^+-i interface) interconnected by a photoconductive i layer. It is assumed that the length l of the i layer is much greater than the length of the interface regions and is a constant value. It is further assumed that the photoconductive i layer is resistive and that the photocurrent J_p is constant for the applied voltage $V < V_{oc}$. The conductivity of this i layer is directly proportional to the number of charge carriers in the i layer. This is basically a "black box" engineering approach. The light enters the cell through the p^+ contact layer and the sign convention is that the short-circuit current is positive.

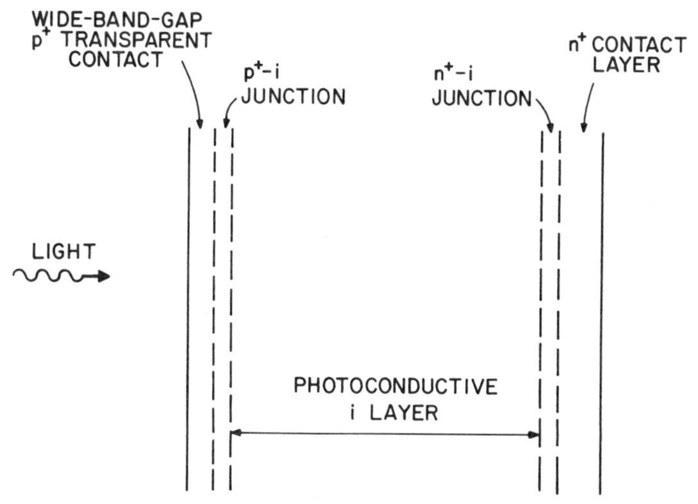

FIG. 1. Schematic drawing of proposed amorphous silicon cell model showing junctions at p^+-i and n^+-i interfaces with intervening photoconductive i layer.

1. CASE I

The total current density J_T through the cell is

$$J_T = J_p - J_d, \tag{1}$$

where J_d is the diffusion current density through the junctions for a given applied voltage across the junctions and J_p the photocurrent density. The diffusion current J_d is related to the voltage at the cell terminals V by

$$J_d = J_0(\exp[(q/\beta kT)(V + R_s J_T + R_c J_T)] - 1), \tag{2}$$

where R_s is the photoconductive resistance of the i layer and R_c any current-independent resistance in series with the cell, such as contact resistance.

Neglecting the -1 term and combining Eqs. (1) and (2) gives

$$V = -R_c J_T - R_s J_T + (\beta kT/q)\ln(J_d/J_0). \tag{3}$$

The energy band diagram for open-circuit condition, where $J_T = 0$, is assumed to have a configuration shown in Fig. 2a and for the short-circuit condition the configuration is shown in Fig. 2b. An equivalent circuit for this model is shown in Fig. 3.

The value of R_s is expressed as

$$R_s = V_s/|J_d| = V_s/|J_p - J_T|. \tag{4}$$

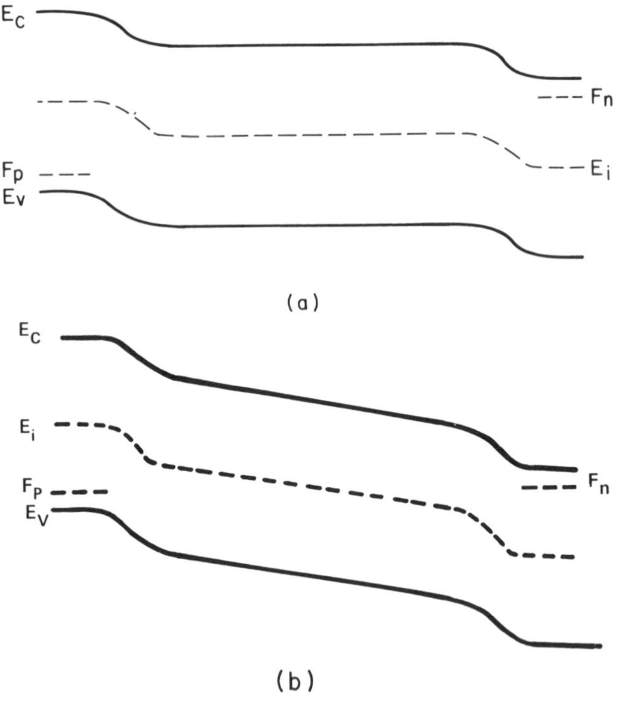

FIG. 2. Band diagram of two-junction model with photoconductive i layer at (a) open circuit and (b) short circuit.

At $V = V_{oc}$, where $J_T = 0$,

$$R_s = V_s/J_p, \tag{4a}$$

and at $J_T \approx J_p$, where almost all the charge carriers are extracted from the i layer, $R_s \to \infty$.

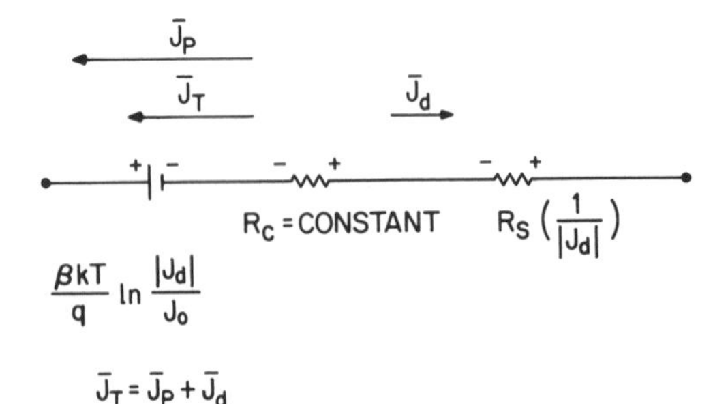

FIG. 3. Equivalent circuit of amorphous silicon solar cell showing voltage source, photoconductive resistance, and fixed series resistance.

Combining Eqs. (3) and (4)

$$V \simeq -R_c J_T - \frac{V_s}{J_p/J_T - 1} + \frac{\beta kT}{q} \ln\left(\frac{J_p - J_T}{J_0}\right), \tag{5}$$

or

$$V = -R_c J_T - \frac{V_s}{J_p/J_T - 1} - \frac{\beta kT}{q} \ln\left(1 - \frac{J_T}{J_p}\right)^{-1} + V_{oc}. \tag{6}$$

To obtain a meaningful comparison of Eq. (6) with experimental data and to measure V_s, consider the slope dV/dJ_T at $V = V_{oc}$ and $J_T = 0$, which is

$$\left.\frac{dV}{dJ_T}\right|_{V=V_{oc}} = R_c + \frac{V_s}{J_p} + \frac{\beta kT}{J_p q}. \tag{7}$$

Measurement of V_{oc} as a function of the log of J_{sc} gives a value for β, and a plot of dV/dJ_T at $V = V_{oc}$ as a function of J_p^{-1} will give a value for V_s. However, experimentally it is more convenient to measure J_{sc} and derive V_s from a plot of dV/dJ_T at $V = V_{oc}$ versus J_{sc}^{-1}, such as shown in Fig. 4 for a test cell.

The relationship in Fig. 4 is linear, with the intercept on the y-axis equal to

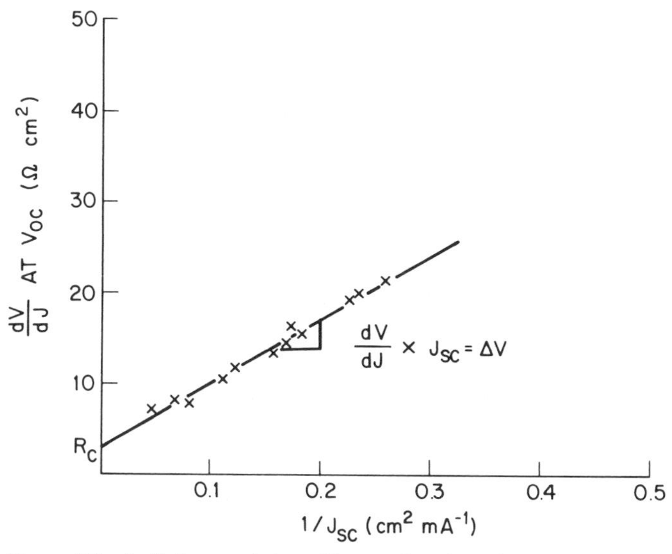

FIG. 4. Slope of the $I-V$ characteristic at V_{oc} as a function of the reciprocal short-circuit current. The relationship is linear with the slope defined as ΔV and the intercept on the y-axis is a current-independent resistance in series with the two junctions, y-intercept $= 2.9\ \Omega\ \text{cm}^2$, slope $= 71$ mV.

R_c and the slope is (Swartz, 1982)

$$\left[\left(\frac{dV}{dJ_T}\right)_{V=V_{oc}} - R_c\right] J_{sc} = \Delta V. \tag{8}$$

The quantity ΔV is related to V_s by

$$\Delta V = \left(V_s + \frac{\beta kT}{q}\right)\frac{J_{sc}}{J_p} \tag{9}$$

and V_s' is defined by

$$V_s' = \Delta V - \frac{\beta kT}{q} = V_s\frac{J_{sc}}{J_p} - \frac{\beta kT}{q}\left(1 - \frac{J_{sc}}{J_p}\right). \tag{10}$$

To simplify further analysis it is assumed that $R_c = 0$. A geometric representation of V_s, V_s', ΔV, and dV/dJ_T at $V = V_{oc}$ in relation to the I–V characteristic is shown in Fig. 5 for the simplified case of $R_c = 0$. The value for J_{sc}/J_p is determined as a function of V_s by setting the terminal voltage in Eq. (6) equal to zero, and, with the values of β and V_{oc} experimentally derived, solving the transcendental equation. With application of the results to Eq. (10), V_s' is plotted as a function of V_s as shown in curve A, Fig. 6, where $\beta = 1.56$ and $V_{oc} = 850$ mV.

For the same parameters used in curve A, Fig. 6, Eq. (6) is solved to determine the I–V characteristic for a range of V_s values. The FF values derived from the I–V curves are plotted as a function of V_s', as shown in curve A, Fig. 7. To compare measured data with the theory in Fig. 7, the fill factor of each cell is measured at several light intensities and extrapolated to

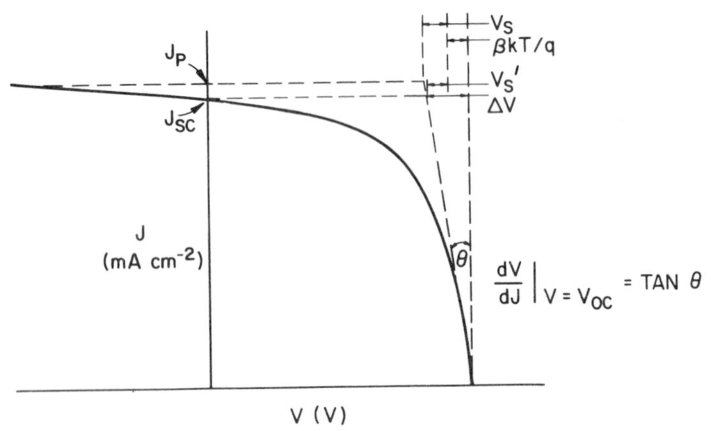

FIG. 5. Geometric representation of the parameters, V_s, V_s', ΔV, and dV/dJ_T at $V = V_{oc}$ with respect to an I–V characteristic for the simplified case of $R_c = 0$.

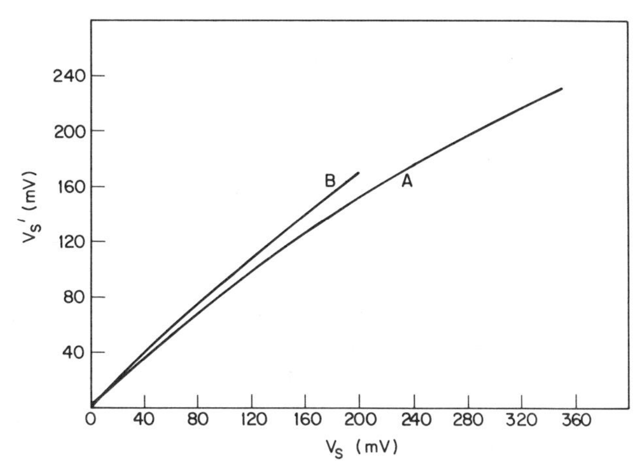

FIG. 6. The parameter V_s' as a function of V_s for the case of electron back-diffusion into the p^+ layer (curve A) and the case of an electron mirror at the p^+-i interface (curve B). $V_{oc} = 850$ mV, $\beta = 1.56$, $R_c = 0$.

zero intensity to eliminate the fill factor reduction caused by power lost in R_c, which in the calculation has been given a zero value. This fill factor at zero light intensity, FF(0), is plotted as a function of V_s' in Fig. 7 for a series of cells that have a wide-band-gap carbon-doped p^+ layer. With the light entering through the p^+ layer, the cell performance of a portion of this series was degraded by illumination with simulated AM1 sunlight for periods

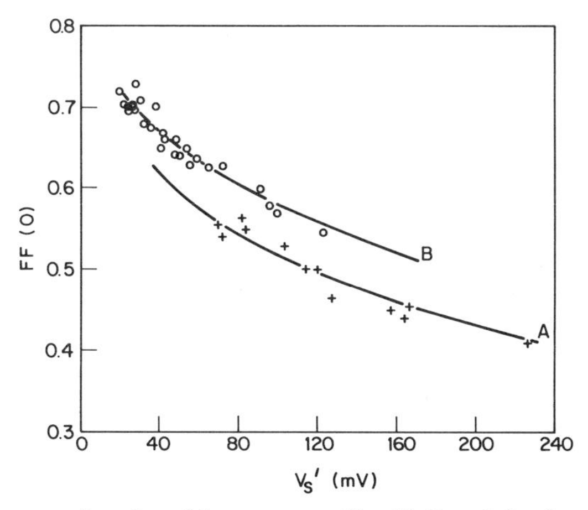

FIG. 7. Fill factor as a function of the parameter V_s', with $R_c = 0$, for electron back-diffusion into p^+ layer (curve A) and electron mirror at p^+-i interface (curve B). Data points are for measured fill factor extrapolated to zero illumination, FF(0), as a function of V_s' for undegraded cells (circles) and cells degraded by extended illumination (crosses).

greater than 48 hr. The theoretical curve A derived from Eq. (6) agrees with the FF(0) measurements on the light-degraded cells. However, the FF(0) measurements on cells, which were taken before the light degradation or on cells that had never been degraded by illumination, are much greater than predicted by Eq. (6) for given values of V_s'. The measured "shunt" resistance dV/dI at $V = 0$ for these undegraded cells is twice the value predicted by Eq. (6) for a given V_s'.

2. Case II

The quantum efficiency at wavelength of 400–500 nm on the a-Si cells with a wide-band-gap p^+ layer is typically 75–80% for the best-efficiency cells (Catalano *et al.*, 1982). This is in contrast to quantum efficiencies of 40–50% for cells without the wide-band-gap p^+ layer (Swartz, 1982) or for cells that were illuminated with AM1 light for an extended time (B. Faughnan, private communication). It was shown that the low quantum efficiency at blue–violet wavelengths may be the result of back-diffusion of minority carriers into the top contact layer; that is, electrons diffuse back into the p^+ layer or holes diffuse back into the n^+ layer (Gutkowicz-Krusin, 1981; Gutkowicz-Krusin *et al.*, 1981). It is assumed that a lack of back-diffusion of the electrons into the wide-gap p^+ layer results in the high quantum efficiency measured on the a-Si cells with the carbon-doped p^+ layer. Photoemission studies (Evangelisti *et al.*, 1984) on a-Si_xC_{1-x}:H/a-Si:H heterojunctions support this assumption by indicating that the potential energy of the valence band edge across the heterojunction is not changed. This gives rise to a large conduction band discontinuity at the heterojunction, which prevents electrons from diffusing into the p^+ layer.

To incorporate the reflections of electrons from the p^+–i interface into the theory, consider the effect of the electron mirror on the photoconductivity of the i layer. It is assumed that the electrons are totally reflected back into the i layer when the internal electric fields are highest at $J_d \to 0$. The electrons are not reflected and back-diffusion exists when the internal fields are low at $J_T = 0$ and $J_d = J_p$. With the assumption that the electrons carry most of the current and that $l_d \geq l$, where l_d is the electron diffusion length and l is the length of the i layer, the resistance of the i layer is

$$R_s = \int_0^l \frac{dx}{\sigma}, \tag{11}$$

where

$$\sigma = nq\mu_n, \tag{12}$$

n is the electron density, and μ_n is the electron mobility. With an AM1 illumination most of the carriers entering the cell through the p^+ layer are generated within 0.1 μm of the p^+-i interface with half the carriers diffusing toward the n^+ contact and half toward the p^+ contact. Without electron reflection at the p^+ interface (Sze, 1981)

$$nq = J_\mathrm{d}\tau_n/l. \tag{13}$$

For the case of electron reflection at the p^+-i interface

$$nq = 2J_\mathrm{d}\tau_n/l. \tag{14}$$

For the case where $l_\mathrm{d} < l$, the holes and electrons diffuse into the i layer for a distance l_d and the electrons are drifted throughout the remaining i layer thickness by the internal field. Thus the electrons are distributed over the distance l and the electron density is still given by Eq. (14) for electron reflection at the p^+-i interface and Eq. (13) without reflection. Thus with back-diffusion at $J_\mathrm{T} = 0$

$$R_\mathrm{s} = l^2/J_\mathrm{d}\mu\tau = V_\mathrm{s}/J_\mathrm{d} \tag{15}$$

and with an effective mirror as $J_\mathrm{d} \rightarrow 0$

$$R_\mathrm{s} = \frac{l^2}{2J_\mathrm{d}\mu\tau} = \frac{V_\mathrm{s}}{J_\mathrm{d}}. \tag{16}$$

As J_T varies from 0 to J_p, V_s is reduced to one-half its value at V_oc. The exact relationship between V_s and J_T is not known. As an approximation the reduction in V_s is given a square root dependence on the ratio $J_\mathrm{T}/J_\mathrm{p}$, which was found to be a good fit to the data. Thus Eq. (6) becomes

$$V = -R_\mathrm{c}J_\mathrm{T} - \frac{V_\mathrm{s}(1 - 0.5\sqrt{J_\mathrm{T}/J_\mathrm{p}})}{J_\mathrm{p}/J_\mathrm{T} - 1} - \frac{\beta kT}{q}\ln\left(1 - \frac{J_\mathrm{T}}{J_\mathrm{p}}\right)^{-1} + V_\mathrm{oc}. \tag{17}$$

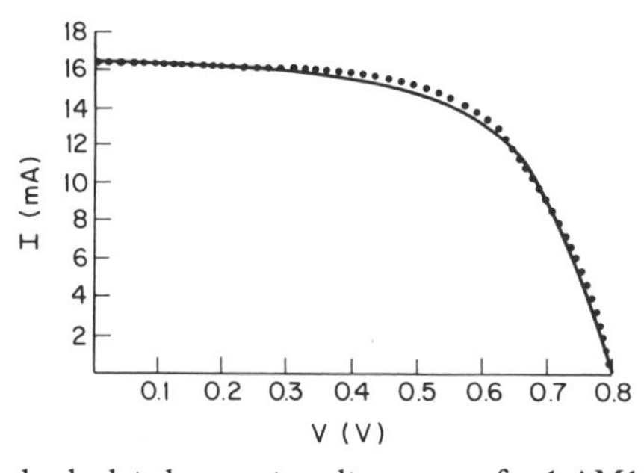

FIG. 8. Measured and calculated current–voltage curve for 1-AM1 illumination. Parameters, which are derived from cell data and used in calculation, are $V_\mathrm{oc} = 800$ mV, $J_\mathrm{sc} = 15.1$ mA cm^{-2}, $V_\mathrm{s} = 44$ mV, $\beta = 1.56$, and $R_\mathrm{c} = 4.44$ Ω cm^2. Dotted curve, measurement; solid curve, theory.

The calculated $I-V$ curve from Eq. (17) and the measured $I-V$ curve of a particular cell are in excellent agreement, as shown in Fig. 8. The values of R_c, J_{sc}, β, V_{oc}, and V_s' used in the calculation are taken from the measured cell data. The combination of Eqs. (17) and (10) with the input parameter values of $R_c = 0$, $\beta = 1.56$, and $V_{oc} = 850$ mV gives the fill factor as a function of V_s', as shown in curve B, Fig. 7, which is in very good agreement with the FF(0) versus V_s' data taken from the series of undegraded cells with a wide-band-gap p^+ layer. A plot of V_s versus V_s' for the electron mirror model described by Eq. (17) is shown in curve B, Fig. 6.

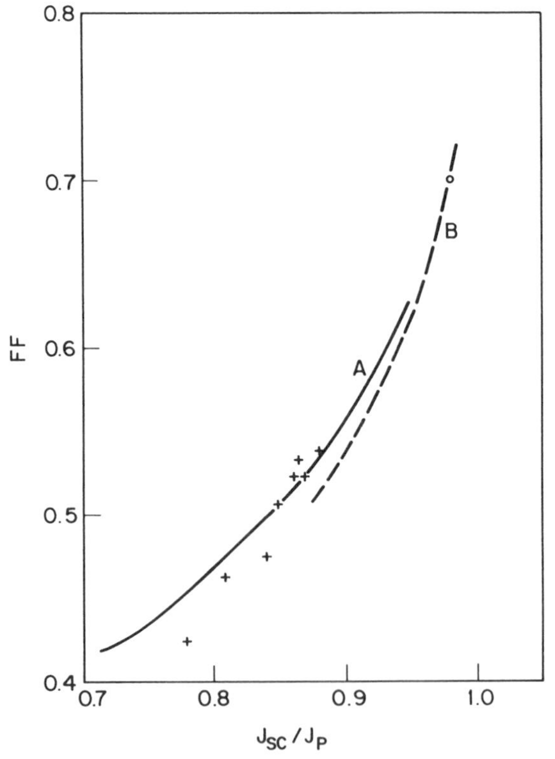

FIG. 9. Fill factor as a function of J_{sc}/J_p for electron back-diffusion into p^+ layer (curve A) and electron mirror at p^+-i interface (curve B). $R_c = 0$, $\beta = 1.56$, and $V_{oc} = 850$ mV. Data points are for a single cell in which J_p was determined from curve B with the values of FF and J_{sc} measured on the cell before the cell performance was degraded by forward biasing. The range of FF and J_{sc} values were obtained by annealing the degraded cell at 80°C for several hundred hours.

III. Interrelation of η, FF, and J_{sc}/J_p

The $I-V$ characteristic described in Eq. (17), which includes the electron mirror effect, and that described in Eq. (6), which does not, can be applied to a single cell by considering the cell before and after extended illumination. For both equations a relationship between the fill factor and the short-circuit current can be established by keeping all parameters except V_s constant. A plot FF verses J_{sc}/J_p for $R_c = 0$, $\beta = 1.56$, and $V_{oc} = 850$ mV is shown in Fig. 9, curve B, for the electron mirror present and curve A for back-diffusion included. A single undegraded cell with a fill factor of 0.7 is plotted on curve A to give a value for J_{sc}/J_p at AM1. Here J_{sc} had been measured and thus J_p is determined. The cell is degraded by forward biasing for 24 hr at 20 times the short-circuit current. The $I-V$ characteristic of the cell is measured and then the cell is annealed at 80°C for several hundred hours with periodic $I-V$ measurements made at room temperature during the course of the anneal. With the value of J_p at AM1 illumination previously deter-

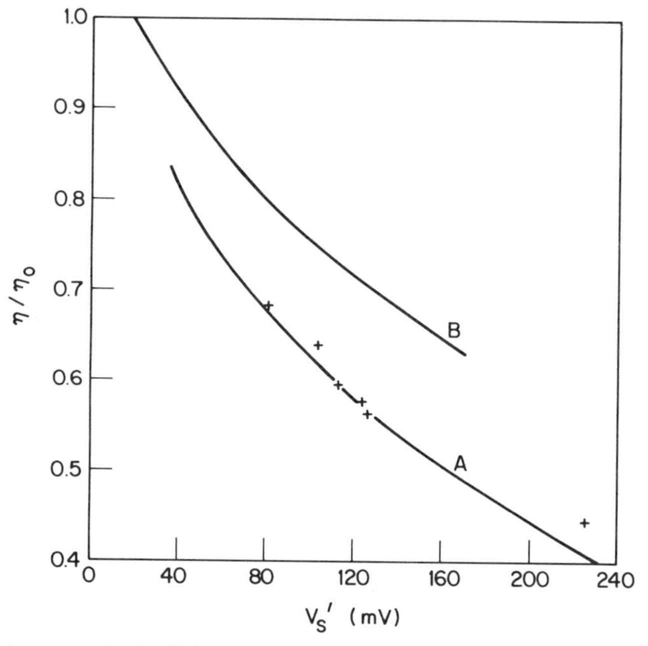

FIG. 10. Normalized cell efficiency as a function of V_s' for electron back-diffusion into p^+ layer (curve A) and electron mirror at p^+-i interface (curve B). The value of V_s' for $\eta/\eta_0 = 1$ is arbitrary. $R_c = 0$, $\beta = 1.56$, and $V_{oc} = 850$ mV. Data points are for a series of cells in which η_0 was determined from curve B with the values of η and V_s' measured on the cells before the cells' performances were degraded by illumination.

mined, the FF is plotted against the value of J_{sc}/J_p for each $I-V$ measurement. The data are in agreement with curve A in Fig. 9.

Since a relationship between fill factor and J_{sc} is established for a constant V_{oc}, the normalized efficiency as a function of V_s, the only parameter that is varied, is also determined. A plot of the normalized efficiency, η/η_0 as a function of V_s', is shown in Fig. 10 for the electron mirror included (curve B) and for the back-diffusion case (curve A). The ratio η/η_0 is arbitrarily fixed at unity for $V_s' = 19$ mV. A series of undegraded cells with carbon-doped p^+ layers in which V_s' and η had been measured were plotted on curve B in Fig. 10 to determine η/η_0 and therefore η_0. The cells were then degraded by illumination under 1 AM1 for periods greater than 48 hr. The values of V_s' and η were remeasured and plotted in Fig. 10. The data follows curve A very closely, which gives support to the theoretical relationship between FF and J_{sc} and indicates the destruction of the electron mirror effect by illumination of the cell with AM1 light for periods as short as 48 hr. For those cells whose data are represented in Fig. 10, the performance values, before and after light soaking, are given in Table I. The data in Fig. 7 also show the destruction of the electron mirror effect by extended cell illumination. This photoinduced reduction of electron reflection at the junction interface is in agreement with Okamoto *et al.* (1983).

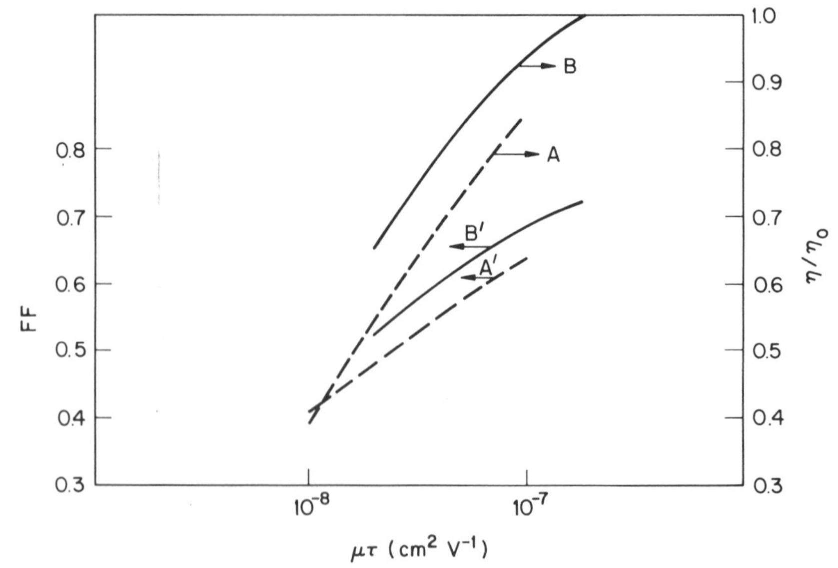

FIG. 11. Normalized efficiency and fill factor as a function of the electron $\mu\tau$ product for electron back-diffusion into p^+ layer (curve A and A′) and electron mirror at p^+-i interface (curve B and B′). $R_c = 0$, $\beta = 1.56$, $V_{oc} = 850$ mV, and i-layer thickness $= 0.6$ μm.

TABLE I

MEASURED CELL PERFORMANCE PARAMETERS BEFORE AND AFTER PERFORMANCE DEGRADATION BY EXTENDED ILLUMINATION

Cell number	Illumination under AM1 time (hr)	At AM1 J_{sc} (mV cm^{-2})	At AM1 V_{oc} (V)	At AM1 η (%)	FF(0)	ΔV (mV)	V'_s (mV)	β
W20907-25-04	0	12.1	0.80	6.85	0.729	68	27	1.58
W20907-25-04	48	10.8	0.79	4.0	0.464	179	127	1.94
C09232-2-04-02	0	14.3	0.759	6.7	0.64	91	50	1.54
C09232-2-04-02	48	11.2	0.763	3.5	0.41	267	226	1.59
C09272-4-24-04	0	14.6	0.80	8.16	0.70	66	23	1.66
C09272-4-24-04	48	13.6	0.78	4.95	0.50	160	114	1.79
C10152-3-02-04	0	13.9	0.79	7.55	0.70	71	26.5	1.71
C10152-3-02-04	72	12.2	0.782	4.5	0.468	175	124	1.95
C08162-2-02-04	0	12.5	0.821	6.3	0.676	81	36	1.74
C08162-2-02-04	48	11.6	0.80	4.6	0.564	122	81	1.56
A07302-1-24-04	0	16.3	0.807	7.93	0.70	79	38	1.56
A07302-1-24-04	48	12.9	0.78	5.5	0.53	160	104	2.15

By referring back to the $I-V$ relationship in Eqs. (6) and (17) and R_s expressed in terms of $\mu\tau$ in Eq. (14), the fill factor and normalized efficiency, as shown in Fig. 11, are determined as a function of the electron $\mu\tau$ product. These relationships shown in Fig. 11 could be tested by utilizing recent work by Faughnan, Moore, and Crandall in which the electron collection length in the cell's i layer at $J_T = J_{sc}$ are determined from quantum efficiency measurements at various bias potentials applied to the cell (Faughnan *et al.*, 1984). The collection length at $V = 0$ is a product of $\mu\tau$ times the internal electric field and the internal field may be determined by the theory from the potential drop across R_s at $J_T = J_{sc}$. Fill factor and efficiency data as a function of the $\mu\tau$ product extracted from the electron collection length before and after extended cell illumination can be used to test this proposed model.

IV. Summary

A model of the amorphous silicon solar cell that includes two junctions interconnected by a photoconductive i layer is used to formulate a closed-form solution for the current–voltage characteristic. The model is extended to include a top contact layer with a wide energy gap that serves to reduce the back diffusion of electron carriers into the contact layer. Reflection of those carriers into the i layer reduces the resistivity of the i layer and thereby increases the output current and fill factor. The fill factor derived from the closed-form solution is related, with and without electron back-diffusion, to the voltage parameter V_s', which is defined in Eqs. (4) and (10). With back-diffusion, the FF relationship to V_s' agrees with the measurement of FF and V_s' on cells that were illuminated for extended time periods. The relationship, without back-diffusion, agrees with the data from cells before they were subjected to extended illumination. The model predicts a relationship between the short-circuit current and the fill factor and therefore, with a constant V_{oc}, the dependence of a normalized efficiency on the parameter V_s or V_s' is established. By illuminating a number of cells and thereby changing the measured value of V_s' and eliminating the electron mirror effect at the p^+-i interface, the predicted dependences of η on V_s' and J_{sc} on FF are shown to be in agreement with measured data. As an addendum to the model the parameter V_s is related to the electron $\mu\tau$ product so that fill factor and normalized efficiency are also related to $\mu\tau$. Future work, which may involve collection length and diffusion length measurements of a-Si:H cells, is required to test this added portion of the model.

ACKNOWLEDGMENTS

I wish to thank J. Dresner and A. Catalano for supplying the deposited amorphous silicon and R. D'Aiello, who oversaw the fabrication of the cells whose performance data were given in this paper. I wish to thank R. Crandall for some helpful discussions. I also wish to thank G. Nostrand for writing the computer software used in the course of the measurements and H. Gervasoni for making the cell measurements.

REFERENCES

Catalano, A., D'Aiello, R. V., Dresner, J., Faughnan, B., Firester, A., Kane, J., Schade, H., Smith, Z. E., Swartz, G., and Triano, A. (1982). *Conf. Rec. IEEE Photovoltaic Spec. Conf.* **16,** 1421.

Crandall, R. S. (1981). *RCA Rev.* **42,** 449.

Crandall, R. S. (1982). *J. Appl. Phys.* **53,** 3350.

Dalal, V. (1980). *Sol. Cells* **2,** 261.

Dalal, V., and Alvares, F. (1981). *J. Phys. Colloq. Orsay, Fr.* **4,** C4-491.

Debney, B. T. (1978). *Solid State Electron. Devices, Spec. Issue* S-15.

Evangelisti, P., Forini, P., Giovannella, C., Patella, F., Perfetti, P., Quaresima, C., and Capozi, M. (1984). *Appl. Phys. Lett.* **44,** 764.

Faughnan, B. W., and Crandall, R. S. (1984). *Appl. Phys. Lett.* **44,** 537.

Faughnan, B. W., Moore, A. R., and Crandall, R. S. (1984). *Appl. Phys. Lett.* **44,** 613.

Gutkowicz-Krusin, D. (1981). *J. Appl. Phys.* **52,** 5370.

Gutkowicz-Krusin, D., Wronski, C. R., and Tiedje, T. (1981). *Appl. Phys. Lett.* **38,** 87.

Hack, M., and Shur, M. (1982). *Conf. Rec. IEEE Photovoltaic Spec. Conf.* **16,** 1429.

Miyamoto, H., Konagai, M., and Takahashi, K. (1981). *Jpn. J. Appl. Phys.* **20,** 1691.

Okamoto, H., Kida, H., Nonomura, S., Fukumoto, K., and Hamakawa, Y. (1983). *J. Appl. Phys.* **54,** 3236.

Reichman, J. (1981). *Appl. Phys. Lett.* **38,** 251.

Rothwarf, A. (1981). *Conf. Rec. IEEE Photovoltaic Spec. Conf.* **15,** 690.

Swartz, G. (1982). *J. Appl. Phys.* **53,** 712.

Sze, S. M. (1981). "Physics of Semiconductor Devices," Chapter 2. Wiley, New York.

CHAPTER 4

Electrophotography

Isamu Shimizu

IMAGING SCIENCE AND ENGINEERING LABORATORY
TOKYO INSTITUTE OF TECHNOLOGY
YOKOHAMA, JAPAN

I. Introduction

Several years have passed since the use of a-Si:H as a photoreceptor for electrophotography was first proposed. Following the dramatic success in the large-scale fabrication of a-Si:H solar cells, from the industrial point of view, remarkable progress has been made in a-Si:H drums for copying machines or for laser line printers.

After a brief introduction about processing in electrophotography, some advantages expected from the a-Si:H photoreceptor will be discussed in comparison with other competitive materials. To satisfy the requirements of "a charge depletion device" during image processing, attempts have been made at designing both the device structure and the material for this purpose. Consequently, an acceptable field of 40–50 V μm^{-1} has been obtained in a-Si:H photoreceptor.

The photoinduced discharge (PID) will be discussed within the framework of evaluating the photoresponse of a-Si:H photoreceptor. The results of the study for the fast PID behavior lead us to conclude that the surface of a-Si:H has significant influence either on the photoresponse or on the image quality. A typical device structure will be proposed for practical use.

Finally, some problems of a-Si:H photoreceptor will be briefly discussed from the manufacturing point of view.

ISBN 0-12-752150-X

II. Advantages Expected from a-Si:H as Photoreceptor of Electrophotography

It is well known that electrophotography, a simple technique for producing hard copies, has been used in copying machines, laser line printers, etc. Furthermore, electrophotography has been put to practical use for nondestructive tests and for medical diagnoses. In electrophotography, the formation of images depends on a physicial effect, the electrostatic force, while chemical reactions are utilized in silver halide photography. Accordingly, electrophotography prevails in the field of office automation as a technique for making images with easy control of treatment. Its simplicity and speed are suitable for copying machines.

The fundamental processes of duplicating images by a copying machine are schematically illustrated in Fig. 1. First of all, electric charges are deposited homogeneously by a corona discharge on the surface of the photoreceptor made of a photoconductive film deposited on a conducting substrate [sensitization (1)]. The surface charges are partly neutralized by illuminating the photoconductor with the light from an original subject [imagewise exposure (2)], and consequently an electrostatic latent image is formed on the surface of the photoreceptor. A visual image is reproduced from the latent image by pouring charged pigment particles (toner) on the photoreceptor [development (3)]; thus the image is formed by attachment and detachment of toner particles by coulombic force. The toner image is transferred to a plain paper and fixed by fusing [transfer and fixation (4)]. The photoreceptor is ready for the next copying after eliminating the electrostatic image by illumination and brushing to remove residual toner [cleaning (5)]. These procedures are repeated in making each copy at the velocity of 12–15 copies min^{-1} or 120 copies min^{-1} in a slow or a fast copying machine, respectively.

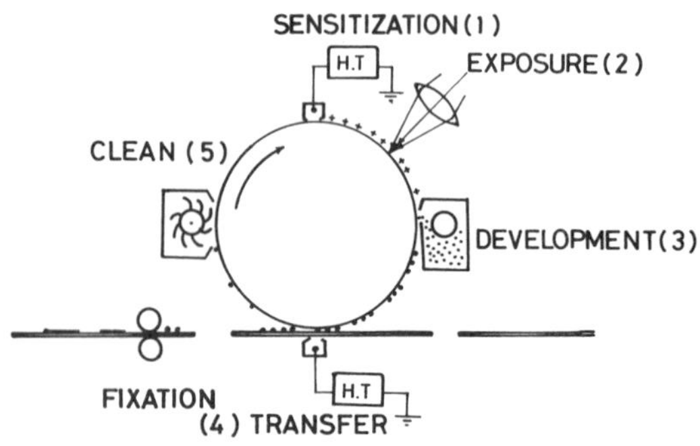

FIG. 1. Duplication steps in electrophotography.

The desirable characteristics of photoconductive films for use in electrophotography are as follows:

(1) High charge acceptance of 10^{11}–10^{12} ions cm^{-2} and a decay time longer than the processing time (t_p sec).
(2) High surface potential (>200 V) sufficient for development with toner.
(3) Excellent photoresponse in forming the electrostatic latent image. The response spectrum must be well matched to the light source.
(4) Durability against heat and chemicals.

Mechanical strength is also an important factor.

Accordingly, photoconductive films for the photoreceptor are believed to need an electrical conductivity (σ_R) lower than $10^{-13}\ \Omega^{-1}\ cm^{-1}$ because the dielectric relation time (t_r sec) given by the equation $t_r = \varepsilon(\varepsilon_0)/\sigma_R$ must be longer than the t_p. Here, $\varepsilon(\varepsilon_0)$ is the permittivity.

With respect to the photoelectric properties, optical absorption $\alpha(\lambda)$ (per centimeter), carrier generation efficiency η, drift mobility μ_d ($cm^2\ V^{-1}\ sec^{-1}$) and lifetime τ sec are important parameters to evaluate the photoconductive films. The a-Si:H is characterized by a rather high drift mobility (Tiedje *et al.*, 1981) and short lifetime (trapping lifetime). Amorphous selenium has similar characteristics (Abkowitz, 1979). In these films, furthermore, conduction by either electrons or holes takes place in the extended states, involving multiple trapping, which is different from other materials used as photoreceptors, i.e., a-As_2Se_3 (Pflister and Scher, 1975) or organic films (Mort and Pfister, 1982), in which hopping transport is dominant. In the latter materials a high photoconductivity gain is obtained due to their long carrier lifetime. In a-Si:H, on the other hand, high drift mobility plays the major role in providing a high carrier range. The mobility-type will be more favorable than the lifetime-type materials with respect to the quick photoresponse that is required of the photoreceptor for copying machines or laser line printers operating at high velocity.

The a-Si:H has some fascinating properties in addition to its excellent photoelectric nature as follows:

(1) high mechanical strength arising from its tetrahedral bond structure as illustrated in Table I, which shows its Vickers strength compared to that of other materials.
(2) heat resistance which releases machines from some design restrictions.

In electrophotography, the photoreceptor must have a wide area equivalent to the size of the final copy because the image on the photoreceptor is directly transferred to the paper. The rf glow-discharge technique has been

TABLE I

MECHANICAL STRENGTH (VICKERS STRENGTH)

Materials	Vickers strength ($kg\ mm^{-2}$)
a-Si:H	1500–2000
$a\text{-}As_2Se_3$	150–200
a-Se	30–50
a-Se(Te)	50–80

established as a suitable method for preparing homogeneous large-area a-Si:H films. So far, we have attained desired results for a-Si:H photoreceptor within the framework of a fundamental study.

III. Material Design for a-Si:H Photoreceptor

As described in the previous chapter, a-Si:H prepared under the optimum condition for minimizing its defects has electrical conductivity in the range of 10^{-8}–10^{-9} Ω^{-1} cm^{-1} at room temperature. A small amount of boron doping ($B_2H_6/SiH_4 \sim 10$ ppm) shifts the Fermi level toward the middle of the gap and results in the minimum $\sigma_R \sim 10^{-11}$ Ω^{-1} cm^{-1}. The dielectric relaxation time t_r less than 0.1 sec deduced from this σ_R value had first led us to a pessimistic conclusion about the potential of a-Si:H photoreceptors. In fact, neither positive nor negative charges, when deposited by a corona discharge, could be retained sufficiently long on the surface of a-Si:H deposited directly on a metal-coated substrate. On the other hand, we obtained a fairly high surface potential (about 200 V) in both polarities on a-Si:H (~ 5 μm thick) grown on a conductive substrate coated with a thin SiO_x film about 100 nm thick (Shimizu *et al.*, 1980). Moreover, the decay rate was sufficiently long (>20 sec), which was longer by more than two orders of magnitude than the t_r estimated from the σ_R value. This discovery led us to an idea that charges emitted at the interface with the electrodes played a major role in dissipating the surface charges on a-Si:H in the dark by thermal emission. The thermal emission rate was expected to be a determining factor in a device with blocking contacts. This encouraged us to develop a-Si:H photoreceptors (Shimizu *et al.*, 1981a).

The photoreceptor of electrophotography is a "charge depletion device" in which replenishing carriers from the electrodes must be completely prohibited. Since the a-Si:H is a typical ambipolar material in which both carriers, i.e., electrons and holes, are allowed to move, great care must be paid to the contact with either the conductive electrode or the free surface termed the "corona electrode." In the early patents on a-Si : H photoreceptors proposed independently by Siemens (Kempter, 1980) and Canon (Hirai *et al.*, 1981), we can find a description of the requirement of a blocking layer. A *p–n* junction and some insulating films, i.e., SiO_x or Al_2O_3, were adopted as the blocking layers. These proposals are suggestive but unsatisfactory for practical purposes.

The potential barriers built up at the contacts, i.e., a *p–n* or a Schottky-type junction, are not sufficient for sustaining the surface charges in the range of 10^{11}–10^{12} ions cm^{-2}. Blocking by insulating films, on the other hand, tends to collect charges at the contact, which often becomes a source of instability of the photoreceptor during repeated duplication. The blocking layer must satisfy another requirement, i.e., that all carriers in the bulk be allowed to flow over the blocking layer. Two types, a homophase and a heterophase blocking layer, were proposed for prohibiting carrier injection at the contact of an a-Si:H photoreceptor with a conducting electrode (Shimizu *et al.*, 1981b). In the first type (homophase), *n*- or *p*-type a-Si:H doped with foreign elements, i.e., phosphorus or boron, is employed, whereas a very thin insulating film (a-SiN_x:H, a-SiO_x; 15–30 nm thick) is used in the latter case.

The blocking behaviors are schematically illustrated in Fig. 2. The blocking performance in the homophase type is interpreted in terms of the short schubweg ($\mu\tau E$) of the minority carrier in the doped layer. The evidence that the blocking ability depends greatly on either the doping level or the thickness supports this explanation (Oda *et al.*, 1981a). The photoinduced carriers in the bulk pass easily through the doped layer because the current flow is in the forward direction.

The metal–insulator—semiconductor (MIS) structure is employed in the heterophase blocking where the thickness of the insulator used is the key factor in satisfying the blocking requirement. Differently from the solar cell with MIS structure, in which an insulating film tens of angstroms thick is used to avoid the back-diffusion of photoinduced carriers (Wronski *et al.*, 1981), the photoreceptor of electrophotography has necessarily a rather thick insulating film to block carrier transport.

Now let us turn to the photoelectric properties of the photoconductive a-Si:H. Apart from devising a-Si:H photoreceptor with the film prepared under optimal conditions, extensive investigations have been made in preparing a-Si with a σ_R value lower than 10^{-13} Ω^{-1} cm^{-1} in order to have a sufficient charge retentivity. For this purpose, two essentially different approaches were considered: (1) preparing wider-gap materials to decrease the free-carrier density and (2) suppressing the drift mobility of carriers by increasing the number of traps. So far, no sufficient photoelectric property for the photoreceptor has been attained with single-layer wide-energy-gap

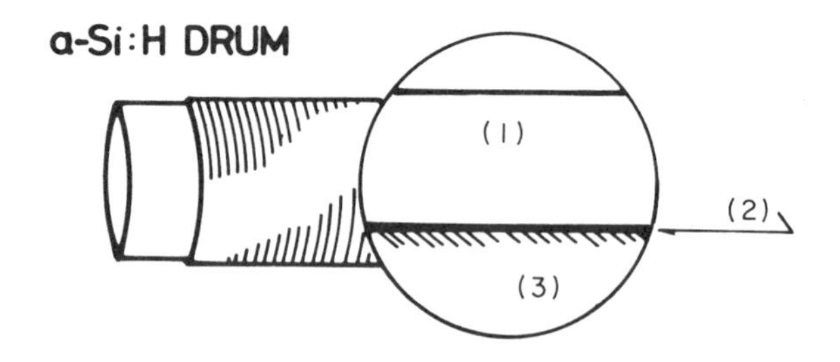

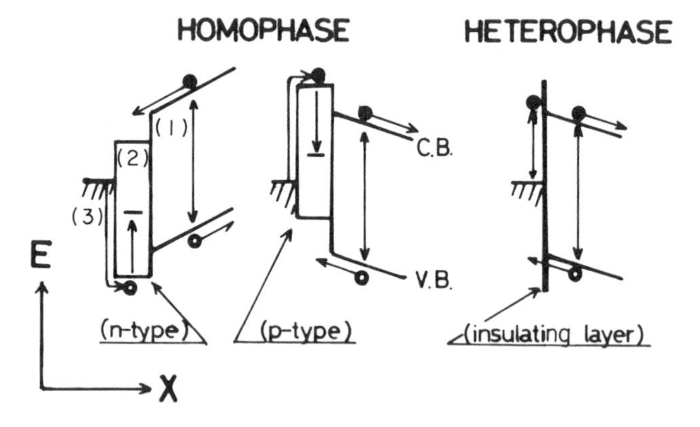

FIG. 2. Structure of a-Si:H photoreceptor and its energy diagram indicating the blocking behaviors. (1) Photoconductive layer, (2) blocking layer, (3) conducting substrate.

alloys, i.e., a-SiN_x:H (Kurata *et al.*, 1981) or a-SiC_x:H (Tawada *et al.*, 1981), in spite of intensive studies. The band-gap energy of a-Si:H increases with an increase in hydrogen content. Remarkable photoresponse was attained in an a-Si:H image pickup tube, in which a-Si:H with a gap energy of about 2.0 eV was prepared by rf sputtering in a gaseous mixture of Ar and H_2 as demonstrated by Imamura *et al.* (1979). The high-resistivity films containing an excess amount of hydrogen are also attainable by the glow-discharge technique when the films are grown at a high deposition rate. On the other hand, care must be taken to avoid degradation of photoconductivity, mechanical strength, and homogeneity. The a-Si:H film prepared from higher silanes (Si_nH_{2n+2}, $n = 2$ or 3) at fast growth rate (over 15 μm hr^{-1}) showed a low σ_R value ($\sim 10^{-12}$ Ω^{-1} cm^{-1}), maintaining high photoconductivity (Ogawa *et al.*, 1981) and appearing as a promising material from a manufacturing point of view.

Another approach, categorized as item 2, where high resistivity is attained by depressing the drift mobility, has been extensively explored by adding a small amount of neutral impurities, i.e., oxygen or nitrogen, into the a-Si:H. Yamamoto *et al.* (1981) have devoted extensive efforts to preparing the a-Si drums in the practical size used in copying machines by glow discharge of SiH_4 including small amounts of O_2 and B_2H_6. In Fig. 3, the doping feasibility is shown for a-Si:H including 0.01 wt. % oxygen in the film, as the change in the electrical conductivity in the dark or under illumination plotted as a function of the doping level, denoted as the volume ratio of phosphine (or diborane) to silane. A remarkable difference is seen with respect to doping feasibility between the oxygen-doped and the oxygen-free film (dashed line) (Wakita *et al.,* 1982). Accordingly, the doping efficiency of phosphorus and boron in controlling the Fermi level becomes fairly small in the presence of oxygen.

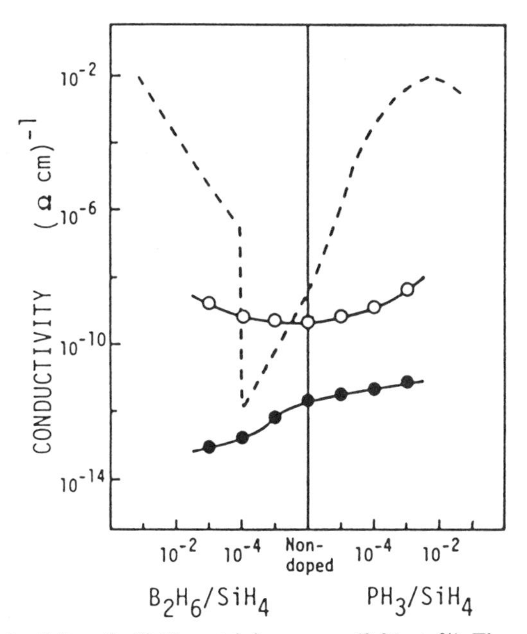

FIG. 3. Doping feasibility of a-Si:H containing oxygen (0.01 wt. %). The conductivity in the dark (●) and under illumination (○) (6328.4 nm, 200 μW cm^{-2}) are plotted as a function of gaseous ratio of dopants to silane. For comparison, the dashed line shows the dark conductivity for a-Si:H without oxygen. [From Wakita *et al.* (1982).]

Mixing both elements, i.e., nitrogen and diborane, into silane was reported by Watanabe *et al.* (1982) as another approach for preparing a high-resistivity film, in which the lowest σ_R value of $10^{-14}\ \Omega^{-1}\ cm^{-1}$ was attained. Different roles, namely, (a) expanding the energy gap and (b) shifting the Fermi level, are considered for the nitrogen incorporated in a-Si:H, depending on its coordination number. The fourfold coordinated nitrogen behaves as a donor in a-Si:H (b) and the threefold coordinated nitrogen contributes to imparting to the film a wide energy gap (a). In the high-resistivity film for the photoreceptor prepared from a gaseous mixture of SiH_4, N_2, B_2H_6, and H_2, the amount of nitrogen incorporated is not sufficient for expanding the energy gap. Consequently, we classified this film into approach 2. It had been established that the drift mobility of electrons was dramatically decreased by mixing oxygen into a-Si:H (Wakita *et al.*, 1982). This evidence supports the idea that the high resistivity resulting from doping with a small amount of foreign element is mainly attributable to the reduction of the carrier transport due to an increase in the trapping density. A minimum amount of doping is preferable for suppressing just the dark electrical conductivity of a-Si:H; otherwise one of the great advantages of a-Si:H, its excellent photoconductivity, would be lost.

We can find various types of photoreceptors in the patent literature (Tabei *et al.*, 1980; Nakagawa *et al.*, 1981) dealing with a-Si:H and organic materials; however, details are lacking.

IV. Photoinduced-Discharge (PID) Characteristics

After the corona sensitization, an imagelike pattern of illumination is projected to make an electrostatic latent image on the photoreceptor. The surface potential V_S (volt) decreases due to the neutralization of surface charges by the photoinduced carriers traveling through the film. Typical PID curves, where V_S is plotted as a function of time, are shown in Fig. 4a. Here "intrinsic" a-Si:H, 10 μm thick, deposited on a blocking layer (*n*-type a-Si:H, 0.2 μm) is used. Hereafter, the doping levels will be indicated in a shorthand form, P-10 or B-10, which denotes the volume ratio of the dopant gas against silane in ppm, such as PH_3/SiH_4, 10 (ppm) or B_2H_6/SiH_4, ~10 (ppm), respectively. The time derivatives (dV_S/dt) calculated from the PID curves are plotted as a function of V_S in log–log plots (see Fig. 4b), which relate to the photocurrent J_p ($A\ cm^{-2}$) versus voltage V_S (volt) for the photoreceptor because the current flux during the PID can be given by the equation $J_p = Cd|V_S|/dt$, assuming a constant capacitance C (coulomb per centimeter). The most distinctive feature in these PID curves is the small residual voltage V_R (volt) independent of the wavelength and intensity of illumination. We can see in the log J_p–log V_S curves that the photocurrent

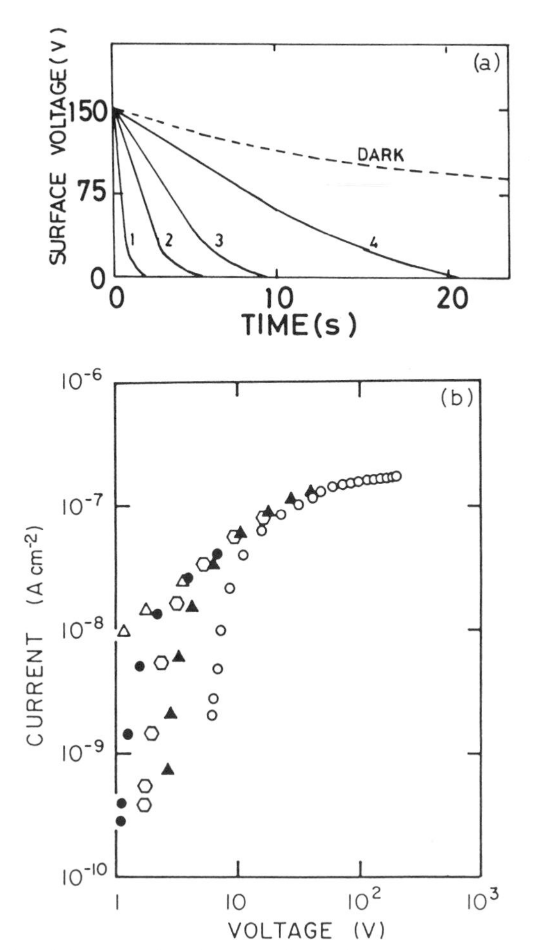

FIG. 4. Photoinduced-discharge (PID) curves of a-Si : H photoreceptor (multiple-layer type). (a) Surface voltage plotted as a function of exposure time for the following light intensities at 600 nm: (1) 1.6 μW cm^{-2}, (2) 0.32, (3) 0.16, and (4) 0.08. (b) The photocurrent (estimated from the time derivative of the PID curves) versus surface voltages for a-Si : H photoreceptor (a-Si : B, 10 ppm, 10 μm thick; exposure: 600 nm, 2.0×10^{12} photons cm^{-2} sec^{-1}). [From Oda *et al.* (1981b).]

saturates with increasing voltage. The saturated value is exactly proportional to the light flux impinging on the photoreceptor (F photons cm^{-2} sec^{-1}). The observed photoconductivity gain ($G = J_p/eF$) in the saturated region is about 0.8 over the whole visible spectrum (see Fig. 5). By taking into account surface reflectance, we can conclude that the maximum gain ($G \sim 1$) is attained in the a-Si : H photoreceptor. The rapid decrease in the gain beyond 670 nm is attributed to the drop in the optical absorption of

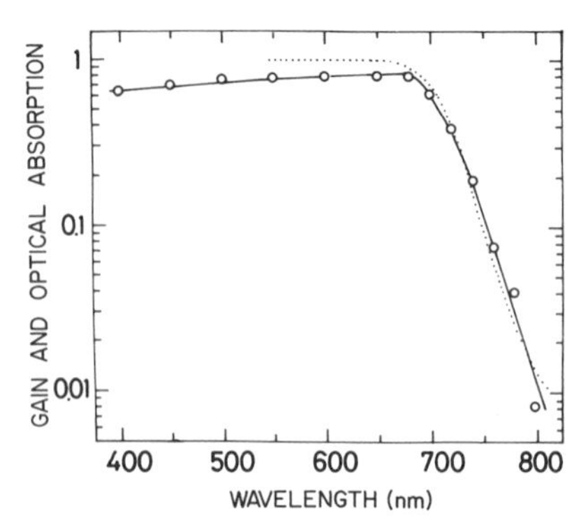

FIG. 5. Spectral sensitivity of a-Si:H photoreceptor. The dotted curve shows the optical absorption of a-Si:H film (10 μm thick).

a-Si:H. Turning again to the PID curves, the rate of the photoinduced discharge decreases gradually with a decrease in the surface voltage in spite of constant illumination. A range-limited current, given by Eq. (1) (Hecht's rule), is a plausible illustration for the photocurrent in this region,

$$G = \frac{Cd|V_S|/dt}{eF} \sim \eta \frac{\mu\tau E}{L} [1 - \exp(-L/\mu\tau E)]. \tag{1}$$

Here L is the thickness of the photoreceptor. In the $J_p - V_S$ curve for hole transport, when the photoreceptor is illuminated with strongly absorbed light after sensitizing the surface with a positive corona, the photocurrent decreases with decreasing V_S more sharply than deduced from Hecht's rule. Both the carrier generation efficiency η and the drift mobility μ_d (which depends on the electric field) can be ruled out as major causes of the rapid reduction of the discharge rate or low electric fields. The η values of unity had been established in a-Si:H, independent of the photon's energy $h\nu$ when an electric field higher than 10^4 V cm^{-1} is applied (Silver *et al.*, 1980), which is different from the carrier generation process, including geminate recombination as interpreted for, a-Se or other amorphous semiconductors (Pai and Enck, 1975). The modification of the effective electric field in the film by the space charge resulting from the PID is the most plausible explanation for the reduction in the PID rate. The $\mu\tau$ value of 10^{-7}–10^{-8} cm^2 V^{-1} was obtained for hole transport of a-Si:H (B-10) by fitting Eq. (1) to a set of PID curves with varying initial surface potential to eliminate the effect of the space charge (Oda *et al.*, 1981b). This $\mu\tau$ value for holes is smaller by two orders of magnitude than that for electrons. Accordingly, care must be taken in preparing the film to maintain this level ($> 10^{-8}$ cm^2 V^{-1}) at least for the hole transport if we expect an excellent photoelectric performance in the a-Si:H photoreceptor.

Now we discuss the PID performances of photoreceptors made of high-resistivity films. In Fig. 6, typical PID curves are illustrated for the a-Si:H drum made of a film 44 μm thick containing oxygen. A sufficient amount of charge with either polarity can be maintained on the surface of the film freshly deposited onto an Al substrate. Nakayama *et al.* (1982) had put their a-Si:H drum to a practical test inside a copying machine and presented their conclusions from laboratory-level measurements of copying life and image quality. Further critical evaluation should be undertaken for the a-Si:H photoreceptor from an industrial viewpoint. The performances are summed up in Table II for the two types of photoreceptor.

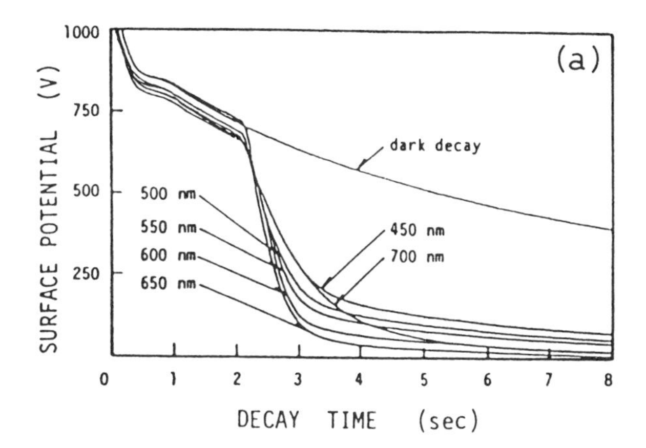

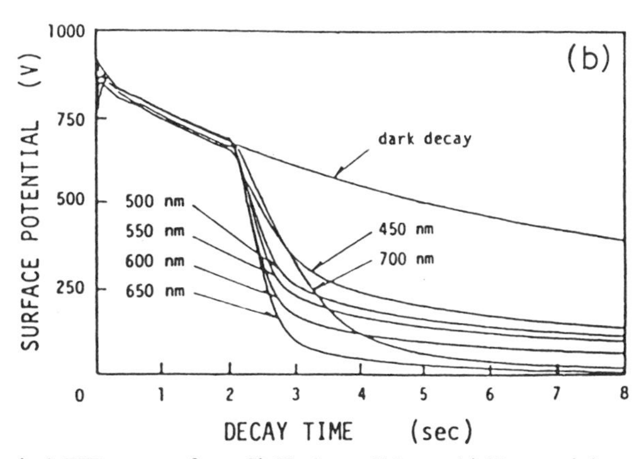

FIG. 6. Typical PID curves for a-Si:H drum (44 μm thick) containing oxygen. Light (0.3 μW cm^{-2}) of various wavelengths impinges on the photoreceptor after sensitizing with (a) positive and (b) negative corona. [From Nakayama *et al.* (1982).]

TABLE II

PERFORMANCE OF a-Si:H PHOTORECEPTORS

Structure	Multilayer p^+ (n^+) a-Si:H/ a-Si:H (intrinsic)/ a-SiN$_x$:H (a-SiC$_x$:H)	Single layer a-Si:H (:B:O)
Charge acceptance	40–50 V μm^{-1}	30–40 V μm^{-1}
Sensitization	Positive (negative)	Both polarities
Photosensitivity (erg cm^{-2})	0.25–1.0	0.8–1.0
Spectral photosensitivity	<670 nm	<650 nm
Dynamic range (mW cm^{-2})	10^{-4}–more than 10	—
Repeatability[a]	Fairly good	Fairly good

[a] The PID rate is exactly proportional to the photon flux.

With respect to the reconstruction of images, the surface plays an important role in the photoreceptor. The surface states secure the landing sites of charges brought by various ions, i.e., O_3^-, O_2^-, H^+ (H_2O), or N_2^+producted by the corona discharge. Both the lateral diffusion of the charges on the surface of the photoreceptor and their injection into the bulk must be prevented to have superior performance in duplicating copies. In spite of such an important functional role from the electronic point of view, the surface of the photoreceptor is treated in copying machines roughly as follows:

(1) Some active chemicals are adsorbed on the surface during sensitization with the corona discharge,
(2) the charged pigments are poured over the surface during the developing process, and
(3) the surface is touched directly with papers and with a brush for transferring images onto the paper and for cleaning the surface to ready it for the next duplication.

Consequently, controlling the surface properties in the a-Si:H photoreceptor should be a key concern because it is well known that the surface of a-Si:H is extremely sensitive to adsorbates (Tanielian, 1982). Our knowledge of the surface properties, however, is insufficient for solving these difficulties, and recently a diagnostic investigation has been made from a practicial point of view. Several proposals have been made to avoid instabilities related to the surface. A very thin insulating film (a-SiN$_x$ or a-SiC$_x$, 100 nm thick) provided on the top of a-Si:H is effective to maintain the surface charges produced by the corona. This falls in the category of heterophase blocking used at the interface to the conducting electrode. The thickness of the insulating film must be limited to less than 100 nm; otherwise, charges collected at the top layer produce a residual potential.

For the photoreceptor made of high-resistivity film containing oxygen, no treatment is required for the blocking property either at its free surface or at the contact to the electrode, implying that the electronic sensitivity is lost either at its surface or in the bulk.

Now let us turn to the sensitive surface of the a-Si:H photoreceptor, employing a film prepared under optimal conditons. A distinctive residual potential V_R (volt) was found in a fast PID process when the surface was illuminated with an intense and strongly absorbed light. The V_R value, as shown in Fig. 7a, depends greatly on either the polarity of the charge carrier (i.e., electron or hole) or the doping levels of boron in the film (Oda *et al.*, 1982). In the undoped film (B-0), for example, the V_R of half of its initial potential can be seen for hole transport while there is no V_R for electron transport. The opposite trends are observed in a-Si:H doped with a slight amount of boron. In the film a-Si:H:F, prepared from a gaseous mixture of SiH_4 and SiF_4, the same behavior as in a-Si:H can be observed with regard to the residual potential. Consistency has been established in the photoelectric properties of these two films. It should be noted that a surprisingly small amount of boron causes a dramatic change in V_R. This residual potential is observed in the fast but not in the slow PID process. In Fig. 7b, the PID curves are illustrated for an undoped film when it is illuminated with a monochromatic light of various wavelengths. The V_R value resulting from illumination with a strongly absorbed light is suddenly reduced to zero when 650-nm light impinges on the same photoreceptor. It is unlikely that the carrier range, equivalent to the V_R as shown in Fig. 7, of either holes or electrons is changed by doping with such a small amount of boron. Moreover, with optical absorption by a-Si:H at 650 nm, the carrier generation occurs within 1 μm of the surface, which is thin compared to the thickness of the photoreceptor (about 10 μm). This evidence leads us to conclude that the residual potential observed in the fast PID process cannot be explained in terms of the range-limited current but rather is related to the space-charge layer formed in the vicinity of the surface. If this were so, not only the residual potential but also the acceptance potential under the corona should be related to this surface effects. Although the origin of the space-charge layer is an open question, the charge transfer between the surface states involving some adsorbates and the bulk may plausibly result in a space-charge layer. The remaining problems observed in repeating the duplication with the a-Si:H photoreceptor were the instability of the charge acceptance depending on the ambient humidity and also the smearing of images that happened occasionally. Both are considered to be strongly related to the

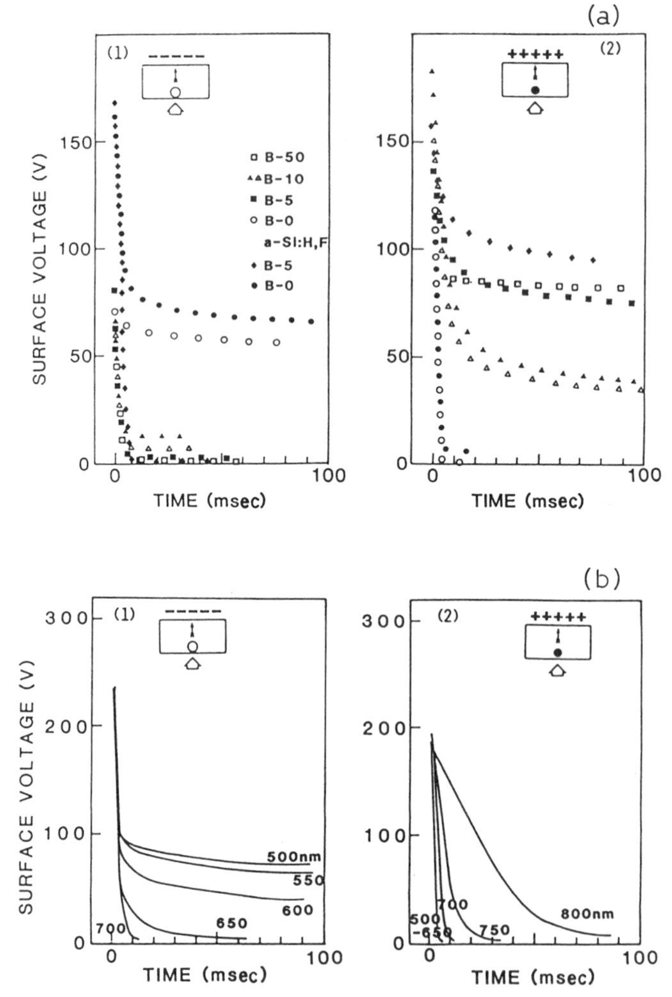

FIG. 7. PID curves in fast region for (a) a-Si : H photoreceptors with various doping levels of boron under 514.5-nm Ar laser light of 0.15 μW cm^{-2} and (b) undoped (99a) a-Si : H,F photoreceptors under illumination with monochromatic light (W/amp with monochromator at 1.1×10^{15} photon cm^{-2}) of various wavelengths. Hole transport (left) and electron transport (right) are shown after illumination with strongly absorbed light (514.5 nm). [From Oda *et al.* (1982).]

surface of a-Si : H and to involve adsorbates on the surface (Shimizu *et al.*, 1981a).

The extremely high quality of the electrostatic latent image on the surface of a-Si : H was established by reconstructing the optical image on a television frame using an image pickup tube. The materials and the device structure of the target were essentially the same as those of the photoreceptor except for their ambient conditions—in vacuum for the image pickup tube and in air for electrophotography. There were 800 television lines clearly resolved on the a-Si:H (B-10) 1-in.-diameter target and no deterioration was found in the image quality despite the long-term operation. In the a-Si:H photoreceptor, on the other hand, either the acceptance voltage or the image quality often changed greatly, especially when operating under high-humidity conditions. These evidences lend support to the idea that the surface of the a-Si:H photoreceptor must be desensitized.

One solution has been successful in improving the stability, namely, a photoconductive film with a wide energy gap (about 1.8–2.0 eV) and a thickness of 1 μm, corresponding to the width of the space-charge layer, is deposited on the top of the a-Si:H. The structure of the photoreceptor is shown in the inset in Fig. 8. According to this structure, the band bending resulting from the charge transfer at the surface is expected to limit an increase in its surface conductivity and carrier injection is blocked by the wide energy gap. In the present state of the art, the photoelectric properties of materials with a wider energy gap are not sufficiently good for making a single-layer photoreceptor; therefore we recommend their use only as a top layer with a limited thickness. Films of a-SiN_x:H or a-SiC_x:H prepared from gaseous mixtures of SiH_4–NH_3 or SiH_4–CH_4, respectively, are appli-

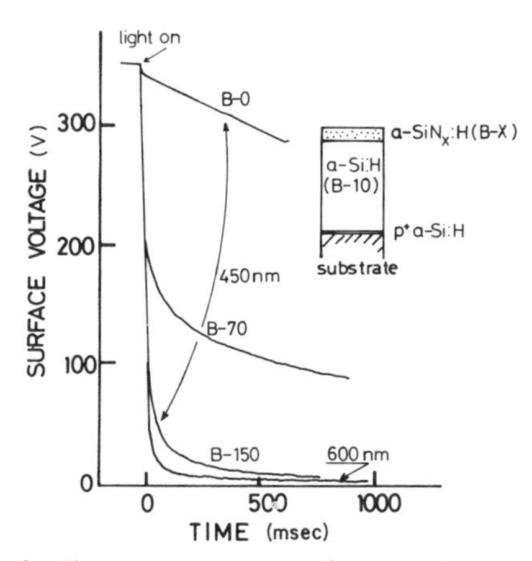

FIG. 8. PID curves of a-Si:H photoreceptors with top a-SiN_x:H layer. The wider gap materials, i.e., a-SiN_x:H including various concentrations of boron (B-0, undoped; B-70 and B-150), were deposited on the top of the a-Si:H (B-10, 6 μm thick). Illumination at 450 or 600 nm was used.

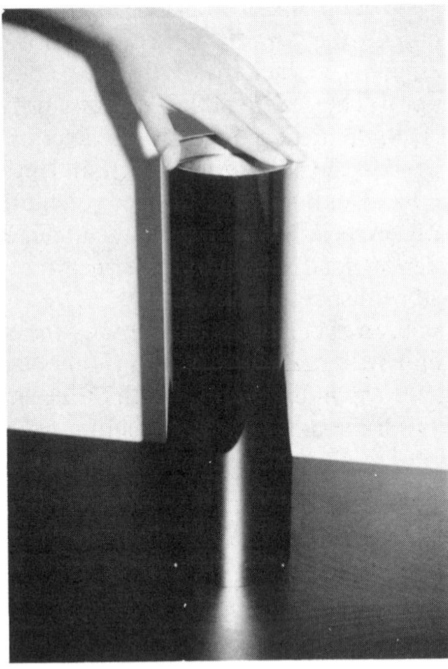

FIG. 9. Photograph of a-Si:H drum used in a copying machine. (Courtesy of Canon Corporation.)

cable to this purpose. In Fig. 8, typical PID curves are shown for the photoreceptor with an a-SiN_x:H top layer. An excellent stability has been established in repeated operation, without sacrificing its photoelectric performance.

Finally, we give a short account of the characteristics of the a-Si:H photoreceptor regarding image quality. The quality of the final image is greatly affected by the development with charged pigments and by the processes of forming the electrostatic latent image. It is quite difficult to evaluate independently the different contributions to the quality of the image on the photoreceptor. Generally, these studies have been made using actual copying machines and consequently only the images on the final copies are evaluated. In Figs. 9 and 10, a photograph of an a-Si : H drum and a copy obtained with the photoreceptor are presented. With respect to the resolution, contrast, and homogeneity, the images duplicated by the a-Si : H photoreceptor are highly rated. Furthermore, the a-Si : H photoreceptor has successfully undergone practical examination for durability and operational stability at the industrial level.

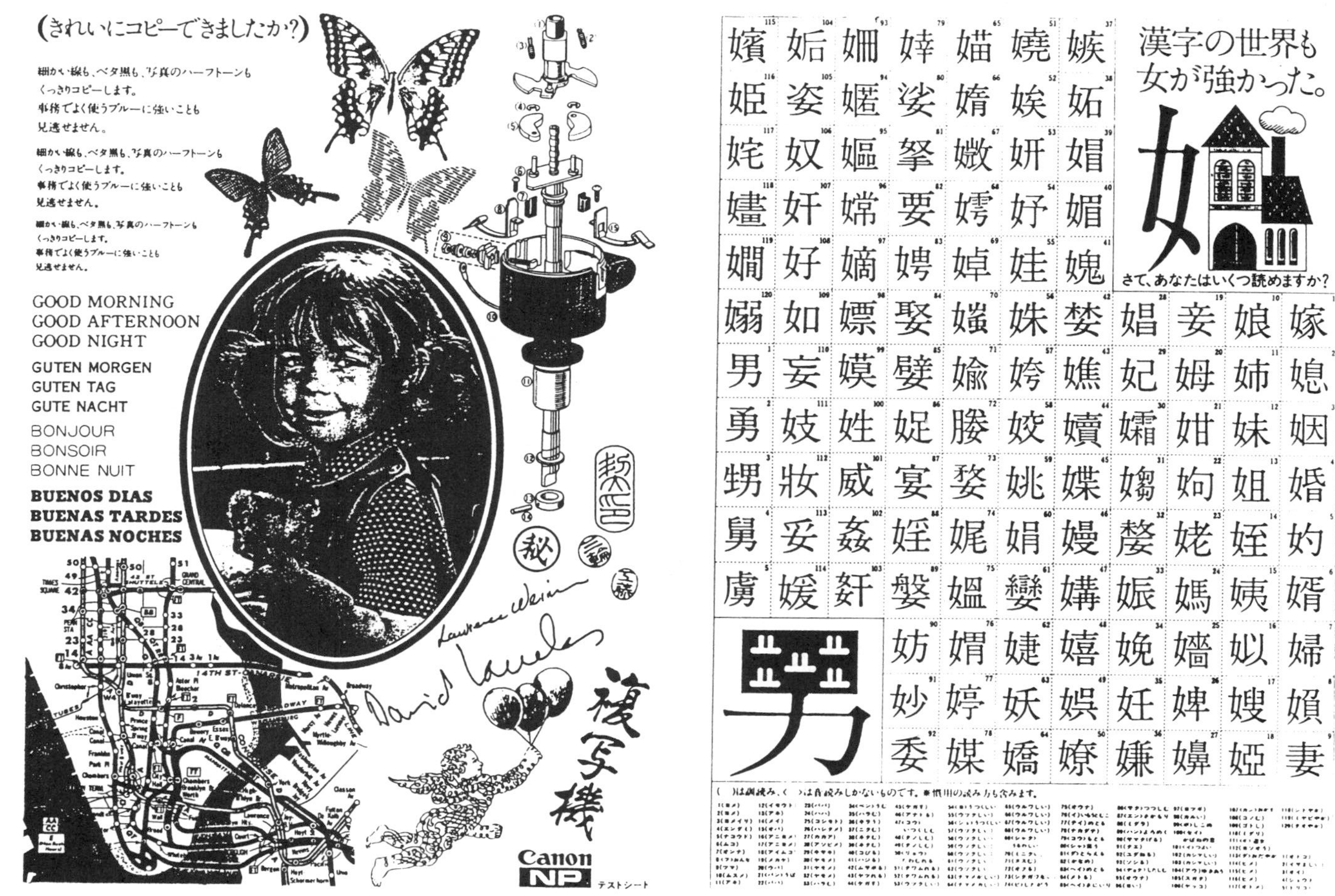

FIG. 10. A copy duplicated by the drum after making hundreds of thousands of copies, here reduced by 44%. (Courtesy of Canon Corporation.)

V. Problems Remaining Today

Several problems have remained unsolved for the a-Si:H photoreceptor from an industrial point of view. The establishment of the technology for mass production is one of the most important and serious problems. A film more than 10 μm thick is required for the photoreceptor; therefore rapid growth of a high-quality a-Si:H is essential to allow low cost and high productivity. After intensive worldwide studies, a-Si:H can now be prepared by glow discharge of SiH_4 at a growth rate of 4–5 μm hr^{-1} without sacrificing any photoelectric property; but this is not sufficiently fast for the commercial production of photoreceptor. The a-Si:H drums were produced at a growth rate of 6–7 μm hr^{-1}, sacrificing slightly its photoelectric properties. Ogawa *et al.* (1981) have demonstrated that a-Si:H prepared from Si_2H_6 at a deposition rate exceeding 20 μm hr^{-1} exhibited adequate photoelectric properties for use as the photoreceptor. Further investigation of large-scale production techniques is necessary.

Another problem in material preparation is the spectral tuning to the light source. Thus, some diode lasers emit near infrared light (~800 nm). The application of a laser diode as the light source is a recent trend in line printers employing electrophotography. As shown in Fig. 5, the photoresponse of a-Si:H photoreceptor, drops dramatically at wavelengths longer than 670 nm due to poor optical absorption. A new material with an energy gap narrower than that of a-Si:H is desired to fit the spectral sensitivity of the light source. The alloys a-$SiGe_x$:H and a-$SiSn_x$:H have been extensively investigated for this purpose. However, the photoelectric properties of these materials are not sufficient and need further improvement.

References

Abkowitz, M. A. (1979). *In* "The Physics of Selenium and Tellurium" (E. Gerlach and P. Grosse, eds.), p. 210. Springer-Verlag, Berlin and New York.

Hirai, Y., Komatsu, T., Nakagawa, K., Misumi, T., and Fukuda, T. (1981). U. S. Patent 4,265,991.

Imamura, Y., Ataka, S., Takasaki, Y., Kusano, C., Hirai, T., and Maruyama, E. (1979). *Appl. Phys. Lett.* **35,** 349.

Kempter, K. (1980). U. S. Patent 4,225,222.

Kurata, H., Hirose, M., and Osaka, Y. (1981). *Jpn. J. Appl. Phys.* **20,** L811.

Mort, J., and Pfister, G. (1981). *In* "Electronic Properties of Polymers" (J. Mort and G. Pfister, eds.), Chapter 6. Wiley (Interscience), New York.

Nakagawa, K., Komatsu, T., Hirai, H., Misumi, T., and Fukuda, T. (1981). Japanese Patent 56-1943.

Nakayama, Y., Sugimura, A., Nakano, M., and Kawamura, T. (1982). *Photogr. Sci. Eng.* **26,** 188.

Oda, S., Saito, K., Tomita, H., Shimizu, I., and Inoue, E. (1981a). *J. Appl. Phys.* **52,** 7275.

Oda, S., Saito, Y., Shimizu, I., and Inoue, E. (1981b). *Philos. Mag. B* **43,** 1079.

Oda, A., Terazono, S., and Shimizu, I. (1982). *Sol. Energy Mater.* **8,** 123.
Ogawa, K., Shimizu, I., and Inoue, E. (1981). *Jpn. J. Appl. Phys.* **20,** L639.
Pai, D. M., and Enck, R. C. (1975). *Phys. Rev. B* **11,** 5163.
Pfister, G., and Scher, H. (1975). *Bull. Am. Phys. Soc.* **20,** 322.
Shimizu, I., Komatsu, T., Saito, K., and Inoue, E. (1980). *J. Non-Cryst. Solids* **35/36,** 773.
Shimizu, I., Shirai, S., and Inoue, E. (1981a). *J. Appl. Phys.* **52,** 2789.
Shimizu, I., Oda, S., Saito, K., Tomita, H., and Inoue, E. (1981b). *J. Phys. Colloq. Orsay, Fr.* **42,** C4-1123.
Silver, M., Madan, A., Adler, D., and Czubatyj, W. (1980). *Conf. Rec. IEEE Photovoltaic Spec. Conf.* **14,** 1062.
Tabei, M., Takeda, K., Kawajiri, K., and Azuma, A. (1980). Japanese Patent 55-166647.
Tanielian, M. (1982). *Philos. Mag. B* **45,** 435.
Tawada, Y., Okamoto, H., and Hamakawa, K. (1981). *Appl. Phys. Lett.* **39,** 237.
Tiedje, T., Cebulka, J. M., Morel, D. L., and Abeles, B. (1981). *Phys. Rev. Lett.* **46,** 1425.
Wakita, K., Nakayama, Y., and Kawamura, T. (1982). *Photogr. Sci. Eng.* **26,** 183.
Watanabe, H., Katoh, K., and Yasui, M. (1982). *Jpn. J. Appl. Phys.* **21,** L341.
Wronski, C. R., Goldstein, Y., Kelemen, S., Abeles, B., and Witzke, H. (1981). *J. Phys. Colloq. Orsay. Fr.* **42,** C4-475.
Yamamoto, N., Nakayama, Y., Wakita, K., Nakano, M., and Kawamura, T. (1981). *Jpn. J. Appl. Phys., Suppl.* **20-1,** 305.

CHAPTER 5

Image Pickup Tubes

Sachio Ishioka

CENTRAL RESEARCH LABORATORY
HITACHI, LTD.
TOKYO, JAPAN

I. Introduction

When we consider the three well-known applications of amorphous selenium, in solar cells, electrophotography (Mort and Pai, 1976), and the image pickup tubes (Weimer and Cope, 1951), it is not surprising that the material that has shown such great promise for practical photovoltaic cells should also be useful in the other two applications (Rose, 1979). Similarly, since hydrogenated amorphous silicon (a-Si:H) has shown itself useful in solar cells, it ought to be regarded as a candidate for the other two uses. However, certain additional properties that go beyond the needs of energy conversion are required if a-Si:H is to be used for image processing. That is why much more time is needed for the development of imaging applications than for solar cell applications.

The first attempt to realize such imaging device uses was made by Imamura *et al.* (1979), and a vidicon-type image pickup tube was proposed. Especially in the field of color television cameras, there have been great expectations that this device will lead to the realization of high-performance imagers.

This section focuses on a-Si:H image pickup tubes and presents a-Si:H property requirements and fabrication and impurity doping techniques, as well as the structure of photoconductive targets. Pickup tube characteristics that have been attained and some of their applications will be described.

ISBN 0-12-752150-X

II. a-Si:H Image Pickup Tube

Figure 1a shows the fundamental structure and operation mechanism for the image pickup tube. Figure 1b is the equivalent circuit. The photoconductive plate in a pickup tube is called a target. The resistivity of the photoconductor in the target must be high enough to be subdivided into an array of picture elements having resistivity r_p and capacitance c_p.

In ordinary image pickup tube operation, the transparent electrode is biased positively with respect to the cathode. When the low-velocity scanning beam lands on the surface of the photoconductive target, the surface is charged negatively and brought to cathode potential.

Since the effective resistance of the photoconductor is much higher than the electron-beam resistance R_b, the surface potential of the photoconductor can be regarded as being almost the same as that of the cathode. When the photoconductor is illuminated, r_p is decreased by the photogenerated carriers, and the surface potential is increased. This change is accumulated and stored during the scanning period. At the next electron-beam scanning,

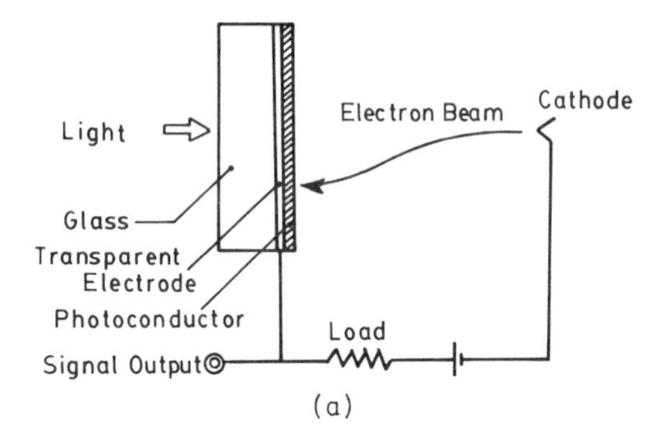

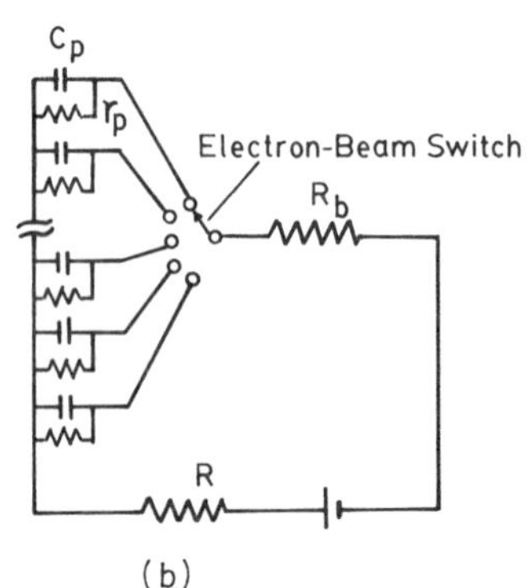

FIG. 1. (a) The fundamental structure and operation mechanism of the image pickup tube. (b) The equivalent circuit of the image pickup tube.

the surface is recharged to the cathode potential. This additional beam current landing on the target corresponds to the light intensity at each picture element and is detected through the load resistance R as a signal current.

This storage type of operation can produce signals at low light levels approximately 10^5 times larger than would nonstorage operation, in which each element in the equivalent circuit switches only when the beam scans the element in question.

If c_p is too small, the potential of the photoconductor surface rises to that of the transparent electrode within the electron-beam scanning period, and the signal storage procedure is stopped at that point. Therefore a time constant

$$c_p r_p = \varepsilon\varepsilon_0 \rho > T_f \tag{1}$$

is needed, where T_f is a scanning frame period ($\frac{1}{30}$ sec). When c_p is too large, on the other hand, it takes a few frames to read out the whole stored charge, since the signal-reading time constant is the product of c_p and R_b. Consequently, the decay lag characteristics become poor.

From these points of view, the total capacitance of the photoconductor for the imaging target should be around 1000 pF. The photoconductor resistivity should then be large to satisfy Eq. (1). For instance, in the case of an a-Si:H target 26 mm in diameter and 5 μm thick, a resistivity of 10^{14} Ω cm is required. High resistivity is important for achieving high resolution.

In summary, to apply a-Si:H to image pickup tubes, high photoconductivity and resistivity high enough for long storage of image signals are required simultaneously. Although a-Si:H is believed to have the potential to fulfill these conditions, the resistivity of a-Si:H is not high enough by itself. Furthermore, as was explained before, the illuminated side of the target is positively biased in the normal image pickup tube operation. Accordingly, photogenerated holes, which generally have a very small $\mu\tau$, become very important (μ is drift mobility and τ is carrier lifetime). Additional requirements that are vital from the point of view of image pickup tube application are listed below.

(1) *Low dark current.* Dark current, the signal current when there is no illumination, not only defines the minimum detectable limit of the signal level, but also causes shading phenomena and instability during temperature changes. The dark current level needs to be reduced to less than 1–2 nA.

(2) *Photoresponse properties.* Photoresponse properties are very important properties for imaging devices, especially in broadcasting use. The amount of decay lag can be defined as the ratio of the residual signal current

at the third field (50 ms) to the constant photosignal current. The lag must be less than a few percentage points.

(3) *Spectral photoconductivity.* Each image device has a desirable spectral photosensitivity relevant to its purpose. For color television cameras, in particular, this range of photosensitivity should cover the whole visible spectrum and extend very little into the infrared region.

Trials of a-Si:H fabrication and improvement of target structure will be discussed in the next section in the context of the requirements that have just been described.

III. Properties of a-Si:H

As was explained in several chapters of Volume 21A, many a-Si:H preparation methods are known. In this section, sputtering will be discussed with the aim of clarifying a-Si:H properties (Imamura *et al.*, 1980), and it will be seen that electrical and optical properties can be controlled over a considerable range by manipulating reaction parameters (Paul *et al.*, 1976). The sputtering system details are described in Volume 21A, Chapter 4, by Moustakas.

A polycrystalline silicon plate of purity 5–9's was sputtered in a mixed atmosphere of argon and hydrogen. Figure 2 shows the relation between the H_2 partial pressure of the reactive gas atmosphere and optical band-gap values (E_g^{opt}). Optical band gaps are calculated from plots of $(\alpha h\nu)^{1/2}$ versus

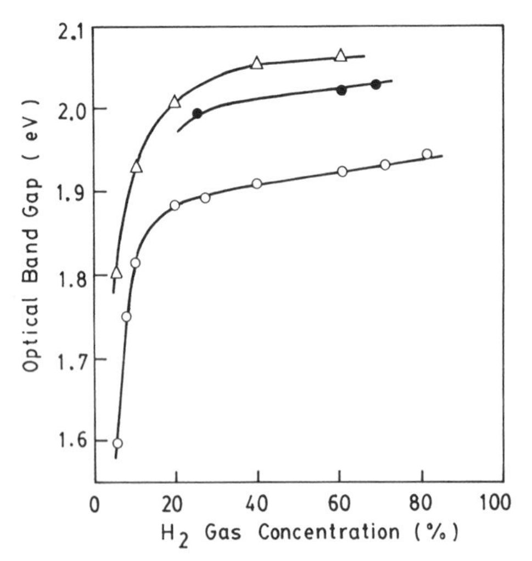

FIG. 2. Relation between the partial pressure of the sputtering gas and optical band gap E_g^{opt}. T_{sub} = 110°C (△), 160°C (●), and 250°C (○).

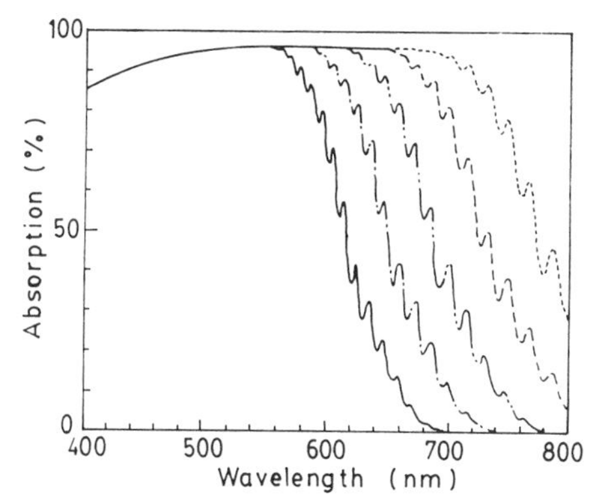

FIG. 3. Absorption spectra for a-Si:H having various optical gaps E_g^{opt} (eV): ———, 2.0; —·—, 1.9; —··—, 1.8; -------, 1.7; ·····, 1.6.

$h\nu$. These results show that a-Si:H with an E_g^{opt} of 1.9–2.0 eV and high resistivity are obtained when the H_2 ratio of the mixed sputtering gas is 40% or more.

When the substrate temperature is low, E_g^{opt} also becomes large. The resulting photosensitivity, however, is not good enough for image pickup tubes. Hydrogenated amorphons silicon prepared by glow discharge with low substrate temperature also shows poor sensitivity. The substrate temperature during the sputtering reaction should be ~200–250°C.

The absorption spectra for 2-μm-thick a-Si:H films having various E_g^{opt} are shown in Fig. 3. Since the spectral shape of the absorption of a-Si:H coincides with that for the photoconductivity of the film, the E_g^{opt} of a-Si:H for a color imager must be around 1.9–2.0 eV.

Typical deposition parameters for image pickup tube a-Si:H are listed in Table I.

Samples of a-Si:H prepared under these conditions have an E_g^{opt} of 1.9–2.0 eV and contain 10–15% of bonded hydrogen atoms. Infrared absorption measurement of a-Si:H provides information about the Si–H bonding configuration (see Volume 21B, Chap. 4 by Zanzucchi).

TABLE I

PREPARATION CONDITIONS OF a-Si:H

Parameter	Value
Total pressure	$1-5 \times 10^{-3}$ Torr
Hydrogen gas concentration	40% or more
Deposition rate	~3 A sec^{-1}
Substrate temperature	200–250°C

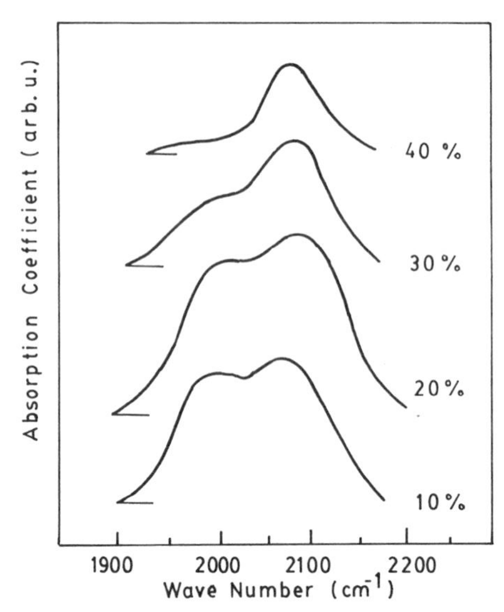

FIG. 4. Infrared absorption spectra of a-Si : H as a function of the H_2 ratio of sputtering gas at Si–H stretching mode.

Figure 4 shows the infrared absorption spectra for a-Si : H at the Si–H stretching mode frequencies with the H_2 partial pressure in the sputtering gas as a parameter. As can be seen in Fig. 4, infrared absorption peaks at 2000 cm^{-1} (monohydride) and 2100 cm^{-1} (dihydride) are both observed under low H_2 gas ratio sputtering conditions. At an H_2 ratio of 40%, the 2100 cm^{-1} peaks becomes dominant.

These results suggest that a-Si : H of dihydride configuration provides the better characteristics as a photoconductor for the image pickup tube. The result that the dihydride configuration a-Si : H has a large E_g^{opt} is consistent with previous theoretical calculations (Allan and Joannopoulos, 1980). It is interesting that this situation seems to be quite different from results reported for an a-Si : H solar cell, where a monohydride configuration is believed to be favorable.

IV. Blocking Contact Structure of the Photoconductive Target

If the electrodes for the photoconductive targets are ohmic-type contacts, the dark current is at least comparable to the value determined by the bulk resistivity of the a-Si : H itself. In addition, carriers continue to flow into a-Si : H even after the illumination is stopped, until the carriers trapped in a-Si : H are neutralized. Thus, it is hard to obtain a low dark current and a high photoresponse.

On the other hand, if the target electrodes are blocking-type contacts; that is, if carrier injection into the photoconductive target is suppressed while photogenerated carriers in the a-Si:H are allowed to be drawn out at the electrode, only a primary photoinduced current contributes to the signal current from the target (Rose, 1963). Such contacts are effective in reducing the dark current and they obtain a rapid photoresponse, thus satisfying the requirements listed in Part II.

The conductivity of a-Si:H can also be changed through impurity doping. As a consequence, a blocking structure can easily be incorporated in an $n-i-p$ diode structure (Shimizu *et al.*, 1980). By using such photodiodes as the target, and under reversed-bias operation, dark current levels of less than a few nanoamperes can be obtained. However in such a structure, n- and p-layer resistivity drops to such a low value that high resolution is difficult to attain. Moreover, it is difficult to separate hole–electron pairs generated inside these layers, thus resulting in decreased photosensitivity.

To avoid these problems, other types of blocking contacts have been proposed (Oda *et al.*, 1981, Hatanaka *et al.*, 1982; Ishioka *et al.*, 1983). In these cases, a wide-gap transparent thin film such as SiO_2 or Si_xN_{1-x}, is used as a hole-blocking layer in place of the n layer. A chalcogenide, such as Sb_2S_3 or $As_2Se_{1.5}Te_{1.5}$, or an oxide such as CeO_2 thin film, is then used instead of the p layer.

Figure 5 illustrates an example of the schematic structure for an a-Si:H target. On a glass substrate, a tin oxide transparent electrode layer is deposited by CVD. On this layer, a very thin SiO_2, as a hole-blocking layer, a photoconductive a-Si:H layer, and an electron-blocking Sb_2S_3 layer are added, in that order. The thickness of the a-Si:H layer is 2–4 μm.

The current–voltage characteristics of this SiO_2/a-Si:H/Sb_2S_3 target are shown in Fig. 6. The SiO_2 thickness was chosen to be around 20 nm. The Sb_2S_3 electron-blocking layer is a porous film about 60 nm thick, chosen with the aim of preventing secondary electron emission during the electron-beam scanning. The blocking layers remarkably suppress the dark current to less than 1 nA, up to 3×10^5 V cm^{-1}, which is satisfactory for television

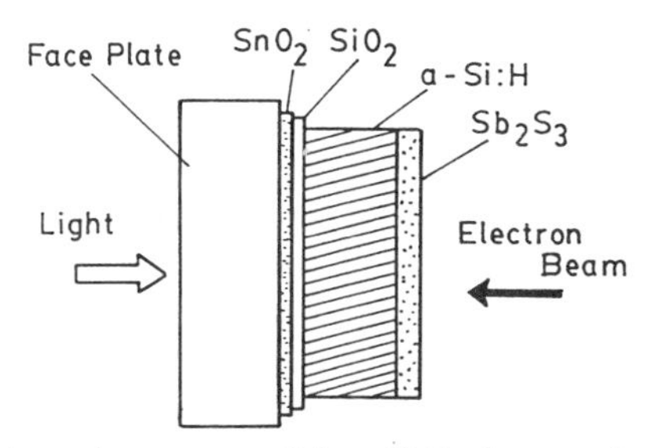

FIG. 5. Schematic structure of the a-Si:H photoconductive target.

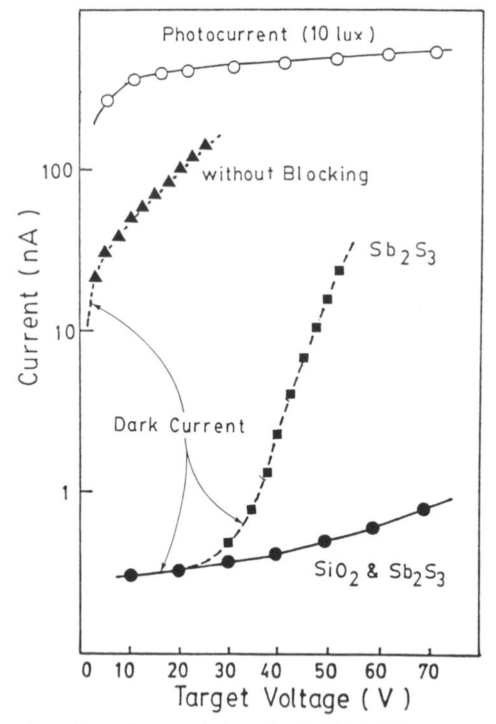

FIG. 6. The current–voltage characteristics of a-Si : H blocking contact structure target.

camera use. The photocurrent reaches the saturation level, which suggests that photogenerated carriers in the a-Si : H are effectively drawn out through the SiO_2 layer. The situation is the same when the SiO_2 thickness is increased to 40 nm.

V. Impurity Doping of a-Si : H

An undoped a-Si : H film is thought to have *n*-type characteristics, while holes travel a longer distance than electrons in a photoconductive target. Therefore, a *p*-type a-Si : H is preferable for the present purpose.

Figure 7 shows the photocurrent–voltage characteristics for a-Si : H targets that have various boron-doping levels (Ishioka *et al.*, 1983). Boron doping was carried out by mixing diborane (B_2H_2) into the sputtering gas. With imaging devices, the dopant amount must not be excessive, for fear of degrading the resolution. The amount of dopant is designated in terms of the diborane gas-mixing ratio.

However, when the incident light is 600 nm, photocarriers are generated

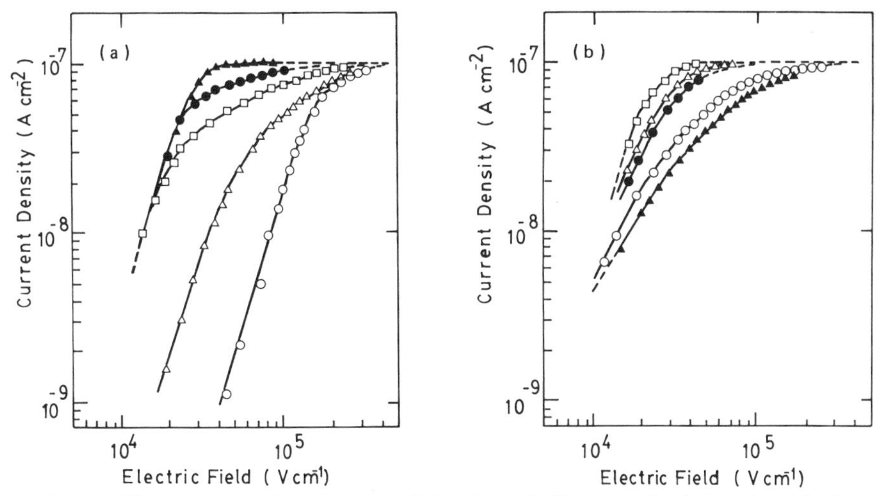

FIG. 7. Photocurrent–voltage characteristics for a-Si:H targets having various doping levels. The wavelength of incident light is (a) 425 nm and (b) 600 nm. Doping levels: ○, 0 ppm; △, 0.5 ppm; □, 5 ppm; ●, 10 ppm; ▲, 50 ppm.

throughout the whole layer of a-Si:H. Therefore, it can be seen in Fig. 7b that both holes and electrons contribute to the photocurrent. The photocurrent of the a-Si:H target increases with increasing target voltage until it reaches the saturation level at voltage V_s. This V_s value can be seen to be proportional to $(\mu\tau)^{-1}$ of a-Si from the results of analysis of primary photocurrent–voltage characteristics (Oda *et al.*, 1983).

Therefore a-Si:H with a low V_s is supposed to have a large carrier mobility and a low trap density. As shown in Fig. 7a V_s decreases monotonically as the content of B_2H_6 increases up to 50 ppm. The V_s controlling the electron current, on the contrary, decreases at a low doping level but increases again when the content of B_2H_6 is 50 ppm. These results indicate that a low doping of B_2H_6, up to a 10-ppm level, serves to decrease the trap density. Hydrogenated amorphous silicon produced by rf sputtering is thought to have a smaller $\eta\mu\tau$ value, but this problem can be improved by a light doping of boron. The same effects are observed in glow-discharge-produced a-Si:H films. The dark current is reported as being at a minimum at the point where B_2H_6/SiH_4 is 10 ppm (Oda *et al.*, 1981).

VI. Characteristics of a-Si:H Image Pickup Tubes

Image pickup tubes are operated at an applied voltage high enough to draw out the saturated level of the photocurrent as shown in Fig. 7. In this region, the photocurrent is determined by the number of electron–hole

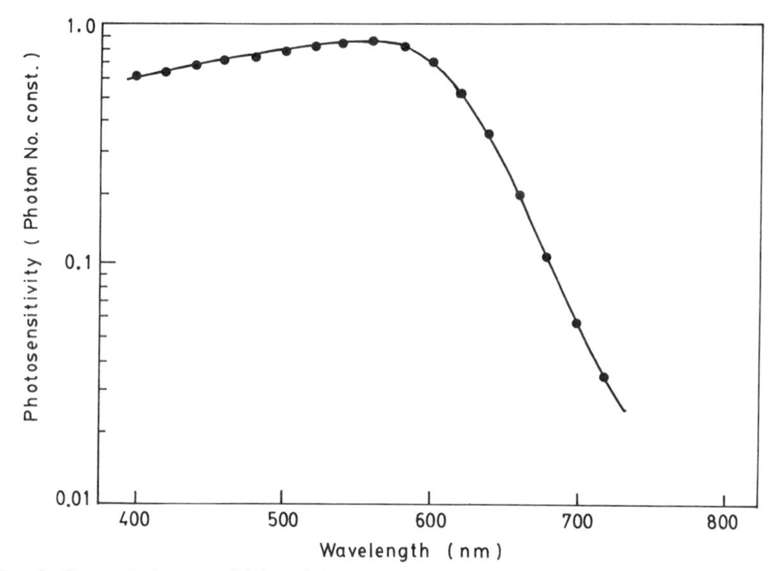

FIG. 8. Spectral photosensitivity of the sputtering-produced a-Si:H target. The target structure is SiO_2 (15 nm)/a-Si:H (2 μm)/Sb_2S_3 (60 nm).

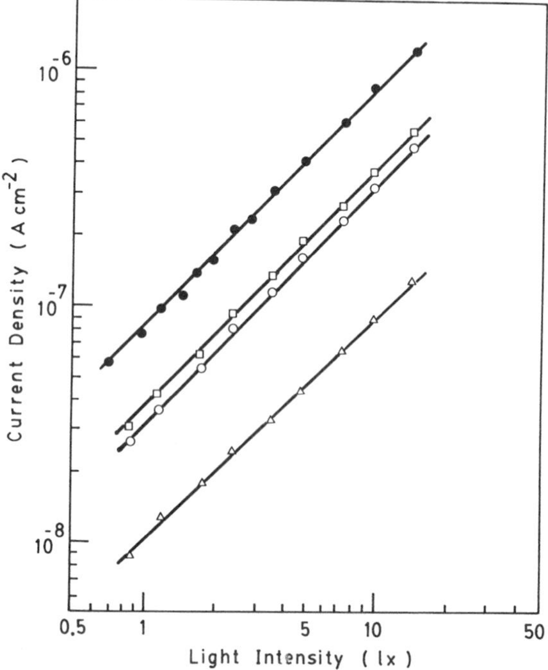

FIG. 9. Relation between the signal current of the a-Si:H target and the incident light intensity for white (●), green (□), red (○), and blue (△) light.

pairs generated in the a-Si:H. Consequently, the carrier generation efficiency can be evaluated if the number of photons absorbed in the photoconductor is known.

Figure 8 shows the spectral photosensitivity of an a-Si:H produced by sputtering (Ishioka *et al.,* 1983). It covers the whole visible range, shows a high quantum efficiency, and extends little into the infrared region. Thus, it is favorable for color imaging. The a-Si:H target produced by glow-discharge CVD, on the other hand, has a narrow band gap of 1.6–1.8 eV, which results in high infrared sensitivity.

Figure 9 shows the relation between the signal current of the a-Si:H target and the intensity of the incident light. The signal current is proportional to the light intensity ($\gamma = 1$), and this provides a great advantage in balancing color signal elements.

The resolution pattern taken from a sputtered a-Si:H target pickup tube is shown in Fig. 10. The horizontal resolution is not affected by boron doping and is more than 600 television lines for a tube 18 mm in diameter. This value is the same as for chalcogenide pickup tubes, which are used in the broadcasting industry (Maruyama *et al.,* 1974).

The decay lag for an undoped a-Si:H target with scanning area of 8.8 × 6.6 mm is about 10%, at 50 msec after turning off the light of a 0.2-nA signal intensity. On the other hand, the decay lag is drastically reduced to less than 4% by a light doping with boron. This value is satisfactory for industrial or

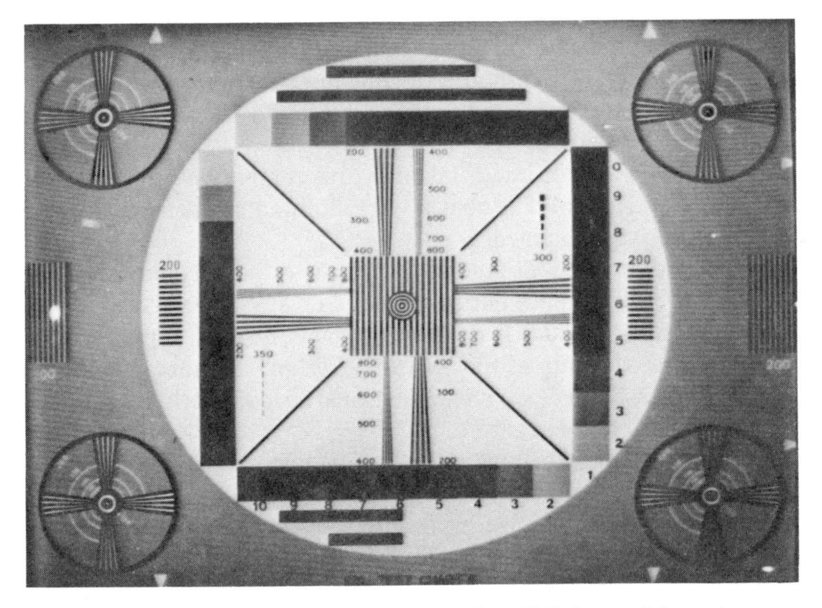

FIG. 10. Resolution pattern reproduced by the a-Si:H image pickup tube.

home-use cameras. Finally, a-Si : H image pickup tubes have no anomalous effects, such as after-image burning or blooming under strong illumination.

VII. Applications for a-Si : H Target

The properties of an a-Si : H image pickup tube with a blocking structure are summarized here:

(1) Spectral photosensitivity is high and favorable for color imaging. Sensitivity in the green region is more than twice as large as that for conventional vidicon tubes.

(2) Resolution is high, and almost no burning or blooming can be seen at any applied voltage. This is a big advantage over solid state imagers.

(3) An a-Si : H film is stable up to 100°C or more. This is also favorable for cameras used for outdoor service.

These properties indicate that a-Si : H image pickup tubes are quite useful over a wide area of applications. The TV camera for an x-ray image intensifier system is an example of such an application (Iamura *et al.*, 1982; Ishikawa *et al.*, 1982).

The peak wavelength of the light emitted from the fluorescent plate of the intensifier is around 520 nm, which matches well the a-Si : H photosensitivity spectrum. The image resolution is measured to be 20% better than in conventional intensifier tubes.

Furthermore, by applying a color-filter-integrated substrate to the a-Si : H target, a single-tube color-imaging camera was successfully fabricated (Ishioka *et al.*, 1983). Stripe-shaped organic filters of cyan and yellow color combinations are employed, and signals can be obtained using a frequency division multiplex method.

Figure 11 shows a color image taken by this camera system (here reproduced in black and white). The objects were illuminated at only 50 lux, but a bright image was still obtained.

Hydrogenated amorphous silicon is thought to be very promising, not only for image pickup tubes but also for solid state imagers such as "two-story" area sensors (Tsukada *et al.*, 1979). That is, an a-Si : H photoconductive layer is placed on top of the Si scanner. Many advantages have been reported for the a-Si : H "two-story" sensor over conventional solid state imagers (Baji *et al.*, 1982).

Although a-Si : H has been proved to be a very good photoconductor for image pickup tubes, some problems remain to be solved. As has already been mentioned, the decay lag in a-Si : H tubes is presently 4%, and it is necessary to reduce this lag, especially in order to apply the technology to broadcasting use. The decay lag as determined by the capacitance of the

FIG. 11. Color image (reproduced here in black and white) reproduced by a-Si:H single-tube camera.

photoconductive layer is calculated as being about 2%. The difference is thought to be caused by a trapping level inside the a-Si:H layer. Thus, a-Si:H fabrication techniques should continue to be improved to decrease the density of harmful states in the a-Si:H layer. To achieve high performance of a-Si:H image pickup tubes, a new operation method with a different target structure has been proposed (Kusano *et al.,* 1983).

Regarding device reliability, the stability of a-Si:H under such conditions as high light intensity, high temperature, or operation in a vacuum should be confirmed. During fabrication, in addition, decreasing the deposition temperatures to 150°C or less would also be useful if a-Si:H were deposited directly on a filter-integrating substrate.

REFERENCES

Allan, D. C., and Joannopoulos, J. D. (1980). *Phys. Rev. Lett.* **44,** 43.

Baji, T., Yamamoto, H., Matsumaru, H., Koike, N., Akiyama, T., Sasano, A., and Tsukada, T. (1982). *Jpn. J. Appl. Phys.* **21-1,** 269.

Hatanaka, Y., Yamagishi, K., So, H., and Ando, T. (1982). *Proc. Meet. TV Eng., 1982* p. 69 (in Japanese).

Imamura, Y., Ataka, S., Takasaki, Y., Kusano, C., Hirai, T., and Maruyama, E. (1979). *Appl. Phys. Lett.* **35,** 349.

Imamura, Y., Ataka, S., Takasaki, Y., Kusano, C., Ishioka, S., Hirai, T., and Maruyama, E. (1980). *Jpn. J. Appl. Phys.* **19-1,** 573.

Imamura, Y., Hirai, T., Maruyama, A., and Nobutoki, S. (1982). *Proc. Meet. TV Eng., 1982* p. 71 (in Japanese).
Ishikawa, K., Ikeda, M., and Okabe, K. (1982). *Proc. Meet. Jpn. Radiol. Technol. 38th, Tokyo 1982* p. 299 (in Japanese).
Ishioka, S., Imamura, Y., Takasaki, Y., Kusano, C., Hirai, T., and Nobutoki, S. (1983). *Jpn. J. Appl. Phys.* **22-1,** 461.
Kusano, C., Ishioka, S., Imamura, Y., Takasaki, Y. Shimonoto, Y., Hirai, T., and Maruyama, E. (1983). *Int. Electron Device Meet of Tech. Dig.*, 509.
Maruyama, E., Hirai, T., Goto, N., Isozaki, Y., and Sidara, K. (1974). *Proc. Int. Conf. Amorphous and Liquid Semicond., 5th,* p. 581. Taylor and Francis, London.
Mort, J., and Pai, D. M. (1976). *In* "Photoconductivity and Related Phenomena," p. 421. Amer. Elsevier, New York.
Oda, S., Tomita, H., and Shimizu, I. (1983). *In* "Amorphous Semiconductor Technologies and Devices" (Y. Hamakawa ed.), p. 113. Ohmsha, Tokyo.
Oda, S., Saito, K., Tomita, H., Shimizu, I., and Inoue, E. (1981). *J. Appl. Phys.* **52,** 7275.
Paul, W., Lewis, A. J., Connell, G. A. N., and Moustakas, T. D. (1976). *Solid State Commun.* **20,** 969.
Rose, A. (1963). "Concepts in Photoconductivity and Allied Problems." Wiley (Interscience), New York.
Rose, A. (1979). *Phys. Today* **12,** 20.
Shimizu, I., Oda, S., Saito, K., and Inoue, E. (1980). *J. Appl. Phys.* **51,** 6422.
Tsukada, T., Baji, T., Yamamoto, H., Hirai, T., Maruyama, E., Ohba, S., Koika, N., Ando, H., and Akiyama, T. (1979). *Tech. Dig.—Int. Electron Devices Meet.* p. 134.
Weimer, P. K., and Cope, A. D. (1951). *RCA Rev.* **12,** 314.

CHAPTER 6

The Development of the a-Si : H Field-Effect Transistor and Its Possible Applications

P. G. LeComber and W. E. Spear

CARNEGIE LABORATORY OF PHYSICS
THE UNIVERSITY OF DUNDEE
DUNDEE, SCOTLAND

I. Introduction

The fundamental work on amorphous semiconductors during the 1970s established the considerable applied potential of hydrogenated amorphous silicon (a-Si : H) prepared by the decomposition of silane in a glow-discharge plasma. A significant contribution to this development was the early field-effect experiments of the Dundee group (Spear and LeComber, 1972; Spear, 1974; Madan *et al.*, 1976; Madan and LeComber, 1977), which provided information on the density-of-states distribution $g(E)$ in this material. In particular the work drew attention to a unique property of glow-discharge a-Si : H—the remarkably low $g(E)$ throughout most of the mobility gap, which has made it into an electronically viable material for a growing range of applications.

On the basis of the field-effect studies the authors and colleagues at the Royal Signals and Radar Establishment (RSRE), Malvern, proposed in 1976 the use of a-Si : H field-effect devices in the addressing of liquid crystal matrix displays, as an alternative to the thin-film CdSe transistors which had

ISBN 0-12-752150-X

been developed by Brody *et al.* (1973). They reasoned that an elemental covalently bonded material such as a-Si : H should have distinct advantages over the more complex II–VI compounds as far as ease of preparation, reproducibility, and stability are concerned. The design and characteristics of an a-Si : H insulated gate FET suitable for driving liquid crystal displays were first published by LeComber *et al.* (1979), and experiments at RSRE showed that such a device could switch liquid crystal elements satisfactorily. Subsequent developments in our laboratory concerned with the a-Si : H FET and its possible applications have been published in a number of papers (Snell *et al.*, 1981a,b; French *et al.*, 1983; Mackenzie *et al.*, 1983) and reviewed at the Grenoble Conference (LeComber *et al.*, 1981). The work has aroused considerable interest and during the last three years a growing number of industrial and university laboratories in Japan, Europe, and the United States have entered the field.

The main motivation for using a-Si : H as an FET material is that it is the only material suitable for a large-area array of such devices. Although in present devices the gain factor is small, many applications appear possible. In this chapter we shall consider the design, fabrication, and performance of a-Si : H FETs and also discuss important features such as stability, reproducibility, and radiation hardness of the devices; some recent studies of the ON-state are included in Part VII. The final sections deal with the possible applications. The most prominent of these, the use of a-Si : H FETs in addressable liquid crystal displays, forms the subject of the following article by Ast and will not be discussed here. Instead, we shall review some exploratory work on the application of the FETs to simple logic circuits and addressable image sensors that has been carried out at Dundee and at the Tokyo Institute of Technology.

As mentioned above, the field-effect experiments on a-Si : H have been of basic importance to the subject of this article, and we shall therefore begin with a brief discussion of the principles underlying this experimental method.

II. The Field Effect in Glow-Discharge a-Si : H

In the experiment, the localized state distribution is investigated by displacing it with respect to the Fermi level E_f. This is done by means of an external electric field, applied in a direction normal to the surface of the film to be studied. Figure 1 shows two arrangements that have been used in our work. In Fig. 1a the a-Si : H film is deposited directly onto a thin ($\simeq$ 170-μm) quartz substrate Q, which also acts as the dielectric for applying the external field. The other surface contains the narrow gate electrode G, which is carefully aligned with the gap between the source and drain electrodes S

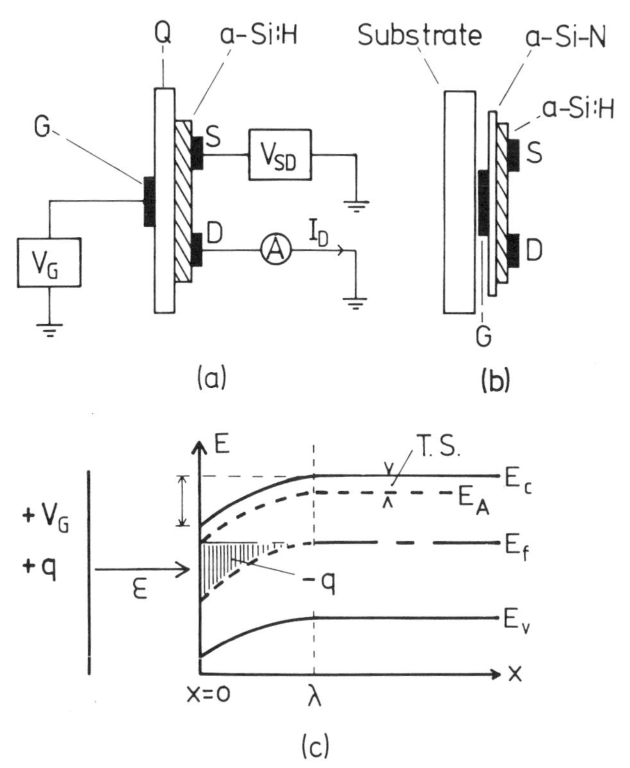

FIG. 1. (a and b) Specimen geometries used in the field-effect experiments. S, D, and G represent the source, drain, and gate electrodes; Q is a thin quartz dielectric. (c) Band diagram showing the formation of an electron accumulation layer near the surface between $x = 0$ and $x \simeq \lambda$. The electric field ϵ from the positive gate electrode induces a charge $-q$ in the a-Si:H. The tail-state distribution TS between E_c and E_A is likely to limit the band bending at the surface.

and D, deposited on the free surface of the a-Si:H film. The measurement consists of applying a relatively small constant voltage V_{SD} between the source–drain contacts and recording the specimen current I_D as voltages V_G, of up to 10 kV, are applied between the gate electrode and the free surface of the a-Si:H, which will remain essentially at ground potential. Measurements are made for both positive and negative V_G.

The main disadvantage of using the quartz substrate as the dielectric is the necessity for large gate voltages. This can be overcome by the thin-film insulator shown in Fig. 1b. A film ($\lesssim 1\ \mu m$) of amorphous silicon nitride (a-SiN) is deposited by the glow-discharge process onto a substrate which carries the evaporated gate electrode G. The a-Si:H is deposited directly onto this SiN film and source and drain electrodes are then evaporated as before. In this way much higher fields can be applied with comparatively low voltages, significantly extending the range of measurements.

As an example, Fig. 2 shows some of the first I_D versus V_G curves for a-Si:H (Spear and LeComber, 1972). The arrangement of Fig. 1a was used in the experiments but with a thin glass dielectric instead of the quartz substrate Q. Although the use of glass is likely to complicate the analysis (Madan *et al.*, 1976) it does not affect the general conclusions that can be drawn from these early results. The striking feature of the curves is the remarkable increase in I_D by over three orders of magnitude for two of the specimens, which occurs when V_G is made positive. As illustrated in Fig. 1c, this means that the extended electron states above E_c are pulled toward the Fermi level E_f in a region that extends from the dielectric/a-Si:H interface ($x = 0$) to $x \simeq \lambda$ in the a-Si:H film; λ depends on V_G and $g(E)$ and lies generally between 100 and 1000 Å. The high electron density in this accumulation layer forms a conducting path between the S and D electrodes, producing the large increase in I_D.

The charge $-q$ induced capacitively in the a-Si:H consists mainly of electrons condensed into localized gap states which have been moved below E_f when V_G is applied. Clearly, for a given V_G, the increase in I_D will be

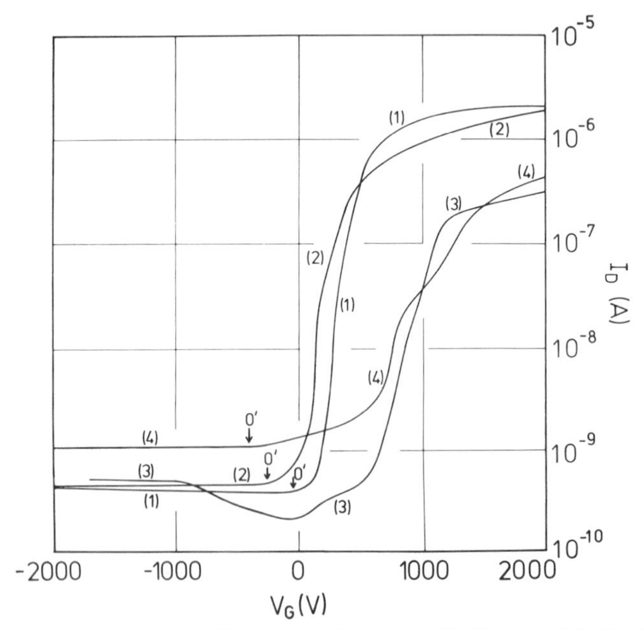

FIG. 2. Examples of early field-effect curves of I_D versus V_G (Spear and LeComber, 1972). (1) and (2) refer to specimens prepared at a substrate temperature of about 300°C; (3) and (4) were deposited at 200°C. O′ denotes the approximate flat-band position.

largest for specimens with the lowest $g(E)$ in the energy range probed. This applies to curves (1) and (2) in Fig. 2, coresponding to specimens deposited at a substrate temperature of about 300°C. It is also likely that for these low $g(E)$ specimens the leveling out of I_D around 10^{-6} A is caused by E_f approaching the onset of the rapidly rising localized tail-state distribution at E_A, as indicated in Fig. 1c. This limitation to I_D will be discussed further in Part VII. Curves (3) and (4), on the other hand, were obtained from specimens deposited at 200°C, which undoubtedly possess a higher overall density of gap states. Figure 2 shows that the sign of the majority carrier, as well as useful general information of $g(E)$ as a function of preparation conditions, can be obtained directly from the experimental results, without detailed analysis.

With increasing negative V_G, one passes through the flat-band position denoted by O′ in Fig. 2. An electron-depletion layer will then be formed giving little further change in I_D. At higher negative V_G, hole conduction through the inversion layer at the surface has been observed in FETs without n^+ contacts.

III. Design and Fabrication of the a-Si:H FET

Two important points were established at an early stage of the present development. The first is that conventional photolithographic techniques, widely used in the semiconductor industry, can be applied successfully to a-Si:H thin-film devices. The second equally relevant point is that the a-Si:H FETs perform adequately at voltages below 15 V, so that they are compatible with modern integrated-circuit voltage levels.

Figure 3a shows a section through an individual a-Si:H device and Fig. 3b illustrates the design of the FET in an array of transparent conducting indium-tin-oxide (ITO) electrodes on glass (Snell *et al.*, 1981a). The first step in the fabrication consists in etching the required ITO pattern; a chromium film is then evaporated and etched to provide the gate electrodes G, about 20 μm in width. The prepared substrate is now coated with 0.3-0.4-μm-thick layers of amorphous silicon nitride (SiN) and undoped silicon; near the end of the run a thin (~ 150-Å) highly doped n^+-Si film is formed. The deposition in the rf glow discharge plasma is a continuous process, controlled by the gas mixture flowing through the reaction tube. Addition of ammonia to the silane produces the SiN dielectric layer, whereas 3000 vppm of phosphine added at the final stage leads to the highly doped n^+ film.

The unwanted Si is etched away and contact holes, denoted by A in Fig. 3, are etched through the silicon for connecting the drain contact to the ITO pads. The top electrode configuration, S and D, is then formed from the

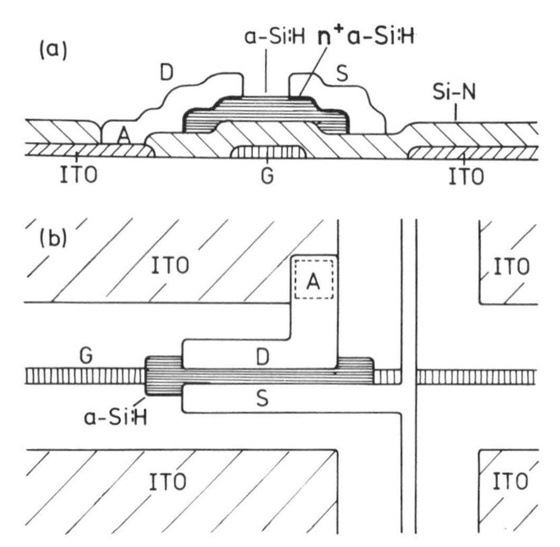

FIG. 3. Design of a-Si : H FET. (a) Section through device. (b) FET in part of matrix array. ITO, indium–tin oxide squares; A, contact hole etched through SiN film. [From Snell *et al.* (1981a).]

evaporated aluminum top metallization, and the final step consists in etching away the conducting n^+ film between S and D. In some devices source–drain lengths have been reduced to about 4 μm; the channel widths are 200–500 μm, depending on the application.

A systematic assessment of the preparation technique has recently been carried out in our laboratory, aimed at optimizing the ON current and the stability of the FETs (Mackenzie *et al.,* 1983). The main factors investigated were the silane–ammonia ratio for the deposition of the silicon nitride and the formation of the SiN/a-Si : H interface. The former not only influenced the dielectric properties of the SiN but also played a significant role in the short-term stability of the FETs. Fortunately, the ratio that was found to give optimum stability also gave a relatively pinhole-free SiN film with a large breakdown strength ($>$ 1 MV cm^{-1}). For the particular reactor used, a gas ratio SiH_4–NH_3 of 1 to 4 gave the best results, although we believe this optimum to be dependent on the deposition unit.

The experiments also showed that the time taken for the changeover of the gases in the deposition unit affected the device stability. Clearly, the time involved in the change from the SiH_4–NH_3 mixture to the pure silane used for the a-Si : H will depend strongly on the size of the deposition unit, gas flow rates, pumping speeds, etc. The experiments led to the conclusion that a rapid changeover of the gases produced the most stable devices. In the system studied, with a reaction chamber of a few liters volume, this changeover could be accomplished with fast pumps within 30 sec.

The n^+ underlay was used with the source and drain contacts to provide good electron-injecting properties. This caused the reproducibility of the contacts to increase, and the ON current obtained in this way increased significantly. For example, devices with n^+ contacts consistently produced source-drain currents a factor of about three higher than the best currents obtained without the n^+ layers.

IV. Direct-Current Characteristics

The dc transfer characteristics of an FET device produced under the optimized conditions described above are shown in Fig. 4 (Mackenzie *et al.*, 1983). The thicknesses of the SiN gate insulator and the a-Si film were each 0.3 μm; the source-drain width and length were 500 μm and about 4 μm, respectively. Figure 5 gives the output characteristics of a similar device. These FETs represent a significant achievement in a-Si:H device technology: ON-currents are typically seven orders of magnitude greater than the OFF-current at $V_G \simeq 0$; source-drain currents of 20 μA are obtainable with applied voltages in the 10-15-V range, and fractions of a milliampere can be obtained if supply voltages are increased to 30-40 V. It should be noted that under the latter conditions the n^+ contacts supply current densities of several hundred amperes per square centimeter to the conducting channel!

To assess the present state of the art it is of interest to compare a-Si:H

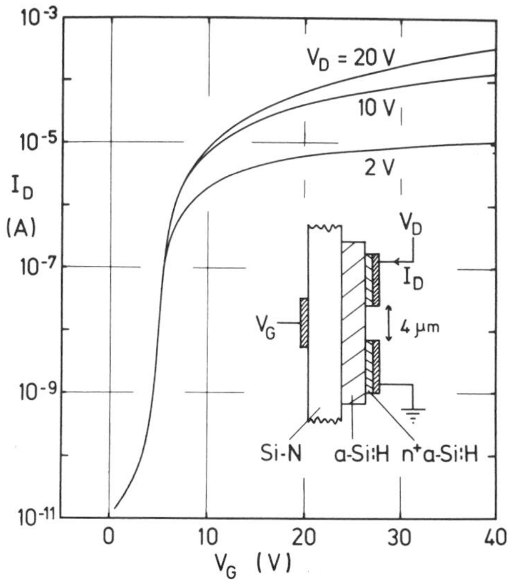

FIG. 4. Transfer characteristics of a-Si:H FET element. The drain current I_D is plotted against the gate voltage V_G for three-drain potentials V_D. [From Mackenzie *et al.* (1983).]

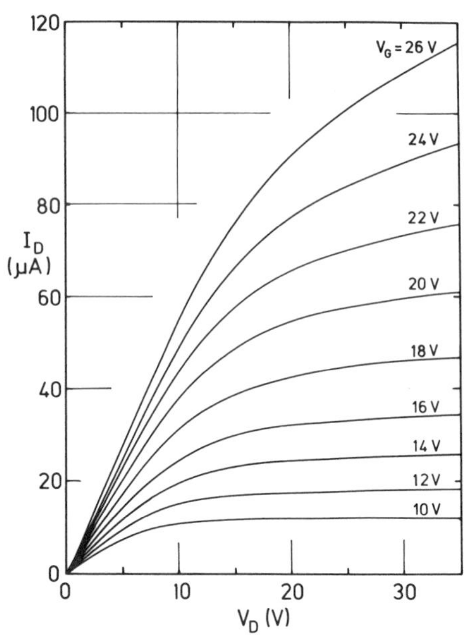

FIG. 5. Output characteristics of a-Si : H FET element ($L = 4\ \mu$m). [From Mackenzie *et al.* (1983).]

FET characteristics published by a number of laboratories engaged in this development. In Fig. 6 some of the most recent results are shown. We follow Powell *et al.* (1981) in plotting the sheet conductance G_S against the field at the insulator/a-Si : H interface in order to allow for different device geometries and gate dielectrics. The interface field has been calculated from $\epsilon_i V_G/\epsilon_{Si} d$, where ϵ_i and ϵ_{Si} are the relative permittivities of the gate insulator and of the a-Si : H, respectively, and d is the thickness of the gate insulator. The curve denoted by D represents the data described in this paper. Curve P is from Powell *et al.* (1981) at the Philips Research Laboratories; the curves marked T are from data published by Matsumura *et al.* (1981) and Hayama and Matsumura (1981) and curves C and F represent recent results from the Canon (Okubo *et al.*, 1982) and Fujitsu (Kawai *et al.*, 1982) laboratories, also in Japan.

Curve X represents some new results from the Xerox Corporation laboratories (Tuan *et al.*, 1982) for a double-gate device. In this FET the a-Si : H layer is sandwiched between two SiN insulating films so that essentially two electron accumulation channels (one at each a-Si : H – SiN interface) can be produced by the application of the appropriate gate voltages. It can be seen that at present these structures give the same maximum sheet conductance as the best single-gate devices. It is also interesting to note that in double-gate

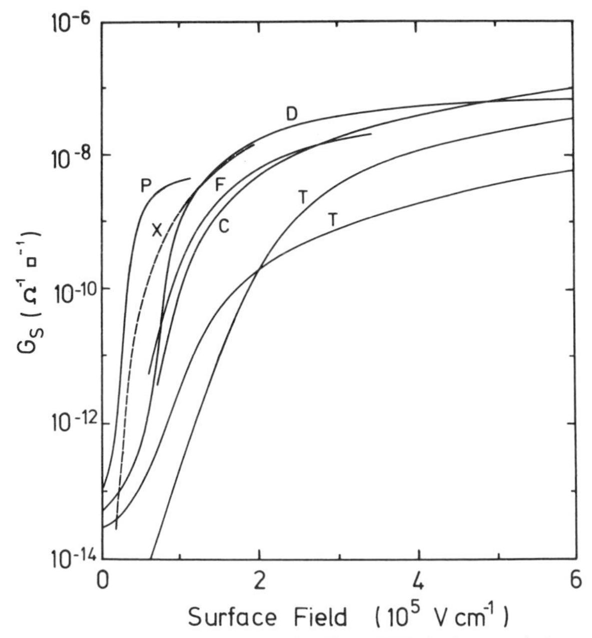

FIG. 6. A comparison of the performance of a-Si:H FET devices made by various research groups. D, Dundee; P, Philips; F, Fujitsu; X, Xerox (double gate); C, Canon; T, Tokyo. [From Mackenzie *et al.* (1983).]

operation the observed current is more than the sum of the currents measured when either gate is operated independently. Powell and Pritchard (1983) suggest that this arises from the interaction of the two electron-accumulation layers. In an earlier paper (Mackenzie *et al.*, 1983), we proposed that this effect may possibly be associated with some modification of the injecting properties at the source–drain electrodes. Both these explanations could play a part, with the former more important at the lower current levels and the latter at higher currents.

In their recent paper Powell and Pritchard (1983) presented calculations of the field-effect sheet conductance G_s that included the effects of any interfacial states and any fixed charge at both surfaces of the a-Si:H film in a single-gate FET. They conclude that in devices 0.3 μm thick and with a bulk density of states of about 10^{17} cm^{-3} eV^{-1} at E_f, the gate field will have some effect on the surface charge distribution on the opposite side of the a-Si:H film which carries the source–drain electrodes. This rather unexpected effect could modify performance of the FET and suggests the importance of controlling the fabrication and processing of both surfaces in the FET production.

V. Dynamic Performance

To assess the dynamic operation of the FETs, the performance of recent devices was tested under conditions approaching those in a liquid crystal (LC) panel (Mackenzie *et al.*, 1983). The circuit is shown in Fig. 7 together with the waveforms applied to drain and gate. The 10-pF capacitor C_{LC} in the circuit simulates the LC cell capacitance in an addressable display panel or alternatively a stray capacitance in other applications. The voltage V_D applied between drain and ground is alternated between -10 V and $+10$ V, in this case with a period of about 3 msec. When the gate voltage V_G is switched from -5 V to $+15$ V, the FET turns ON and charges the capacitor C_{LC} toward the drain potential V_D as shown by the voltage pulses V_{LC}. The very high OFF- resistance of the FET means that the accumulated charge on the capacitor is stored for times of the order of hundreds of milliseconds.

An important parameter for addressing the elements in the LC display is the gate pulse width required to charge the capacitor to a particular value of V_{LC}. In Fig. 8 the rms value of V_{LC} is plotted against the width of the applied gate pulse for a-Si:H FETs with the indicated source–drain lengths. The data for $L = 40$ μm is taken from earlier work (Snell *et al.*, 1981a), and the considerable improvement in speed obtainable with present devices is clearly evident. For the 4-μm devices, the voltage rises to 4 V rms within a gate pulse width of about 5 μsec. Even increasing L to 10 μm only requires a gate pulse width of about 10 μsec to reach the same V_{LC}. This voltage level should be suitable for low-threshold twisted-nematic liquid crystal displays,

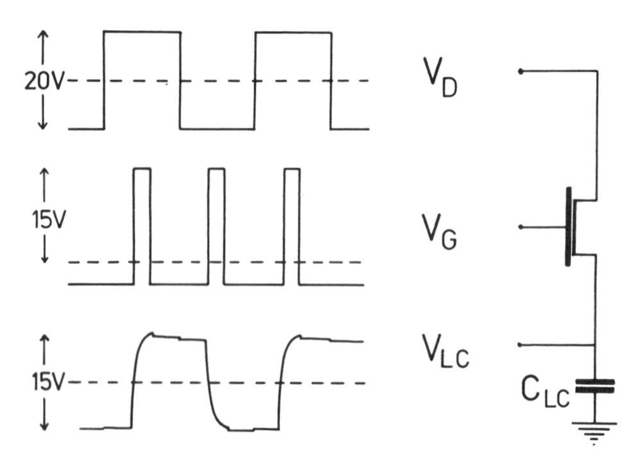

FIG. 7. Circuit and waveforms used to study the dynamic behavior of the FET and the simulated liquid crystal element. V_D, potential applied to drain; V_G, gate pulses; V_{LC}, observed potential across 10-pF capacitor simulating liquid crystal element. The dashed lines indicate 0 V.

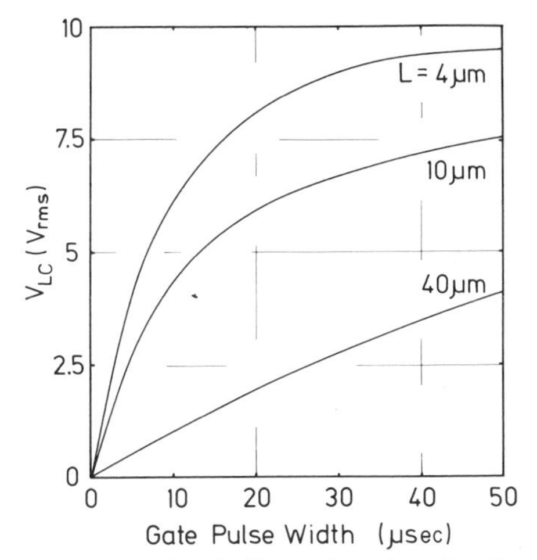

FIG. 8. Voltage level V_{LC} across the 10-pF capacitor plotted against the width of the gate pulse necessary to obtain the value of V_{LC}, using the circuit shown in Fig. 7. $V_D = \pm 10$ V; $V_G = +15$ V. [From Mackenzie *et al.* (1983).]

and with a frame time of 25 msec would indicate that an array consisting of a few thousand lines could, in principle, be addressed by the usual multiplexing procedure.

VI. Reproducibility and Stability

An important factor in the FET development is the reproducibility and uniformity of characteristics in an array of devices. Figure 9 shows the transfer characteristics of a linear array of elements (Snell *et al.*, 1981a). For clarity, successive curves have been displaced along the V_G axis by 10 V. The similarity of these curves demonstrates the uniformity of performance that can be achieved in a given deposition. All elements had $R_{OFF} > 10^{11}$ Ω and threshold voltages were within ±0.4 V of the mean value.

Figure 10 illustrates the short-term dc stability of the devices (Mackenzie *et al.*, 1983). The output characteristics of one FET are shown for 10 successive scanning cycles of V_G from −10 V to +45 V and back. The gate voltage was scanned in both directions at a rate of about 0.5 V sec^{-1}. The traces show remarkably little drift or hysteresis; in fact, the maximum variation in V_G is approximately 0.4 V for a given value of source–drain current.

In an earlier paper (Snell *et al.*, 1981a) we reported on the long-term dynamic stability of a-Si:H FETs. Tests on one specimen have been continued during the last 2 years and Fig. 11 shows photographs of the

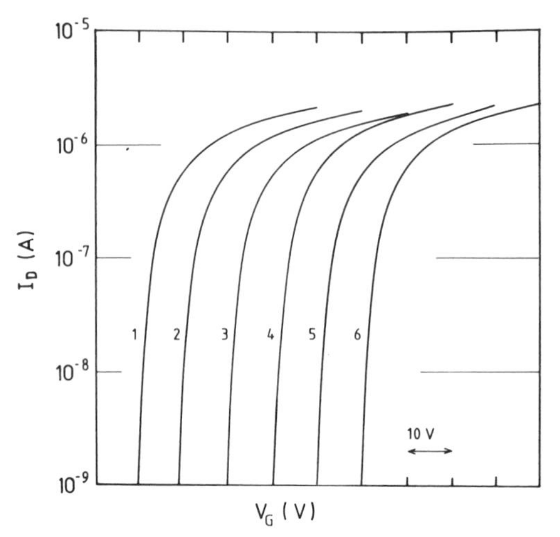

FIG. 9. Transfer characteristics of six FETs in a linear array as a test of device reproducibility. Successive characteristics have been displaced by 10 V along V_G axis. $V_D = 2$ V. [From Snell *et al.* (1981a).]

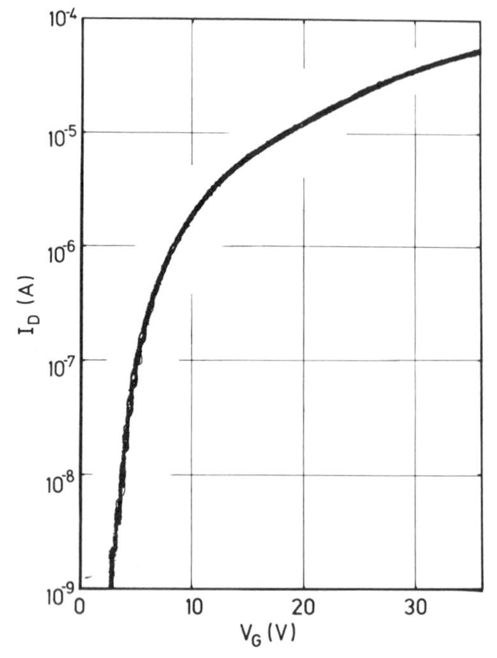

FIG. 10. Multiple transfer characteristics obtained from a single FET by sweeping the gate voltage from −10 to +45 V and back a total of 10 times $V_D = 20$ V; $L = 20$ μm. [From Mackenzie *et al.* (1983).]

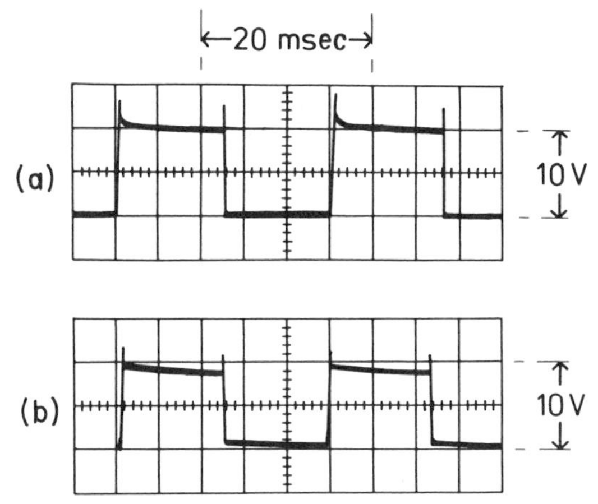

FIG. 11. Comparison of the dynamic performance of an a-Si:H FET in (a) September 1980 and (b) December 1982 after more than 6×10^9 switching operations. [From Mackenzie *et al.* (1983).]

observed waveforms. The potential was measured across a 10-pF capacitor in a circuit similar to that shown in Fig. 7, but one operating at a repetition rate of 80 Hz. The upper trace gives the initial waveform, taken in September 1980, whereas the lower trace shows the pulses observed in December 1982 after some 6×10^9 switching operations. Evidently after 2 years of continuous operation the peak voltage level changed by only 10–15%. This is encouraging especially since this device is totally unpassivated and unencapsulated.

VII. Investigation of the ON-State

As shown in the preceding sections the present FETs have significantly higher ON-currents than previously—larger than can be accounted for by the reduction in source–drain separation alone. To gain some understanding of the reason for this, a series of experiments has been performed to investigate the transport in the conducting channel of the devices (Mackenzie *et al.*, 1983).

The temperature dependence of the ON-current and also of the field-effect mobility μ_{FE} have been studied in a number of optimized devices. Since both these quantities behave in a similar way, we shall concentrate here on the field-effect mobility.

Two methods have been used to determine μ_{FE}. Both rely on the following simple expression for the source–drain current (e.g., Müller and

Kamins, 1977)

$$I_D = (C_i W/L)\mu_{FE}(V_G - V_T - \tfrac{1}{2}V_D)V_D, \tag{1}$$

where W and L refer to the width and length of the source–drain electrodes, C_i is the gate capacitance per unit area and V_T is the threshold voltage. For $V_D \ll V_G$ Eq. (1) simplifies to

$$I_D = (C_i W/L)\mu_{FE}(V_G - V_T)V_D. \tag{2}$$

At low V_D the experimental data for I_D (see Fig. 5, for example) show a linear increase with V_D so that Eq. (2) can be applied in this range. In the second method of determining μ_{FE} the value of I_D at the onset of saturation in the output characteristics is used. According to the simple theory of the device I_D saturates for $V_D > V_G - V_T$.

In the experiments we have used the circuit shown in the insert to Fig. 12 in which V_D is connected directly to V_G. Under these conditions

$$I_D^{1/2} = ((C_i W/2L)\mu_{FE})^{1/2}(V_D - V_T). \tag{3}$$

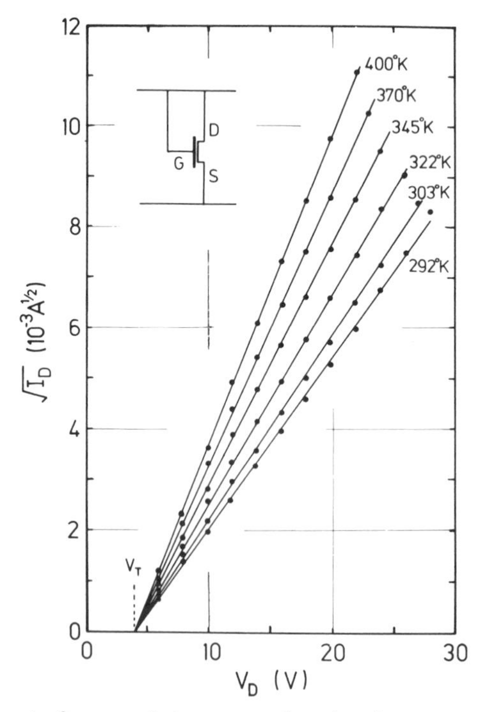

FIG. 12. Square root of source–drain current plotted against source–drain voltage, with gate and drain connected together, for various temperatures. [From Mackenzie *et al.* (1983).]

Equation (3) predicts that a plot of $\sqrt{I_D}$ versus V_D should give a straight line, whose slope can be used to determine μ_{FE} and which has an intercept on the V_D axis at $V_D = V_T$ corresponding to the threshold voltage. Figure 12 shows data plotted in this way for one specimen (Mackenzie *et al.*, 1983), taken at the temperatures indicated. At each temperature a very reasonable straight line can be drawn through the points enabling a value of μ_{FE} to be calculated. Furthermore, the curves all intersect the V_D axis at a value of about 4 V for this particular sample, which is identified as the threshold voltage.

Values of μ_{FE} determined by both the above methods agree to within 10% or better and, at room temperature, were typically 0.2–0.3 $cm^2\ V^{-1}\ sec^{-1}$ in magnitude. The sample represented in Fig. 12 has a room-temperature mobility of 0.31 $cm^2\ V^{-1}\ sec^{-1}$ with an activation energy E_μ of 0.11 eV. Figure 13 shows μ_{FE} versus $10^3/T$ curves for three devices measured under these conditions, and Fig. 14 summarizes the values of E_μ as a function of V_G. Curve a in Fig. 14 represents data from earlier samples that did not employ n^+ contacts at the source and drain contacts. Curve c represents data from the latest optimized FETs and curve b an intermediate stage in this development. At zero gate voltage, all three curves lead to E_μ values of $\simeq 0.7$ eV corresponding to $E_c - E_f$ of undoped a-Si with conduction in the electron extended states. With increasing V_G, $E_\mu \simeq E_c - E_f$ decreases, eventually approaching 0.2 eV for curve a and 0.1 eV for curve c.

These results point to the fundamental reason for the improvement in the

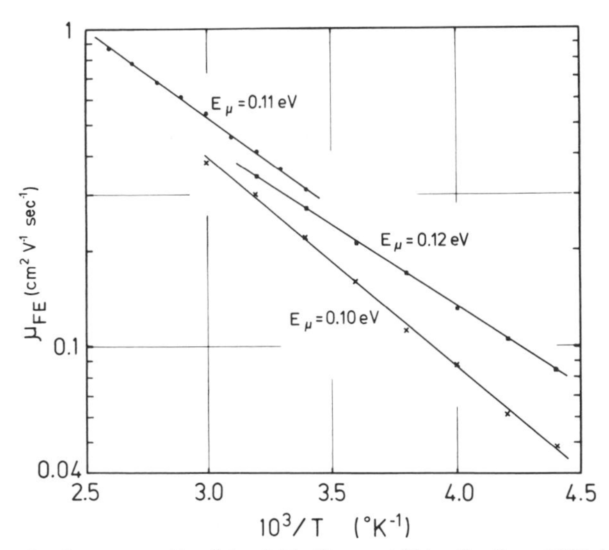

FIG. 13. Activation energy E_μ of the field-effect mobilities for three FET devices. [From Mackenzie *et al.* (1983).]

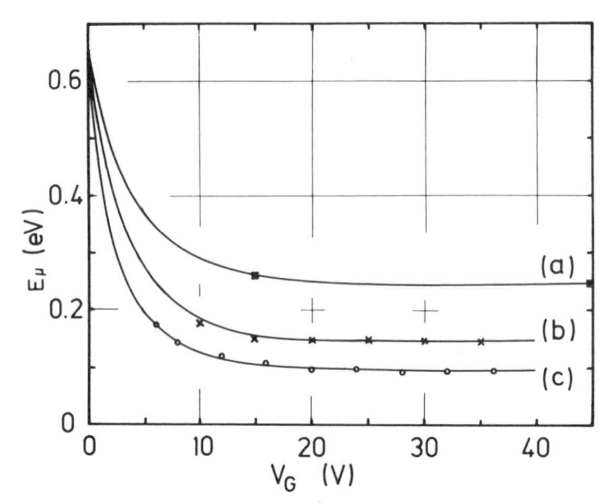

FIG. 14. Variation of activation energy of the source–drain current with gate voltage: (a) early devices without n^+ contacts; (b) intermediate stage of the optimization; (c) optimized devices described in the paper by Mackenzie *et al.* (1983).

ON-current of recent FETs—quite apart from the increase associated with the reduced source–drain length. The depth of the accumulation layer at the interface has been increased by about 0.1 eV, which means that the average carrier density in the conducting channel is now larger by the factor $\exp(0.1\ \text{eV}/kT) \sim 50$ at room temperature. This decrease in $E_c - E_f$ could be associated with several causes. It is possible that the deposition conditions used to form the present devices produce a material with a somewhat increased long-range order. As has been established in our work on the glow-discharge microcrystalline phase (Willeke *et al.*, 1982), even a very limited increase in order will begin to delocalize the electron tail states of the material. Previous work on earlier material established the tail-state width $E_c - E_A$ in Fig. 1) at about 0.2 eV in agreement with the data for curve a of Fig. 14, so that a reduction to 0.1 eV by an optimized preparation technique, coupled with a very low density of interfacial states, would be entirely feasible.

Another possibility may be that the n^+ contacts have removed any previous contact limitations to the supply of excess electrons. The high carrier density could lead to the filling up of the lower lying tail states and allow E_f to move closer toward E_c. With present devices we induce a maximum of about 4×10^{12} electrons cm^{-2} into the a-Si:H conducting channel. If the latter is assumed to extend for about 100 Å at large V_G, then for a linear tail-state distribution of 0.2-eV width having $g(E_c) \sim 10^{21}\ \text{cm}^{-3}\ \text{eV}^{-1}$, a simple estimate suggests that E_f would lie about 0.16 eV below E_c, even at the highest gate voltages. However, in view of the assumptions inherent in this estimate we can not entirely rule out this explanation. In fact it would not be surprising if this and the previous suggestion were both playing a part in reducing the observed activation energies. Whatever the reason, it seems unlikely that any further dramatic improvements in the FET performance could be achieved solely by improvements in the a-Si:H (rather than, for example, by using significantly thinner SiN gate insulators). We therefore believe that the ON-currents reported here are likely to be close to the maximum possible for the a-Si:H. In this connection it is of interest to note the field-effect mobilities measured by the four groups that produce devices with the highest performance, namely, 0.31 $cm^2\ V^{-1}\ sec^{-1}$ by Dundee (Mackenzie *et al.*, 1983), 0.3 $cm^2\ V^{-1}\ sec^{-1}$ by the Canon Research Center (Okubo *et al.*, 1982), 0.2 $cm^2\ V^{-1}\ sec^{-1}$ by the Xerox Palo Alto Research Center (Thompson *et al.*, 1982), and 0.26 $cm^2\ V^{-1}\ sec^{-1}$ by the Philips Research Laboratories (Powell, 1983). These remarkably similar values would support the suggestion that the present layers are approaching an optimum for a-Si.

VIII. Radiation Hardness of a-Si:H FETs

In a recent paper French *et al.* (1983) investigated the effect of γ radiation on the properties of a-Si:H FETs by exposing devices to doses of up to 5 Mrad (Si). The notation implies that this energy is absorbed in the a-Si film.

The effects of ionizing radiation on crystalline MOS structures have been widely studied to assess the reliability of MOS devices in space and nuclear environments. The most important effect is a negative shift of the threshold voltage V_T, which provides a sensitive measure of any electronic changes at the insulator–Si interface. The observed shift ΔV_T arises because the irradiation creates electron–hole pairs in the insulating oxide layer of the crystalline device. The electrons, having a relatively high mobility, are extracted from the oxide, while the majority of the holes become trapped. The latter form a fixed positive charge sheet in the insulator which produces band bending in the silicon, in addition to that caused by any applied voltage on the gate. One would therefore expect ΔV_T to be negative, as is in fact found in most crystalline devices. It has also been found that ΔV_T is increased if during irradiation a potential V_G is applied to the gate electrode. In addition to the shift in threshold, irradiation generally produces some degradation of the transconductance of the FET.

In the present study (French *et al.*, 1983) a series of FETs with closely matched characteristics were irradiated at room temperature using a dose rate of 0.17 Mrad (Si) hr^{-1}. The first group received a γ-ray dose of 1 Mrad, the second of 5 Mrad. Each included devices held during irradiation at V_G values of -8 V, -5 V, 0 V, $+5$ V, and $+8$ V with respect to the source and drain electrodes, which were connected together. The characteristics of all specimens were carefully measured before and after irradiation. It was noted that the Corning 7059 glass substrate showed pronounced darkening due to radiation damage.

Figures 15a and 15b summarize the results for the threshold shifts after the 1-Mrad and 5-Mrad irradiations. The shift ΔV_T is plotted against V_G, the gate voltage applied during the irradiation. In spite of the scatter of experimental points, results for the 1-Mrad dose show a tendency toward a small negative threshold shift for both positive and negative V_G. This is consistent with observations made on crystalline devices, and it is possible that a similar mechanism is responsible in both cases. However, Fig. 15b for a dose of 5 Mrad shows a predominantly positive ΔV_T, suggesting that other mechanisms begin to play a part during the longer irradiation period.

The magnitude of the threshold voltage shift is less than 3 V for all the samples investigated, and similar results were obtained if all electrodes were left floating. This compares well with "radiation-hardened" crystalline MOSFETs having a similar gate oxide thickness. In general, reducing the oxide thickness improves the radiation resistance of the device, and a

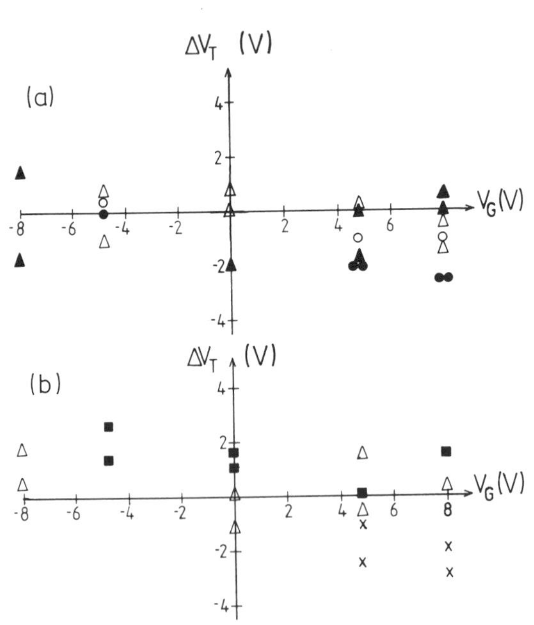

FIG. 15. Change in threshold voltage ΔV_T plotted against gate voltage V_G applied during irradiation for samples given (a) 1-Mrad and (b) 5-Mrad γ-ray doses. Each experimental point is representative of a different device. Samples prepared in the same deposition run are denoted by the same symbol. [From French *et al.*, (1983).]

relationship varying from a linear to a square law dependence is observed between ΔV_T and the oxide thickness. Hu *et al.* (1981) have reported a crystalline Si FET with a 450-nm-radiation-hardened oxide which led to $\Delta V_T = -5$ V after a 0.33-Mrad (Si) exposure to an electron beam while at $V_G = +5$ V. A crystalline device with only 40 nm of hardened oxide gave a threshold shift of 0.3 V after an irradiation of 1 Mrad from a Co^{60} source with a +10-V gate bias (Nordström and Gibbon, 1981).

Figure 16 shows the transfer characteristics of some typical a-Si:H FETs used in the experiments, both before and after irradiation. The postirradiation curves have been shifted by the amount of their ΔV_T (stated for each curve) to facilitate the comparison and show clearly the changes produced by the irradiation. It is evident that no major deterioration in performance has occurred; in fact, the decrease in transconductance was less than 10% in all cases. When the above-mentioned samples were annealed at 130°C, with their gate, source, and drain contacts connected together, the FETs returned to their original transfer characteristics and threshold voltages.

On the basis of the present experiments, it is concluded that a-Si:H FETs are remarkably resistant to the effect of radiation, despite the relatively thick dielectric layer used at present in these devices.

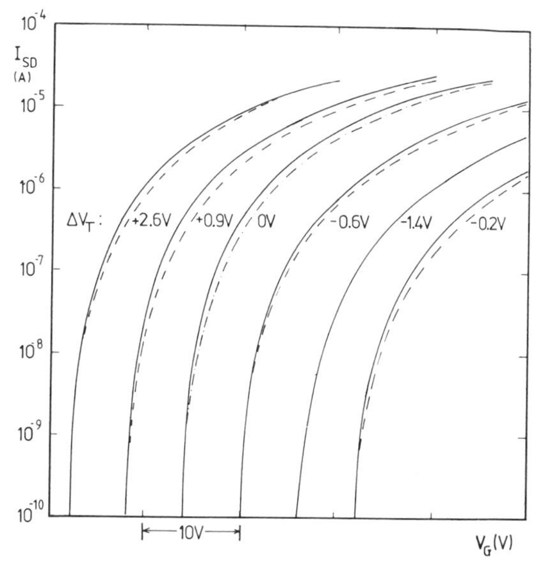

FIG. 16. Transfer characteristics for six FETs before (solid curves) and after irradiation with 5 Mrad (dashed curves) and 1 Mrad (dotted-dashed curves). For clarity the characteristics after irradiation have been shifted along the V_G axis by the $-\Delta V_T$ values shown for each device. $V_{SD} = 15$ V. [From French *et al.*, (1983).]

IX. Some Possible Applications of a-Si:H FETs

As mentioned earlier, probably the most promising application of a-Si:H FETs is their use in addressable liquid crystal displays, which will be discussed in the following chapter. Here we shall review the results of some exploratory experiments on the application of the FETs to a number of logic and image-sensing circuits.

Matsumura, Hayama, and co-workers first showed that an integrated inverter circuit (Matsumura and Hayama, 1980) and an image sensor (Matsumura *et al.*, 1980) can be fabricated from a-Si:H FETs. The devices used in their work were made from a-Si:H deposited in a dc glow discharge and with SiO_2 produced by low-pressure CVD as the gate insulator. Although these FETs were relatively leaky and turned on rather slowly, they demonstrated the feasibility of using the a-Si technology in these circuits. At the same time, work in our laboratory had been concerned with similar applications (Snell *et al.*, 1981b; LeComber *et al.*, 1981), initially aimed at providing integrated a-Si drive circuits for the liquid crystal display panels mentioned above. In the following discussion we shall briefly review the work on the inverters and related devices.

1. Simple Logic Circuits

The inverter circuit described in recent publications by Nara and Matsumura (1982) is shown in Fig. 17. It consists of an *n*-channel enhancement mode driver FET and a *p*-channel load FET produced by a hot-cathode arc-discharge decomposition of silane gas mixtures (Matsumura and

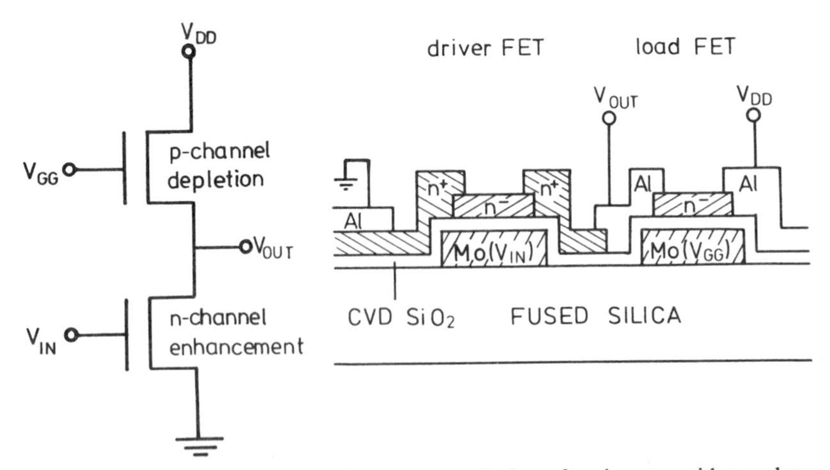

Fig. 17. The equivalent circuit and cross-sectional view of an inverter with a *n*-channel enhancement mode driver FET and a *p*-channel load FET. [From Nara and Matsumura (1982).]

Uchida, 1981). Figure 17 also contains a cross-sectional view of the inverter. The transfer curves obtained from the devices are shown in Fig. 18. The solid lines are the experimental data and the dashed lines were calculated from the FET parameters. Although there are some differences between the curves, both show the steplike behavior associated with the inverter logic.

In the devices investigated in our laboratory (Snell *et al.*, 1981b; Le-Comber *et al.*, 1981) a load resistor is formed by an integrated a-Si:H gap cell. This is simpler and has the further advantage that the a-Si:H resistor requires less space than an FET load. Figure 19 shows a section through the integrated device and also its layout in plan view. The input voltage is applied to the gate of the FET, the source is connected to ground, and the output is taken from the drain connection. To obtain the appropriate value of load resistance, a doped a-Si:H layer, about 250 Å thick and about $3 \times 10^8\ \Omega \square^{-1}$, was deposited. This was etched away from the source–drain gap of the FET after the top Al metallization, leaving doped a-Si:H under the source and drain, which had the added advantage of improving the electrical contact at these electrodes (see Part III).

The dashed curve in Fig. 20 shows the transfer characteristics of such a device having a load of approximately 30 MΩ and measured with a supply potential of 15 V. The inverter logic is clearly seen: the output changes from about 14.5 to 2 V as the input swings from about 5 to 15 V. The full line in

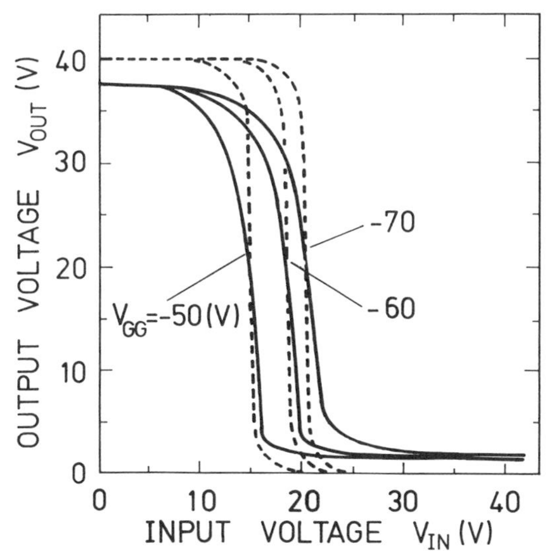

FIG. 18. Transfer characteristics of the inverter shown in Fig. 17. V_{DD} was fixed at 40 V with $V_{GG} = -70$ V, -60 V, and -50 V. The dashed lines are the theoretical curves, the solid lines from experiment. [From Nara and Matsumura (1982).]

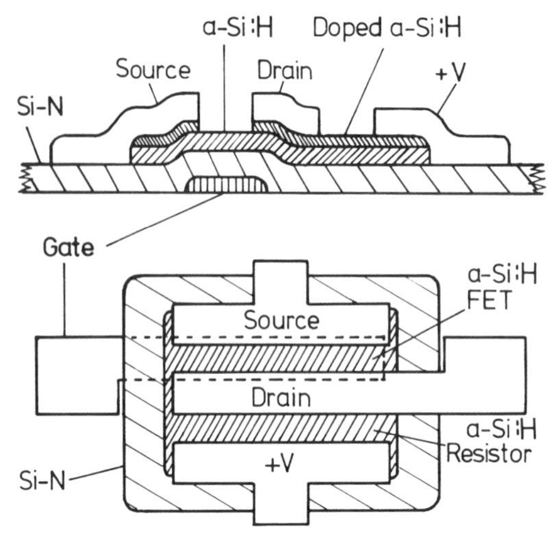

FIG. 19. Section and plan view of the integrated a-Si : H inverter. [From Snell *et al.* (1981b).]

Fig. 20 shows the characteristics of three such devices connected in series. As well as sharpening the characteristics this demonstrates the important point that the output from one device can be successfully used as the input for successive stages.

By simple extension of the basic design given above, logic circuits such as NAND and NOR gates as well as bistable multivibrators have been produced (Snell *et al.*, 1981b; LeComber *et al.*, 1981). One of the primary aims of the work was to investigate the possibility of integrated-drive circuits for the liquid crystal panels, in particular the fabrication of a shift register to

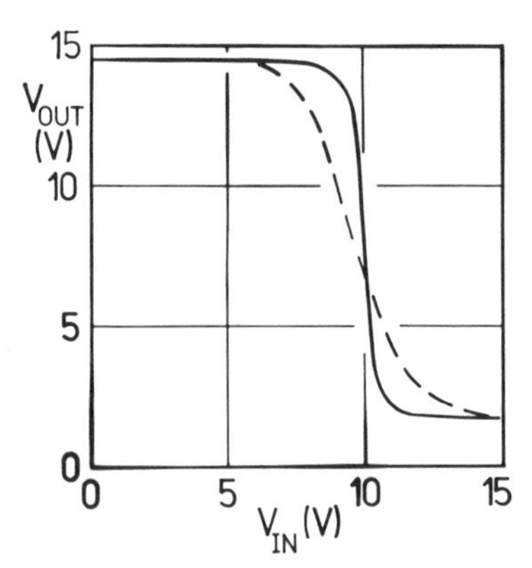

FIG. 20. Characteristics of single a-Si : H (dashed curve) inverter and three inverters connected in series (solid curve). $V_{SD} = 15$ V. [From Snell *et al.* (1981b).]

address sequentially the gate buses. Recently, a four-stage shift register using a-Si:H FETs has been made in which each stage can shift a 20-msec pulse by 10 msec. Nara and Matsumura (1982) have interconnected nine inverters to produce an integrated ring oscillator. The maximum oscillation frequency of this device was 500 Hz with a power-delay product per gate of about 480 pJ and a minimum propagation delay of about 110 μsec.

All the circuits mentioned in this section are slow by crystalline standards, although it must be emphasized that these exploratory a-Si:H circuits were in no way optimized for speed of response. The limitations are briefly discussed in Part X.

2. Addressable Image Sensors

In this section we review the work of Matsumura *et al.* (1980) and our results (Snell *et al.*, 1981b; LeComber *et al*, 1981) on a-Si:H image sensors. The circuit used initially by both groups was that originally suggested by

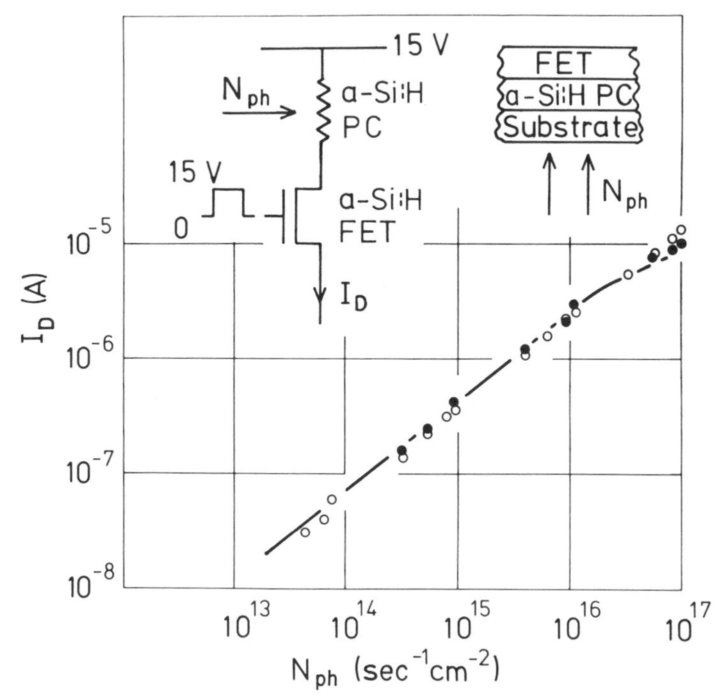

FIG. 21. The output current I_D from an elemental integrated-image sensor as a function of the incident photon flux N_{ph}. The experimental points were obtained when the sensor was addressed individually (●) and as part of an array (○). It is found that $I_D \propto N_{ph}^{0.8}$. The photosensor area was 1.4×10^{-3} cm^2. The full line was calculated from the individual characteristics of the FET and the a-Si:H photoconductor. [From Snell *et al.* (1984).]

Weimer *et al,* (1969) and consisted of an a-Si photoconductor, a thin-film capacitor, and an a-Si:H FET. More recently we have dispensed with the capacitor and investigated linear arrays of 10 sensing elements (Snell *et al.*, 1984), each consisting of an a-Si photoconductor with an a-Si:H FET integrated vertically above it as shown schematically in Fig. 21. The characteristics of one such vertical integrated-circuit image sensor are plotted in Fig. 21, which shows the output current I_D as a function of the white-light photon flux N_{ph} incident on the photoconductor. During the "read" operation the FET was pulsed on with $V_G = 15$ V. The solid line represents the response calculated from the individual characteristics of the FET and the photoconductor and the circles represent the experimental points when the device is operated independently (full circles) or when it is addressed within an array (open circles). The output current I_D varies with the photon flux N_{ph} as $I_D \propto N_{ph}^{0.8}$ over $3\frac{1}{2}$ orders of magnitude of light intensity, providing an excellent gray scale. With an output load resistor of 10 kΩ the rise and fall times of the current pulse were both less than 10 μsec, suggesting that in principle less than 100 msec would be required to read the information from an array consisting of 1000 lines of image-sensing devices.

Further work on this potentially important application is presently being carried out at Dundee and in a number of other laboratories.

X. Limitations of Present a-Si:H FETs

The work described above has clearly demonstrated the feasibility of using a-Si:H FETs in a number of practical circuits. In particular, the devices presently available already offer a viable solution to the problem of multiplexing liquid crystal displays (Snell *et al.*, 1981a; Mackenzie *et al.*, 1983). However, the factor that appears most likely to limit their use in logic and other applications is their frequency response. This property will generally be determined by the time required for the ON-current to charge circuit capacitances, so that an obvious improvement would be achieved with higher ON-currents. The problem can be approached in two ways. First, the device geometry can be optimized, for example, by a reduction in the source–drain separation. Second, one can attempt to optimize the material properties that limit the ON-conductance, namely, the interface state density at the a-Si:H–insulator boundary and the density of states in the a-Si:H. From the discussions in Parts IV and VII of this chapter, it is clear that considerable progress in both these directions has been made. It is now possible to make a-Si:H FETs that can charge a capacitance of 1 pF to a few volts in under 1 μsec, so that operation at frequencies approaching 1 MHz may be feasible in certain applications. The maximum operating frequency of an a-Si FET can be estimated from the "figure of merit" f_m, given by the gain-bandwidth product (Sze, 1969)

$$f_m \simeq \mu_{FE} V_G / 2L^2.$$

With the optimum value of $\mu_{FE} \simeq 0.3$ cm^2 V^{-1} sec^{-1} (see Part VII), $V_G = 15$ V and $L = 4$ μm, a value of $f_m \sim 5$ MHz is obtained. In view of the worldwide effort now being made on a-Si:H development it is likely that arrays of such devices will be used increasingly in applications in which this inherent frequency limitation is relatively unimportant.

References

Brody, T. P., Asars, J. A., and Dixon, G. D. (1973). *IEEE Trans. Electron Devices* **ED-20,** 995.

French, I. D., Snell, A. J., LeComber, P. G., and Stephen, J. H. (1983). *Appl. Phys.* **A31,** 19.

Hayama, H., and Matsumura, M. (1981). *Appl. Phys. Lett.* **36,** 754.

Hu, G. J., Aitken, J. M., and Dennard, R. H. (1981). *IEEE Trans. Nucl. Sci.* **NS-28,** 4102.

Kawai, S., Takagi, N., Kodama, T., Asama, K., and Yangisawa, S. (1982). *SID '82 Digest,* p. 42.

LeComber, P. G., Spear, W. E., and Ghaith, A. (1979). *Electron. Lett.* **15,** 179.

LeComber, P. G., Snell, A. J., Mackenzie, K. D., and Spear, W. E. (1981). *J. Phys. Colloq. Orsay, Fr.* **42,** C4-423.

Mackenzie, K. D., Snell, A. J., French, I., LeComber, P. G., and Spear, W. E. (1983). *Appl. Phys.* **A31,** 87.

Madan, A., and LeComber, P. G. (1977). *In* "Amorphous and Liquid Semiconductors" (W. E. Spear, ed.), p. 377. CICL, University of Edinburgh.

Madan, A., LeComber, P. G., and Spear, W. E. (1976). *J. Non-Cryst. Solids* **20,** 239.

Matsumura, M., and Hayama, H. (1980). *Proc. IEEE* **68,** 1349.

Matsumura, M., and Uchida, Y. (1981). *J. Phys., Colloq. Orsay, Fr. Paris* **42,** C4-671.

Matsumura, M., Hayama, H., Nara, Y., and Isibashi, K. (1980). *IEEE Trans. Electron Devices Lett.* **EDL-1,** 182.

Matsumura, M., Kuno, S., and Uchida, Y. (1981). *J. Phys. Colloq. Orsay, Fr.* **42,** C4-519.

Müller, R. S., and Kamins, T. L. (1977). "Device Electronics for Integrated Circuits," pp. 350ff. J Wiley, New York.

Nara, N., and Matsumura, M. (1982). *Jpn. J. Appl. Phys.* **21,** Suppl. 21-1, 257.

Nordström, T. V., and Gibbon, C. F. (1981). *IEEE Trans. Nucl. Sci.* **NS-28,** 4349.

Okubo, Y., Nakagiri, T., Osada, Y., Sugata, M., Kitihara, N., and Hatanaka, K. (1982). *SID '82 Dig.* p. 40.

Powell, M. J. (1983). *In* "Insulating Films on Semiconductors" (J. F. Verweig and D. R. Wolters, eds.), p. 245. North-Holland Publ., Amsterdam.

Powell, M. J., and Pritchard, J. (1983). *J. Appl. Phys.* **54,** 3244.

Powell, M. J., Easton, B. C., and Hill, O. F. (1981). *Appl. Phys. Lett.* **38,** 794.

Snell, A. J., Mackenzie, K. D., Spear, W. E., LeComber, P. G., and Hughes, A. J. (1981a). *Appl. Phys.* **24,** 357.

Snell, A. J., Spear, W. E., LeComber, P. G., and Mackenzie, K. D. (1981b). *Appl. Phys.* **26,** 83.

Snell, A. J., Doghmane, A., LeComber, P. G., and Spear, W. E. (1984). *Appl. Phys. A,* in press.

Spear, W. E. (1974). *In* "Amorphous and Liquid Semiconductors" (J. Stuke and W. Brenig, eds.), p. 1. Taylor and Francis, London.

Spear, W. E., and LeComber, P. G. (1972). *J. Non-Cryst. Solids* **8/10,** 727.

Sze, S. M. (1969). "Physics of Semiconductor Devices," p. 573. Wiley (Interscience), New York.

Thompson, M. J., Johnson, N. M., Mayer, M. D., and Lujan, R. (1982). *IEEE Trans. Electron Devices* **ED-29,** 1643

Tuan, H. C., Thompson, M. J., Johnson, N. M., and Lujan, R. A. (1982). *IEEE Trans. Electron Device Lett.* **EDL-3,** 357.

Weimer, P. K., Pike, W. S., Sadasiv, G., Shallcross, F. V., and Meray-Horvath, L. (1969). *IEEE Spectrum* **6,** 52.

Willeke, G., Spear, W. E., Jones, D. I., and LeComber, P. G. (1982). *Philos. Mag. B* **46,** 177.

SEMICONDUCTORS AND SEMIMETALS, VOL. 21, PART D

CHAPTER 7

a-Si : H FET-Addressed LCD Panel

D. G. Ast

DEPARTMENT OF MATERIALS SCIENCE AND ENGINEERING
CORNELL UNIVERSITY
ITHACA, NEW YORK

I. Introduction

Liquid crystals were discovered in 1888 by F. Reinitzer, who observed that certain organic crystals, when molten, could form fluid phases with anisotropic optical properties. Eighty years later, the first display using a liquid crystal was built by a group at RCA (Heilmeier *et al.,* 1968a,b). Widespread commercial use of liquid crystal displays, however, became feasible only after stable liquid crystals operating at room temperature were developed (Gray, 1975). Presently, liquid crystal displays (LCDs) are used commercially in applications in which low power consumption is important and only modest amounts of monochromic information need be displayed. The display of larger amounts of information requires the presence of additional electronic devices. To understand why these devices are required, the operating principles of liquid crystal displays will be reviewed.

II. Properties of Liquid Crystals

Liquid crystals consist of rigid rodlike molecules whose interaction tends to align them parallel to each other (for more details, see de Gennes, 1974;

ISBN 0-12-752150-X

Chandrasekhar, 1977). This cooperative behavior results in weak elastic properties. Then, the application of an electric field can easily change the molecular orientation, which is initially fixed by the mechanical boundary conditions. The concomitant changes in the optical properties form the basis of liquid crystal displays (LCD).

Liquid crystals can display different degrees of long-range order, dependent on temperature, chemical composition, and the presence or absence of electric fields. In the nematic phase, the molecular axes point in a common direction, denoted by the director $\hat{\mathbf{n}}$ but the molecular centers are otherwise arranged randomly. Because of the low degree of long-range order, nematic LCs have viscosities typical of ordinary liquids, and displays based on nematic LCs can operate at television frame rates. The most popular nematic-based display, the twisted nematic (TN), will be discussed in more detail below.

In the more ordered smectic phase, the molecules again are aligned in a common direction but are, in addition, arranged into layers perpendicular to $\hat{\mathbf{n}}$. Depending on the degree of order within the layer, several subgroups (smectic A, B, C, etc.) exist; no sharp structural differences can be drawn between some that are highly ordered, e.g., smectic B liquid crystals, and true laminar crystalline solids. As a result of the layered structure, smectic crystals are very viscous. The high viscosity is useful in the construction of memory displays based on the smectic A–nematic phase transition (Lu *et al.*, 1982; Hareng *et al.*, 1982, 1983). Such smectic displays can be multiplexed without the use of electronic devices but have a relatively high power consumption since they use Joule heating to induce the phase change.

In cholesterics, the structure is similar to nematics, but the director rotates in a corkscrewlike fashion along $\hat{\mathbf{n}}$. Electric-field-induced transitions between the cholesteric and nematic phases are used in the dye phase change display discussed below.

Most (but not all) LCDs require that the liquid crystal be uniformly oriented i.e., that the LC exist in the form of a single domain in either the OFF or ON condition. The required alignment can be brought about by imparting to the confining surfaces a unidirectional sense, either by rubbing or by the oblique evaporation of SiO or other compounds (Kahn, 1977). A variety of boundary conditions can therefore be set up, including some that result in a metastable behavior of the enclosed liquid crystal phase.

Many different kinds of LCD displays can be constructed, but only twisted nematic (TN), dynamic scattering, and guest–host (GH) displays are presently of commerical importance.

A twisted nematic display consists of a top electrode and bottom electrode, typically spaced 10 μm apart, whose surface alignments are perpendicular. The director, which is parallel to the surface, therefore rotates by 90° from top to bottom. Light enters the device at the top through a polarizer whose axes are rectilinear to the surface alignment. As the light proceeds through the cell, the electrical field vector, "locked in" to the molecular arrangement, is rotated by 90°, and exits through a second polarizing foil, the analyzer, at the bottom. When an electric field is applied, the molecular structure—except for the boundary layers—becomes aligned perpendicular to the cell. The transmitted light, no longer being rotated, is then blocked by the bottom polarizer. Figure 1 shows a schematic

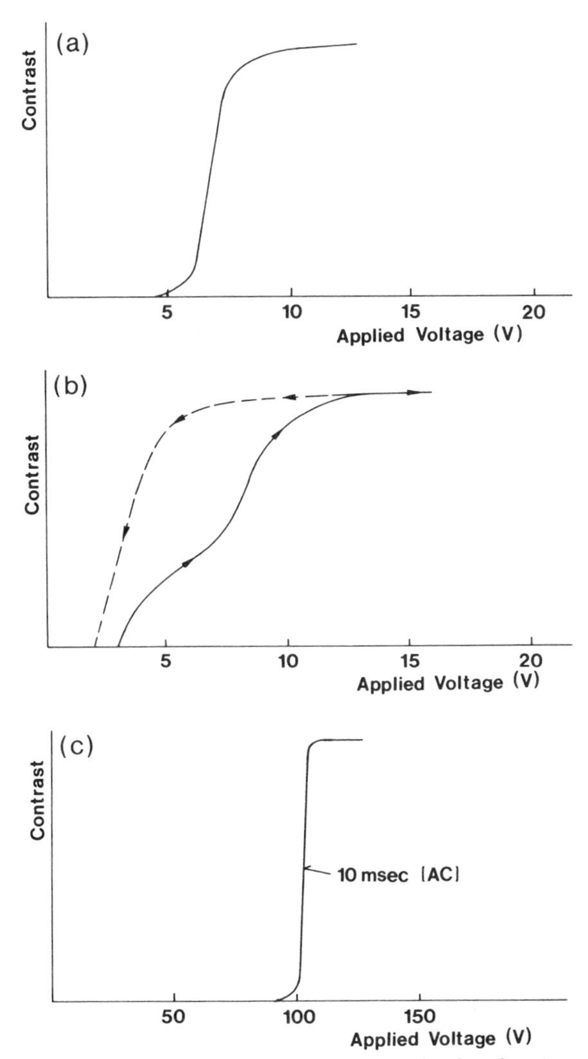

FIG. 1. Schematic diagrams of the contrast versus the applied field for (a) a twisted nematic, (b) a cholesteric guest-host bistable display (Wysocki *et al.*, 1972; Ohtsuka and Sukamoto, 1973), and (c) a bistable LCD (Boyd *et al.*, 1982).

transmission versus applied voltage curve, measured under rms conditions, for such a device.

A TN LCD requires only a modest driving voltage and turns on over a fairly narrow voltage range. The extent of this range depends both on the properties of the liquid crystal and on cell design parameters. A convenient measure for the sharpness of the transition is the ratio of the voltage at which the device is acceptably ON to the voltage at which the device is acceptably OFF. A typical value of this ratio, usually denoted as Q, is 1.1. The width of this transition determines, as discussed below, the upper limit on the number of lines that can be multiplexed without the use of electronic elements.

In the dynamic scattering mode (DSM) LCD, a nematic liquid crystal is also initially oriented by a surface treatment. Application of a dc field results in the transport of ions from one electrode to the other. The transit causes turbulence, which in turn causes local variation in the index of refraction, scattering the incident light (Helfrich, 1969). An alternating field also produces turbulence by a complex mechanism (Dubois-Violette *et al.*, 1971). Dynamic-scattering-mode LCDs typically turn on at 5 V and reach full contrast at about 30 V. The wide transition range facilitates control of gray scales but impedes multiplexing.

In dye-phase change nematic and GH cholesteric displays, contrast arises not from the liquid crystal, but rather from a dissolved pleochroic dye that is ordered relative to the LC. The LC can be in the nematic or cholesteric (planar or focal conical) phase and dielectrically negative or positive; that is, the director aligns, respectively, perpendicular and parallel to the field. This allows a variety of display configurations. Guest-host displays have a wide viewing angle, a good contrast ratio, and—when cholesteric—can be operated satisfactorily without the use of polarizers. Unlike DSM LCDs, cholesteric GH LC displays turn on over a short voltage range (threshold condition). Their driving voltages, which depend on the spacing (pitch) of the cholesteric repeat distance, are typically between 15 and 25 V (see Fig. 1b). They may be kept activated at voltages lower than the driving voltage (see Fig. 1b).

III. Principles of Multiplexing

In a small display it is feasible to individually connect each display element, or pixel, to an external driving circuit. As the number of display elements increases, however, the number of connections quickly becomes prohibitive. For example, a display with a resolution typical for a television set (300 × 300 pixels) would require 90,000 external connections. Matrix addressing reduces this number to 600 connections, a much more feasible number.

A small TN LC display can be multiplexed without the use of additional elements, because a nonlinear relationship exists between contrast and voltage, as shown in Fig. 1a. An addressing scheme such as shown in Fig. 2 can then be employed. Information is written into the display one row at a time. The row to be addressed is pulsed to $+2V$, where V is a suitable voltage, while all other, nonactive rows, are set to zero. Elements along a given row are addressed concurrently, by setting the columns at $-1V$ or $+1V$ depending whether a pixel is to display information or not, respectively. After the row is written, row potential is returned to zero and the next line is addressed. Polarity is reversed between frames to minimize deterioration of the LC by electrolytic processes.

The rms voltage at a pixel can be derived as follows. In a display of, e.g., N rows, each row is addressed a fraction $1/N$ of the frame time. During that time, the voltage on a selected pixel is $3V$. The remainder of the frame time the pixel is exposed to a fluctuating voltage of $+1V$, depending on the information content of other rows in the display. Nonselected elements see $+1V$ over the entire frame period. The scheme, therefore, is known as 3:1 addressing. The ratio R of the rms voltage on selected and nonselected pixels is $\{(3^2/N) + (N - 1/N)\}^{1/2}$. As long as R exceeds Q, TN LCDs can be multiplexed without the use of external threshold devices.

An extension of this analysis to the general case of $A:1$ addressing scheme, where A is an arbitrary multiple of the column voltage, shows (Bigelow *et al.*, 1974; Alt and Pleshko, 1974) that R is given by

$$R = \{[N - 1 + (A + 1)^2]/[N - 1 + (A - 1)^2]\}^{1/2}. \tag{1}$$

The quantity R is maximized when $A = N^{1/2}$, in which case,

$$R = \{(n^{1/2} + 1)/(n^{1/2} - 1)\}^{1/2}. \tag{2}$$

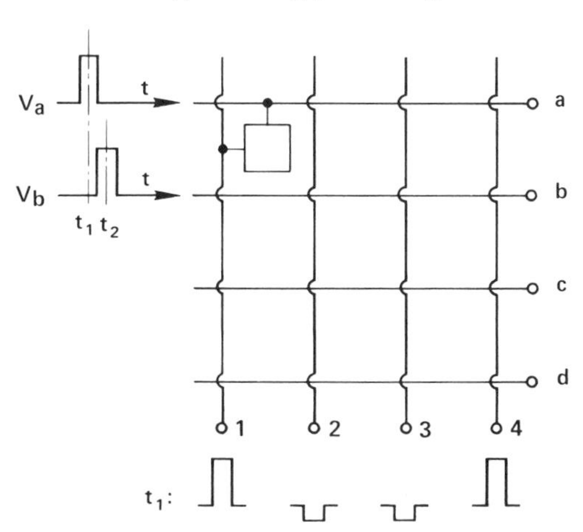

FIG. 2. Addressing scheme in a multiplexed display.

A numerical evaluation of the maximum number of lines that can be multiplexed with such a single-frequency scheme requires specification of the acceptable viewing angle, temperature dependence of the various parameters, and so on; see Kahn and Birecki (1980). The present upper limit appears to be about 25 lines for a direct view, temperature-compensated TN LCD and 91 for a transmission display, but further progress in LC chemistry will increase this number in the future.

The sharper thresholds required for the addressing of larger displays can be brought about by constructing liquid crystal cells with sharper transitions (see Fig. 1c) (Boyd *et al.,* 1982; Wilson *et al.,* 1983), intrinsic threshold control, or by the addition of nonlinear electronic elements, extrinsic threshold control. Liquid crystal displays with gradual turn on such as DSM or GH LCDs always require extrinsic threshold control for matrix addressing.

The number of devices required to multiplex a monochromic television display of moderate resolution is enormous, about 100,000, and color schemes triple this number (see, e.g., Uchida *et al.,* 1982, 1983). To succeed, therefore, a specific technology must demonstrate not only that individual elements meet the (very modest) electrical requirements but also that active matrices, with gate counts comparable to the largest intergrated circuits produced, can be fabricated with acceptable yields.

Compared to competing technologies, the chief advantages of a-Si:H devices are good device uniformity, very low leakage, and the possibility of fabricating the active matrix entirely below 300°C. The low processing temperature permits the use of economical soda-lime glass as a substrate. In addition, a-Si:H shares with some competing technologies (polysilicon, laser-annealed silicon) the following advantages over single-crystal technology: (i) the display size is unlimited (i.e., not tied to available wafer size) and (ii) the substrate is transparent, permitting back illumination of the display. A detailed review of the competing thin-film transistor (TFT) technologies has been given by Lakatos (1982, 1983).

IV. Electrical Specifications of Extrinsic Threshold Devices

1. Introduction

Various matrix addressing schemes are possible with two- and three-terminal devices. For more information, the reader is referred to an early paper of the RCA group (Lechner *et al.,* 1971). Matrix addressing differs from multiplexing in that the LC can be driven at 100% duty factor, maintaining superior optical performance.

2. Two-Terminal Devices

The simplest matrix addressing device that can be implemented with a-Si:H is the back-to-back diode (BTBD) (Ast, 1982a,b, 1983). Interestingly, its use in LCD displays was proposed after the introduction of the more sophisticated a-Si:H-based thin-film transistors (TFT) discussed below (LeComber *et al.*, 1979). The desirable characteristics of a BTBD are (i) a sharp threshold and (ii) a symmetric behavior respective to the applied polarity, important because a dc component impairs the lifetime of the LCD. A back-to-back diode layout as shown in Fig. 3 meets this requirement. The device consists of two Schottky diodes on undoped a-Si:H, which is in ohmic contact with a conductive backplane. This geometry ensures a symmetric electrical behavior of the device since Schottky diodes of equal area are formed on the same side of the a-Si:H layer. The latter is important since for a variety of reasons—nucleation and growth, changing surface potential during deposition, etc.—the top and bottom layers of an a-Si:H film generally have slightly different electronic properties. The current-voltage behavior of such a diode is shown in Fig. 4. For $V > kT$ the reverse current follows the relation:

$$I \propto V^n, \tag{3}$$

where n is the voltage exponent. Breakdown in undoped a-Si:H occurs most probably via thermionic field emission (Snell *et al.*, 1980b). The breakdown voltage is on the order of 20 V, a value suitable for the operation of GH LCDs, and can, if necessary, be adjusted by doping the a-Si:H layer (Snell *et al.*, 1980b).

To obtain a sharp threshold, a high value of n is desirable. Values of $n = 13$ were obtained by the author, and similar values have been reported elsewhere (D. Kaplan, Private communication, 1982; Szydlo *et al.*, 1984). However, the breakdown is not as sharp as in a varistor, a competing technology, in which field exponents as high as 50 can be obtained (Castelberry, 1979). An estimate of the maximum number N_m of rows that can be addressed by a 3:1 addressing scheme incoporating BTBDs can be made by considering charge transfer in selected and nonselected pixels. Assuming that a nonselected device should receive not more than a small fraction k of

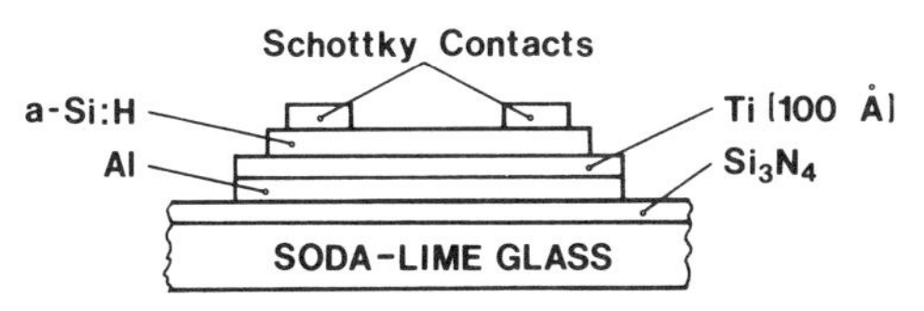

Fig. 3. Schematic layout of an a-Si:H back-to-back diode.

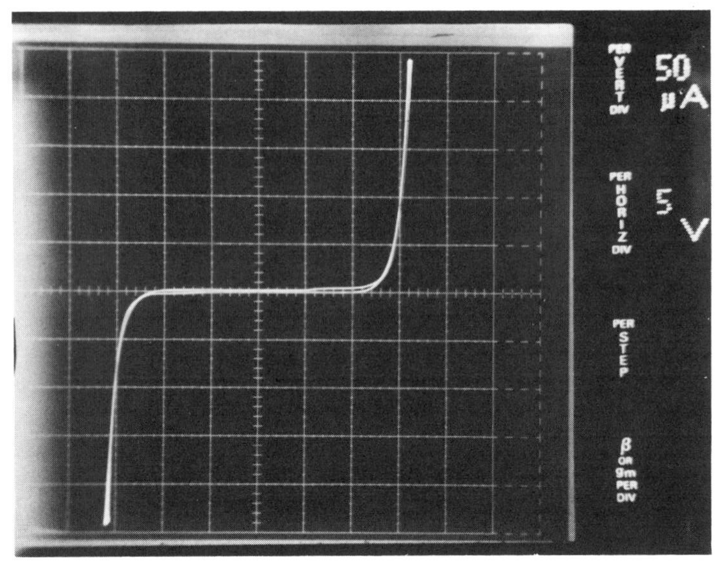

FIG. 4. Electrical characteristics of an a-Si:H back-to-back diode.

the charge necessary to turn on a selected pixel, one obtains to first order:

$$N_m = k \times 3^n. \tag{4}$$

Numerical evaluations, with $k = 0.01$, show that very large arrays ($N_m > 1000$) can be addressed for $n > 10$.

In practical devices, just as in the case of the varistor, N_m is usually limited by the capacitance of the diode, rather than by its static behavior. An evaluation of the diode shows that its minimum size is set by lithography, rather than by current density considerations. A small (2×2-μm) back-to-back diode, fabricated in a 5000-Å-thick amorphous Si film has a capacitance in the order of 10^{-3} pF. The impedance of this capacitance limits the maximum number of lines that can be addressed. It appears that large displays, several hundred lines, can be addressed since the typical capacitance of a 1×1-mm LCD cell is about 10 pF.

The chief advantage of BTBDs is their simple construction. Unlike the TFT, no dielectric layer is required. Since failure of the gate dielectric is the chief cause of low yield, BTBD-controlled matrix displays should have higher yields.

3. Three-Terminal Devices

Figure 5 shows a typical FET LC circuit. The FET acts as a switch regulating charge transfer between the column line and the LC cell. Electri-

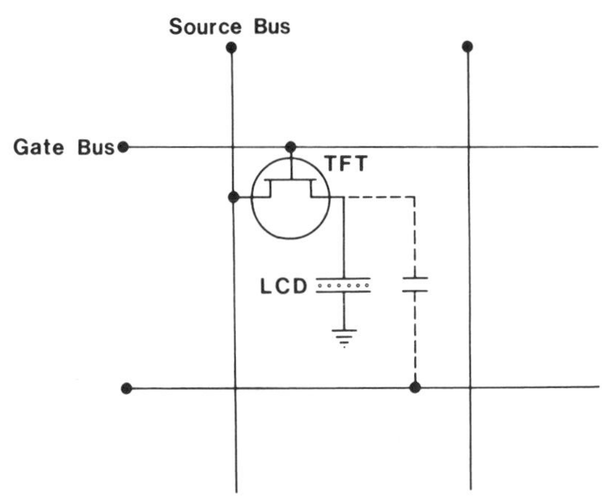

FIG. 5. Matrix-addressed liquid crystal display using an FET as an extrinsic threshold device. The capacitor (dashed) is optional.

cally, the cell behaves like a capacitor (e.g., 10 pF) in parallel with a large resistor (e.g., $10^{10}\ \Omega$). The dotted line indicates the possibility of incorporating an additional storage capacitor. With a-Si:H TFTs, such a storage capacitor is required, if at all, only for very small pixels sizes and its role will not be discussed here. Similar to the multiplexing scheme discussed previously, the gate lines are addressed sequentially, with all gates grounded except those along the selected line in which all TFTs are positively biased. All pixels in a given line are addressed simultaneously by setting the column lines (source lines) either high (selected pixels) or to ground potential (nonselected pixels). To eliminate a dc component, the polarity of the column lines (source lines) is changed after every frame.

Amorphous Si:H has an effective-field drift mobility about 0.2 $cm^2\ V^{-1}\ sec^{-1}$. To gain some perspective on how this material might fit into active matrix display applications, the requirements on a TFT designed to drive a 1-mm^2 GH LCD pixel will be discussed. In the linear regime, the transistor can be modeled as a (gate-voltage-controlled) resistor, and the minimum ON resistance $R_{\min}$ is approximately given by the estimate

$$R_{\min} = R_D(U - V_C)/(V_C N), \tag{5}$$

where R_D is the resistance of the cell (about $10^9\ \Omega$), U the supply voltage, V_C the minimum voltage to turn on the GH LCD, and N the number of lines in the display. Taking typical values, $U = 10$ V, $V_C = 3$ V, $R_D = 10^9\ \Omega$, one calculates that the ON resistance of a transistor in a large ($N = 1000$) display must be below $2.3 \times 10^6\ \Omega$. An exact solution of the time-dependent voltage operating one LCD pixel is given by

$$V(t) = \{U/(1 + (R_{\text{min}}/R_{\text{D}}))\}\{1 - \exp[-(1 + (R_{\text{min}}/R_{\text{D}}))t/(R_{\text{min}}C_{\text{D}})]\}, \tag{6}$$

where C_D is the capacitance of the cell. For the numbers given above, Eq. (6) yields $R_{min} = 2.85 \times 10^6\ \Omega$, which is close to the approximate solution given by (5). Thus one may conclude that the transistor in the linear regime should have an ON resistance of less than about $10^6\ \Omega$.

In most cases, however, the FET will saturate and operate over the major part of the charging cycle of the LCD as a current-limiting device. In this case, the transistor characteristics can no longer be specified by a minimum ON resistance. Rather we now need to specify the minimum source–drain current I_{SD}. $V(t)$ depends on I_{SD} as

$$V(t) = I_{\text{SD}}\, R_{\text{D}}\{1 - \exp(-t/R_{\text{D}}C_{\text{D}})\}. \tag{7}$$

In a multiline display, the time available to charge the LC is small compared to $R_D C_D$ and the equation reduces therefore to

$$I_{\text{SD}} = (V_{\text{C}}N)R_{\text{D}}, \tag{8}$$

which for the preceding numbers, yields $I_{SD} > 3 \times 10^{-6}$ A. Thus the transistor must be able to supply a few microamperes of current.

In the OFF state, the FET appears as a resistor in parallel with R_D. If we arbitrarily specify that the FET should not degrade the R–C product by more than 10%, we obtain for the minimum resistance in the OFF state, R_{OFF}:

$$R_{\text{OFF}} > 9R_{\text{D}}, \tag{9}$$

or about $9 \times 10^9\ \Omega$ for a 1-mm² pixel size.

The preceding estimates are very similar to those obtained by a more involved analysis that considers the various combinations that can occur during successive frames (Luo *et al.*, 1983). By analyzing a similar display ($C = 10$ pF, $N = 500$, 60-Hz refreshment rate), these authors deduce that the FET must deliver a current of about 5 μA, a value also quoted by Lakatos (1982, 1983).

When the FET drives an LCD, two additional factors must be considered. First, as discussed previously the source polarity is changed after every frame to minimize the dc component. However, the transistor operation is not symmetric, since the polarity of the gate voltage stays constant. The resulting imbalance in the transistor transfer characteristics can be compensated by either using a different source pulse height for even and odd frames or suitably biasing the counterelectrode. Experimentally, the asymmetry in the switching characteristics of individual experimental a-Si:H FETs can be investigated by driving an operational amplifier with an input impedance and input capacitance similar to those of a liquid crystal cell (e.g., $10^9\ \Omega$ and 10 pF). The output of the operational amplifier is then connected to an oscilloscope to study the symmetry of the waveform (Luo *et al.*, 1981, 1983; Morin and Le Contellec, 1983). Such a setup is also convenient for studying the long-term stability of a-Si:H transistors.

The asymmetry of the voltage form on the LC decreases with increasing saturation current of the FET since a large saturation current quickly raises the LC voltage to the value of the supply voltage. In such a case, the time-dependent waveform consists essentially of a train of square pulses of alternate polarity, independence of the transfer characteristics of the transistor.

The second consideration is that the transistor must be able to switch sufficiently rapidly. In a 500-line display, operated at 30 Hz, each row is addressed for 76 μsec. This value corresponds to a frequency of about 15 kHz. The maximum operating frequency of an insulated gate FET is given approximately by (Sze, 1969):

$$f_{\max} = \mu(V'/2\pi L^2), \tag{10}$$

where $V' = V_{SD}$ in the triode regime and $V' = V_G - V_T$ in the saturation regime. The symbol L is the gate length, μ the effective channel mobility, V_{SD} the source–drain voltage, V_G the gate voltage, and V_T the threshold voltage. With $V' = 10$ V, $L = 25\ \mu$m, such an estimate yields about 50 kHz for a large device. Transistors with smaller gate length should operate correspondingly faster, and as shown in the preceding chapter, this expectation is confirmed by experiment. In actual devices, trap filling (the number of traps depends on the density of states in the gap) limits the switching time to lower values. In a ring oscillator constructed of nine inverters, the measured delay time was about 110 μsec per gate (Matsumura, 1982). Smaller delays, about 40–50 μsec, were observed on isolated devices (Morin and Le Contellec, 1983) and in a-Si:H TFT matrices (Suzuki *et al.*, 1983). The maximum number of lines that can be addressed with present FETs appears therefore to be about 500–1000 lines at frame rates between 60 and 30 Hz, respectively.

Transistors based on a-Si:H do not follow the Borkan–Weimer equation (Borkan and Weimer, 1963) since the characteristics of an a-Si:H transistor depend on the density of states in the gap. For an exponential density of tail state distribution, appropriate theories were formulated as early as 1975 (Neudeck and Malhotra, 1975, 1976) that have subsequently been developed further (Kishida *et al.*, 1983). If the density of states varies in a nonexponential manner, numerical methods must be used to derive the transistor characteristics. The inverse problem, a derivation of the density of states from field-effect measurements, is discussed in Chapter 2 by Cohen of Volume 21C.

In practice, a-Si : H FETs are nevertheless frequently characterized by means of an effective-field drift mobility, extracted by fitting classical IGFET theory to a a limited range of device performance (this approach is also used in the preceding chapter). The justification for such an approach is that it greatly facilitates a comparison with conventional devices.

With this simplification, one can easily estimate the minimum gate width (W) to length (L) ratio required to drive a 1-mm^2 pixel element. Such an estimate, for $\mu = 0.2\ \text{cm}^2\ \text{V}^{-1}\ \text{sec}^{-1}$, yields $W/L = 10$ for a FET operating in the triode regime, and 17 for a transistor operating in the saturation regime.

Although transparent substrates allow the use of a self-aligned process (Asama *et al.*, 1983), it is difficult to work with gate lengths below 5 μm, especially in large-area devices. Thus L is essentially a fixed number and W is adjusted to satisfy the electrical requirements in the ON state.

The preceding estimates show that a minimum-sized a-Si : H-based thin-film transistor (TFT) driving a 1-mm^2 pixel is still a rather large (5×100-μm^2) device. The devices discussed in the preceding chapter are even larger (4×500-μm^2) with $W/L = 125$. No difficulties are encountered in placing such large devices into a display, as will be seen below, but the large size is undesirable from a yield point of view in that it increases the probability of failure, which scales with the gate area. In comparison, higher mobility materials, such as CdSe and polysilicon, both of which have a mobility of about 40 cm^2 V^{-1} sec^{-1}, could operate the LCD at a W/L ratio of 0.1. However, since W, like L, can not easily be reduced below 5 μm, minimum-sized (5×5-μm^2) FETs of these higher mobility materials are essentially "oversized" by a factor of 10. This oversize is undesirable in that it degrades the resistance in the OFF state. In the large pixels (size, 1 mm^2) considered so far, leakage is not a serious problem, but at higher resolution, e.g., 10 lines per millimeter, the application of high-mobility materials is limited by the low OFF resistance. This comes about because the required ON and permissible OFF currents, which scale with the area, are now reduced by two orders of magnitude. A minimum-sized (5×5-μm^2) transistor implemented in a medium-mobility ($\mu = 40$) material now exceeds its electrically required size by a factor of 1000. Both CdSe and polysilicon transistors will therefore experience difficulties in meeting the required OFF resistance ($10^{12}\ \Omega$ for a 0.1×0.1-mm^2 capacitorless display) unless a dual gate layout, which essentially puts two transistors in series, is adopted (Morozumi *et al.*, 1983). Hydrogenated amorphous silicon transistors have very high OFF resistance, especially if interface trapping occurs. Although this permits the construction of large devices free of leakage, yield considerations dictate that the devices should be kept as small as possible.

The preceding example illustrates that it is actually advantageous to construct TFT for high-resolution displays with a low-mobility material.

This point is not always realized. Amorphous Si:H-based TFTs appear to be well suited for the kind of displays that are now covered by higher resolution CRTs operating at 500–1000 lines over fields between 10 and 25 cm.

The mobility of a-Si:H is insufficient, however, to implement driving and decoding circuits, which must operate at much higher speeds. To integrate those functions into the display, it is necessary to crystallize the a-Si at the locations where the driver circuits are to be located. Work along these lines is being pursued by Morin *et al.* (1982) and by Le Contellec *et al.* (1982), who report mobilities between 15 and 20 $cm^2 V^{-1} sec^{-1}$. Incorporation of the drivers would improve the position of the a-Si:H FET technology relative to competing technologies using higher-mobility materials.

V. Active-Matrix-Addressed Liquid Crystal Displays

4. Introduction

An active matrix display using thin-film transistors (TFTs) was first built at Westinghouse (Brody *et al.*, 1973). This display measured 6 square inches, had a resolution of 120 × 120 lines, and used CdSe TFTs to address TN LC cells. The suggestion to replace CdSe transistors with a-Si:H-based FETs was first published in 1979 (LeComber *et al.*, 1979), but it has been stated that this proposal dates back to 1976 (Snell *et al.*, 1981). The first attempts to drive a separate commercial liquid crystal display element with an a-Si:H FET were reported in the literature in 1981 (Snell *et al.*, 1981). Integrated liquid crystal displays were described by several groups in 1982 (Okubo *et al.*, 1982; Kawai *et al.*, 1982; Ast, 1982a,b, 1983) and in 1983 (Asama *et al.*, 1983; Suzuki *et al.*, 1983). Display with 220 × 240 have been built in Japan (Suzuki *et al.*, 1983) and a 320 × 320 display is being constructed in France (Morin and Le Contellec, 1983). Since it appears certain that even larger displays will appear in the future, it does not seem useful to describe the various displays in detail. Instead, typical considerations concerning the layout of large LCD will be given and some of the options in designing such a display will be discussed.

5. Design Considerations

A development starts typically with the production of individual test devices. Three basic configurations are possible: inverted (gate down), noninverted (gate up), and dual gate (Tuan *et al.*, 1982a,b). Each group, in turn, can be further subdivided according to the nature (e.g., n^+ versus Schottky) and location of the source and drain contacts (e.g., in an inverted FETs the contacts can be at the bottom or the top of the a-Si:H layer). Most critical is the choice of the gate dielectric. The most popular dielectrics are glow-discharge-deposited Si_3N_4 (Mackenzie *et al.*, 1983; Okubo *et al.*, 1982; Ast, 1982a,b, 1983) and SiO_2, deposited by a variety of methods (Kawai *et al.*, 1982; Suzuki *et al.*, 1983; Le Contellec *et al.*, 1982; Ast, 1982a,b, 1983; Morin and Le Contellec, 1983). Wet oxidation of Ta (Lueder *et al.*, 1982), attractive because of its low pinhole density, has not been used with a-Si:H but its use as a gate dielectric for a-Si:H devices should be investigated.

No consensus has yet been reached as to what constitutes the optimum design of display transistors. This is not surprising, since the optimum design is not one that maximizes individual transistor performance, but rather one that simplifies processing and maximizes the number of functioning devices.

After experimenting with inverted, noninverted, and dual-gate a-Si:H TFT prepared with quasi-ohmic Ti–Al source–drain contacts, the author selected inverted TFTs with Ti–Al top contacts (see Fig. 6) for the active matrix.

The use of quasi-ohmic Ti–Al contacts rather than of n^+ contacts represents a compromise between various factors. The use of n^+ contacts increases the source-drain current (see the preceding chapter), but this benefit must be weighed against the additional deposition and etching steps required, as well as the possibility that residual PH_3, lingering in the deposition system, increases the density of states in the active layer of a subsequently deposited device. In general, the choice between various layouts involves trade-offs between the quality of contacts (which for Schottky contacts in turn depends on the deposition sequence, glow-discharge (GD) a-Si:H deposited on metal being more ohmic than metal deposited on a-Si:H), the quality of the dielectric–semiconductor interface (generally better if the semiconductor and the dielectric are deposited in the same deposition system without opening the system to the environment), and the complexity of the processing.

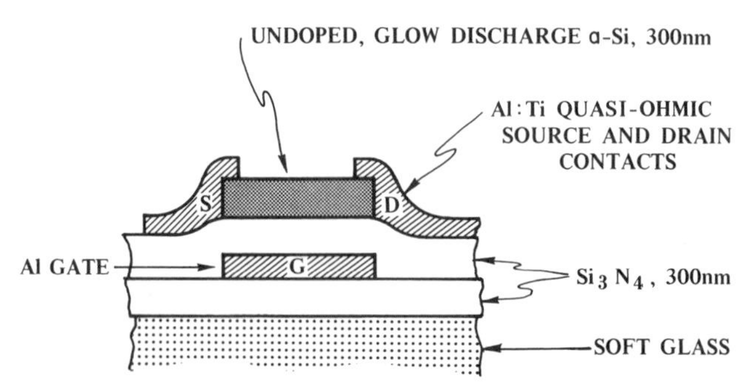

FIG. 6. Schematic layout of the Si_3N_4 gate dielectric transistor. (The SiO_2 gate dielectric transistor is geometrically identical.) $L = 25\ \mu m$, $W = 3400\ \mu m$.

As a rule, those features that are associated with bulk a-Si:H and contacts are not as critical as those associated with the a-Si:H dielectric interface and the quality of the dielectric itself. To evaluate the relative performance of SiO_2 and Si_3N_4, both GD-deposited Si_3N_4 and magnetron-sputtered SiO_2 were used as gate dielectrics in the display described below. Glow-discharge-deposited silicon nitride is a complex material whose properties vary greatly with small variations in the deposition conditions (Stein *et al.,* 1979). The major industrial application of GD Si_3N_4 is as a passivation layer (Sinha *et al.*, 1978), and most studies have therefore been concerned with minimizing interface stresses. Factors that are critical in the fabrication of a-Si:H FETs such as pinhole density, resistivity (Sinha and Smith, 1978), and interface trapping (Fujita *et al.,* 1981), all of which vary with deposition conditions, have not been studied systematically. Preliminary studies indicate that optimum electronic properties require deposition conditions which also minimize pinhole density (see the preceding chapter).

Important factors in the selection of the substrate material are planarity, ease of sealing cost, and the consideration that soda-lime glass substrates, unlike quartz, cannot be processed in a regular silicon line.

The following section will discuss the layout of a display built in 1981 at the Hewlett-Packard Solid State Laboratory. This display includes two innovative features: (i) an edge-free layout of the transistors and (ii) a photoconductive feedback mechanism that maximizes the ON current and minimizes the cumulative trapping at the dielectric – a-Si:H interface.

6. Transistor Layout

A relatively large pixel size of 1×1 mm^2 was chosen to test the suitability of a-Si:H FETs to drive large LC cells. Array size was limited to 26×26 to allow processing with available 3-in. equipment. The transistors were laid out as closed structures surrounding the square drain pad, which simultaneously served as a reflector for the guest – host LC (see Fig. 7a). This layout maximizes the gate width (3800 μm). A more subtle but important advantage of this layout is that it avoids etched edges of the semiconductor running between source and drain. Hydrogenated amorphous silicon is easily damaged during processing, and open edges can act as stripes of more conductive a-Si:H between source and drain electrode, thus degrading the OFF resistivity. Drain and source contacts were made by first evaporating a nominally 300-Å-thick layer of Ti followed by the deposition of a 5000-Å-thick Al layer. The Ti deposition greatly improved the ohmic behavior of the contacts. It also imparted slightly rough texture to the subsequent Al deposition, a desirable feature since this improves the optical performance of the (drain pad) reflector of the GH display.

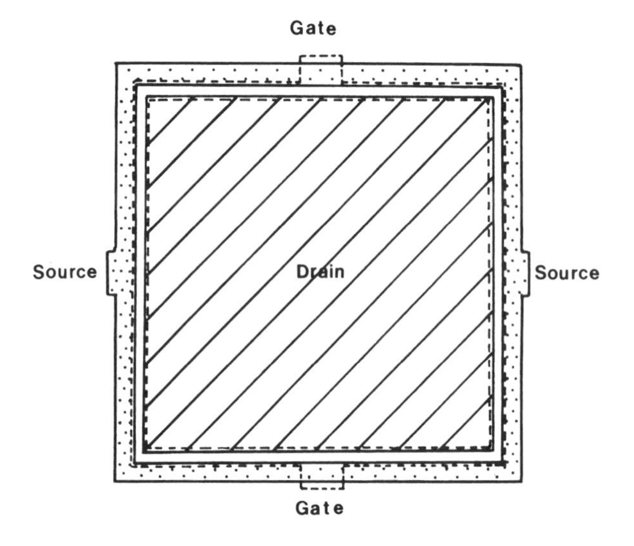

FIG. 7. (Top) Schematic layout of the TFT in the addressing matrix. Cross-hatched area is the drain pad and the dotted area is the source contact. The gap between the source and drain contact is bridged by the underlying gate electrode (broken line) and the a-Si : H semiconductor (not shown). (Bottom) View of the finished matrix. The small dark strip is a-Si : H. The bright central square is the drain pad, doubling as the optical reflector of the guest–host display. Note that the transistor completely surrounds the drain pad.

FIG. 8. Photograph of a completed display under test.

Production of the TFT matrix required only three masking steps: gate definition (liftoff of evaporated Al), patterning of the deposited a-Si:H, and definition of the source–drain contacts (liftoff of evaporated Ti, Al). A modified poly-Si etch was used to pattern the a-Si:H without impairing the underlying Al gate lines. A well-baked photoresist, applied with an adhesion improver, must be used in this processing step. Figure 7 (top) shows a TFT of the completed matrix before encapsulation. The small dark stripe surrounding the large square is the a-Si:H semiconductor. The square itself is the Al drain pad/reflector. Gate length is 25 μm and the W/L ratio is 152. A completed device under test is shown in Fig. 8.

7. Transistor Performance

a. *Instabilities in* a-Si:H *TFTs*

In an amorphous Si:H transistor I_{SD} decreases with time at constant gate voltage because of two effects (i) trapping of electrons in gap states in the amorphous Si and (ii) trapping of electrons in the gate dielectric. These effects have been discussed in more detail elsewhere (Ast, 1982,a,b, 1983).

To assess the stability of TFTs it is important to separate internal (a-Si:H–related) and interface (dielectric-related) trapping. This separation can be brought about by comparing FETs in which the a-Si:H is grown under identical conditions (i.e., in the same deposition run) but in which the gate dielectrics differ. Such side-by-side comparisons were carried out for (i) thermally grown high-temperature gate oxides (using the underlying Si wafer as the gate electrode), (ii) thermally grown high-temperature LPCVD Si_3N_4, (iii) GD-deposited Si_3N_4, (iv) rf-sputtered SiO_2, and (v) thermally evaporated SiO. These comparisons indicate that the short-term decrease ($t < 0.5$ sec) is mostly due to intrinsic trapping. The long-term decrease ($t > 0.5$ sec) is caused by trapping in the gate insulator and is not an intrinsic property of a-Si : H. In the case of GD Si_3N_4 the interface trapping was rather pronounced since the Si_3N_4, because of equipment limitations, was deposited at a relatively low temperature (250°C). The trapping can be reduced by depositing Si_3N_4 at higher temperatures or by using sputtered or GD-deposited SiO_2 as a gate dielectric.

As a result of trapping, the transistor characteristics are time dependent and quite different when measured under dc (Fig. 9) or pulsed (Fig. 10) conditions. Note that in the latter case the GD Si_3N_4 gated transistor switches in excess of 30 μA and its performance becomes comparable to the transistors discussed in the preceding chapter. As discussed below, a limited amount of trapping, as long as it is noncumulative, is tolerable and even useful in display TFTs.

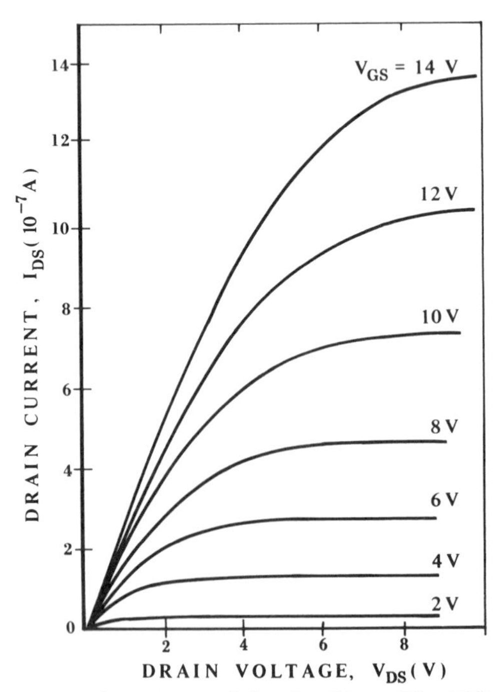

FIG. 9. Dc transistor characteristics. $L = 25\ \mu$m, $W = 3400\ \mu$m.

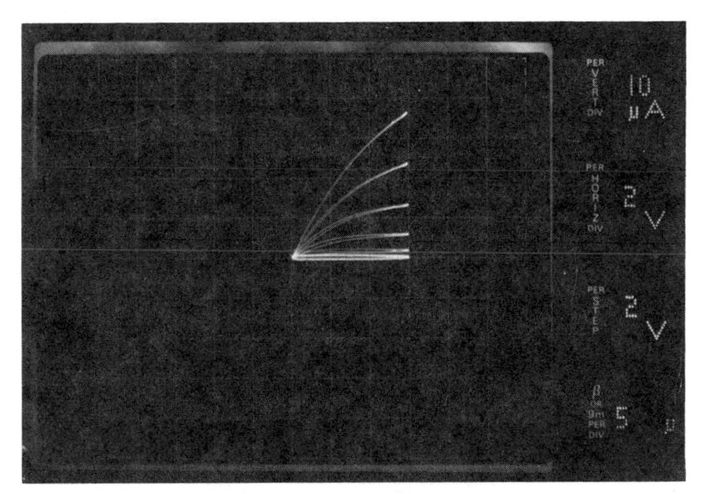

FIG. 10. Transistor characteristics under pulsed conditions.

b. Detrapping by Light- and Field-Induced Instabilities

(1) *Interface Trapping.* A large fraction of the trapped carriers in Si_3N_4 can be released by light. For this reason it is desirable to operate Si_3N_4 gated devices in such a way that they are illuminated in the ON-state. In the OFF-state, of course, the device must be shaded in order to reduce photo-current-induced leakage. Such a selective illumination can be achieved in a guest–host (GH) liquid crystal display in which the GH LC shades the FET. Long-term testing of unshielded 26 × 26 matrix-addressed LC GH displays, fabricated at HP and operated in the dark and illuminated states has shown that the decay of I_{SD} with time is indeed reduced in illuminated devices. This is shown in Fig. 11, which shows the decay of the source–drain current with time in illuminated and dark devices. The scatter in the dark decay curves reflects the fact that the interface between the dielectric and the a-Si:H is sensitive to the preparation conditions (i.e., how the Si_3N_4 layer was terminated and the a-Si:H layer started). In addition to reduced trapping, illumination of the FET in the ON-state increases the current in the ON-state in a superlinear fashion (see below). For this reason we do not recommend a light shield as, for example, described by Okubo *et al.* (1982).

(2) *Staebler–Wronski Effect.* Light (Staebler and Wronski, 1977) and electric fields (Ast and Brodsky, 1978) can induce changes in the electronic properties of a-Si:H. Staebler and Wronski (1977) observed that the photoconductivity and the dark conductivity of undoped a-Si:H decreased slowly during illumination. The fall in the photoconductivity is relatively small

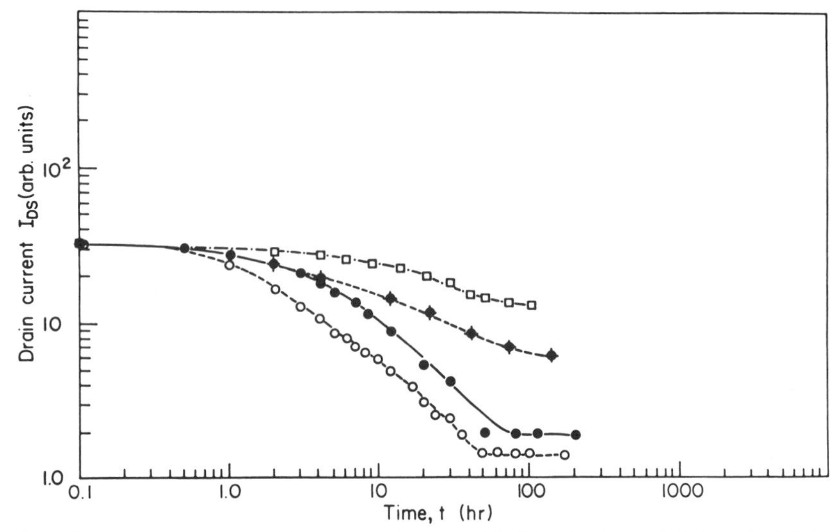

FIG. 11. Long-term decay of source–drain current in illuminated and dark a-Si:H FETs with Si_3N_4 gate dielectric. Pulse width, 500 msec; repeat period, 500 msec; duty cycle, 10%; $V_{SD} = V_G = 13$ V. □, room temperature illuminated; ● and ✦, room temperature dark; ○, 80°C dark.

(about one order of magnitude), but the dark conductivity can be decreased by as much as four to five orders of magnitude. Tests on unshielded, illuminated a-Si:H FETs show a decrease in photoconductivity with time and an increase in the OFF resistance as expected for the Staebler–Wronski effect. Note that both changes are desirable from the operating point of view, especially for unshielded devices. The I_{SD} at a given positive V_G is, however, changed only very slightly since the location of the Fermi level in an ON device depends to a much greater degree on the density of conduction band-tail states than on states in the middle of the gap. In this respect, the behavior of a switched-ON a-Si:H FET is similar to the behavior of phosphorus-doped a-Si:H, which also shows no detectable changes in the photoconductivity with light exposure.

A limited amount of trapping (either intrinsically or extrinsically) is helpful in that it ensures a deep OFF state of the device during the time at which other lines are addressed; that is a (properly adjusted) hysteresis can serve as a memory element over the frame time (Ast, 1982a,b, 1983). This argument is not readily accepted by those who believe that a device must be inherently stable to function in a reliable manner. However, Luo *et al.* (1983) arrived at an identical conclusion in their analysis of the switching performance of CdSe transistors in matrix-addressed LC displays. It is very important that the trapping be noncumulative. Since the duty cycle of a display transistor is very low (about $1/N$, where N is the number of lines), the time available for detrapping is typically two to three orders of magnitude larger than the time available for trapping. For this reason, cumulative trapping in display transistors is low. Individual devices switching simulated loads show only modest changes after many millions of switching operation (see the preceding chapter).

(3) *Superlinear Photocurrents in Illuminated* ON *Devices.* The ON current of an illuminated device is *not* the sum of the photocurrent (measured at $V_G = 0$) plus the channel current measured in the dark at a given V_G. Rather, the photocurrent is increased by about one order of magnitude (see Fig. 12). Technically, this effect is helpful in operating GH liquid crystal displays with large pixel sizes (1×1 mm^2) at low V_{SD} and V_G, which, by amorphous silicon standards, require relatively large drive currents. In this case, the nonlinear photoconductivity contribution together with the varying illumination provided by the GH LC (high when "ON" and low when "OFF") form a *positive feedback mechanism* enhancing I_{SD}. The effect is due to electronic doping and is therefore closely related to the observation that the photoconductivity of phosphorus-doped samples is about two

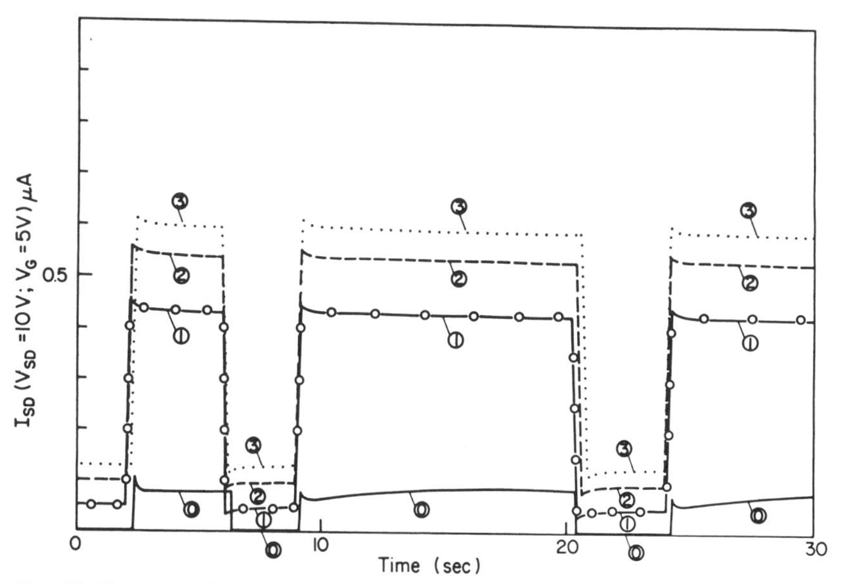

FIG. 12. Source–drain current in an intermittently illuminated a-Si:H FET. (The transistor is an experimental device using thermally grown SiO_2 as a gate dielectric in order to minimize insulator trapping.) (0) ———, dark. (1) -O-, (2) — — —, (3) ....: different illumination levels.

orders of magnitude higher than that of undoped samples (Anderson and Spear, 1977; Kalbitzer *et al.*, 1981).

(4) *Long-term Tests of Operating Display Panels.* Life tests were being carried out at room temperature and at 80°C to study the long-term stability of unshielded devices in contact with GH LC. The devices have been tested for over 1 year without visible degradation.

(5) *Yield.* Yield was higher in Si_3N_4 gated than in magnetron-sputtered SiO_2 matrices. With either gat dielectric, the yield of properly processed devices was mainly controlled by gate leakage. It was found that gate leakage could be greatly reduced by a thorough cleaning of the deposition system prior to the deposition of the gate dielectric (this is one feature that favors inverted TFTs over noninverted layouts), which suggests that part of the problem is related to the fine particles—formed by nucleation in the gas phase—frequently observed in glow-discharge systems ("dust"). Improvement in the yield of functioning matrices is critical since a-Si:H TFTs compete in display applications with polysilicon devices using thermally grown oxide. Further development of a-Si:H TFT active matrix displays should therefore concentrate on improved methods to prepare low-temperature dielectric of low pinhole density. Multichambered deposition systems that separate the deposition of the semiconductor and the dielectric, anodic anodization of Ta, dual-layer dielectrics (e.g., nitride, oxide conversions), and plasma-assisted and OMCVD methods (e.g., ZrO_2) should be explored.

Acknowledgments

The author would like to thank Dr. R. Burmeister for an invitation to the Hewlett-Packard Solid State Laboratory. The support of J. Aronovich, F. Kahn, and the research and development staff at HP Santa Rosa is gratefully acknowledged.

References

Alt, P. M., and Pleshko, P. (1974). *IEEE-Trans. Electron Devices* **ED-21,** 146.
Anderson D. A., and Spear, W. E. (1977). *Philos. Mag.,* **36,** 695.
Asama, K., Kodama, K., Kawai, S., Nasu, Y., and Yanagisawa, S. (1983). *SID* '83, *Dig. Tech. Pap.* p. 144.
Ast, D. G. (1982a). *Proc. US–Jpn. Jt. Semin./Technol. Appl. Tetrahedral Amorphous Solids, Palo Alto, California, 1982.*
Ast, D. G. (1982b). *Conf. Rec. Int. Display Res. Conf., Philadelphia, 1982* p. 152.
Ast, D. G. (1983). *IEEE Trans. Electron Devices* **ED-30,** 532.
Ast D. G., and Brodsky, H. M. (1978) *Conf. Ser—Inst. Phys.* **43,** 1159.
Bigelow, J. E., Kashnow, R. A., and Stein, R. C. (1974). *IEEE Trans. Electron Devices* **ED-22,** 22.
Borkan, H., and Weimer, P. K. (1963). *RCA Rev.* **24,** 153.

Boyd, G. D., Wilson, T., Thurston, R. N., Cheng, J., Storz, F. G., and Westerwick, E. H. (1982). *Conf. Rec. Int. Display Res. Conf., Philadelphia, 1982* p. 136.
Brody, T. P., Asars, J. A., and Dixon, G. D. (1973). *IEEE Trans. Electron Devices* **ED-20,** 995.
Castelberry, D. E. (1979). *IEEE Trans. Electron Devices* **ED-26,** 1123.
Chandrasekhar, S. (1977). "Liquid Crystals." Cambridge Univ. Press, London and New York.
de Gennes, P. G. (1974). "The Physics of Liquid Crystals." Oxford Univ. Press, London and New York.
Dubois-Violette, E., de Gennes, P. G., and Parodi, O. (1971). *J. Phys. Orsay, Fr.* **32,** 305.
Fujita, S., Nishihara, M., Hoi, W.-L., and Sasaki, A. (1981). *Jpn. J. Appl. Phys.* **20,** 917.
Gray, G. W. (1975) *J. Phys. Colloq. Orsay, Fr.* **36,** 337.
Hareng, M., Le Berre, S., Mourey, B., Moutou, P. C., Perbet, J. N., and Thirant, L. (1982). *Conf. Rec. Int. Display. Res. Conf., Philadelphia, 1982* p. 126.
Hareng, M., Le Berre, S., Mourey, B., Moutou, P. C., Perbet, J. N., and Thirant, L. (1983). *IEEE Trans. Electron Devices* **ED-30,** 507.
Heilmeier, G. H., Zanoni, L. A., and Barton, L. A. (1968a). *Proc. IEEE* **56,** 1162.
Heilmeier, G. H., Zanoni, L. A., and Barton, L. A. (1968b). *Appl. Phys. Lett.* **13,** 46.
Helfrich, W. (1969). *J. Chem. Phys.* **51,** 4092.
Kahn, F. J. (1977). *Mol. Cryst. Liq. Cryst.* **38,** 467.
Kahn, F. J., and Birecki, O. (1980). *In* "The Physics and Chemistry of Liquid Devices" (G. J. Sprokel, ed.), p. 79. Plenum, New York.
Kalbitzer, S., Mueller, G., LeComber, P. G., and Spear, W. E. (1981). *Philos. Mag. B* **13,** 439.
Kawai, S., Takagi, N., Kodama, T., Asama, K., and Yangisawa, S. (1982). *SID '82, Dig. Tech. Pap.* p. 42.
Kishida, S., Naruke, Y., Uchida, Y., and Matsumura, M. (1983). *Jpn. J. Appl. Phys.* **22,** 511.
Lakatos, A. (1982). *Conf. Rec. Int. Display Res. Conf., Philadelphia, 1982,* p. 146.
Lakatos, A. (1983). *IEEE Trans. Electron Devices* **ED-30,** 525.
Lechner, B. J., Marlowe, F. J., Nester, E. O., and Tults, J. (1971). *Proc. IEEE* **59,** 1566.
LeComber, P. G., Spear, W. E., and Ghaith, A. (1979). *Electron. Lett.* **15,** 179.
Le Contellec, M., Morin, F., Richard, J., Coissard, P., Morel, M., and Bonell, M. (1982). *SID '82, Dig. Tech. Pap.* p. 44.
Lu, S., Davies, D. H., Chung, C. H., Evanicky, D., Albert, R., and Traber, R. (1982). *Conf. Rec. Int. Display Res. Conf., Philadelphia, 1982* p. 132.
Lueder, E., Kallfass, T., Frasch, W., and Schaible, B. (1982). *SID '82, Dig. Tech. Pap.* p. 186.
Luo, F.-C., Chen, I., and Genovese, F. C. (1981). *IEEE Trans. Electron Devices* **ED-28,** 740.
Luo, F.-C., Pultorak, D., and Freeman, E. (1983). *IEEE Trans. Electron Devices* **ED-30,** 202.
Mackenzie, K. D., Snell, A. J., French, I., LeComber, P. G., and Spear, W. E. (1983). *Appl. Phys.* **A31,** 87.
Matsumura, M. (1982). *Proc. US–Jpn. Jt. Semin. Technol. Appl. Tetrahedral Amorphous Solids, Palo Alto, Ca., 1982.*
Morin, F., and Le Contellec, M. (1983). *Displays* **4,** 303.
Morin, F., Coissard, P., Morel, M., Ligeon, E., and Bontemps, M. (1982). *J. Appl. Phys.* **53,** 3897.
Morozumi, S., Oguchi, K., Yazawa, S., Kodaira, T., Ohshima, H., and Mano, T. (1983). *SID '83, Dig. Tech. Pap.* p. 156.
Neudeck, G. W., and Malhotra, A. K. (1975). *J. Appl. Phys.* **46,** 2662.
Neudeck, G. W., and Malhotra, A. K. (1976). *Solid State Electron* **19,** 721.
Ohtsuka, T., and Sukamoto, M. T. (1973). *Jpn. J. Appl. Phys.* **12,** 22.
Okubo, Y., Nakagiri, T., Osada, Y., Sugata, M., Kitahara, N., and Hatanaka, K. (1982). *SID '82 Dig. Tech. Pap.* p. 40.
Reinitzer, F. (1888). *Monatsh. Chem.* **9,** 421.

Sinha, A. K., and Smith, T. E. (1978). *J. Appl. Phys.* **49,** 2756.
Sinha, A. K., Levinstein, H. J., Smith, T. E., Ouintana, G., and Haszko, S. E. (1978). *J. Electrochem. Soc.* **125,** 601.
Snell, A. J., Mackenzie, K. D., Spear, W. E., and LeComber, P. G. (1980a). *Appl. Phys. Lett.* **24,** 357.
Snell, A. J., Mackenzie, K. D., LeComber, P. G., and Spear, W. E., (1980). *J. Non-Cryst. Solids* **35-36,** 539.
Snell, A. J., Mackenzie, K. D., Spear, W. E., LeComber, P. G., and Hughes, A. J. (1981). *Appl. Phys.* **24,** 357.
Staebler, D. L., and Wronski, C. R. (1977). *Appl. Phys. Lett.* **51,** 3262.
Stein, H. J., Wells, V. A., and Hampy, R. E. (1979). *J. Electrochem. Soc.* **126,** 1750.
Suzuki, K., Aoki, T., Ikeda, M., Okada, Y., Zohta, Y., and Ide, K. (1983). *SID '83 Dig. Tech. Pap.* p. 146.
Sze, S. M. (1969) "Physics of Semiconductor Devices" Chapter 11, p. 573. Wiley (Interscience), New York.
Szyallo, N., Chartier, E., Proust, N., and Magariño, J. (1984). *Appl. Phys. Lett.* **44,** 205.
Tuan, H. C., Thompson, M. J., Johnson, N. M., and Lujan, R. A. (1982a). *IEEE Trans. Electron Devices Lett.* **EDL-12,** 357.
Tuan, H. C., Thompson, M. J., Johnson, N. M., and Lujan, R. A. (1982b). *IEEE Electron Devices* **ED-29,** 1682 (Abstr. only).
Uchida, T., Yamamoto, S., and Shibata, Y. (1982). *Conf. Rec. Int. Display. Res. Conf., Philadelphia, 1982* p. 166.
Uchida, T., Yamamoto, S., and Shibata, Y. (1983). *IEEE Trans. Electron Devices* **ED-30,** 503.
Wilson, T., Boyd, G. D., Thurston, R. N., Cheng, J., Storz, F. G., and Westerwick, E. H. (1983). *IEEE Trans. Electron Devices* **ED-30,** 513.
Wysocki, J. J., Becker, J. H., Dir, G. A., Madrid, R., Adams, J. E., Haas, W. E., Leder, L. R., Mechlowitz, B., and Saeva, F. D. (1972). *Proc. SID* **13,** 114.

CHAPTER 8

Solid-State Image Sensor

S. Kaneko

MICROELECTRONICS RESEARCH LABORATORIES
NEC CORPORATION
KAWASAKI, JAPAN

I. Introduction

An a-Si:H photosensor array has been successively applied to a long linear image sensor and high-performance area image sensor. A photodiode in which a-Si:H is sandwiched between ITO and metal electrodes is used because of its short photoresponse time. In this structure, blocking contacts at both electrodes are necessary to prevent carrier injection from the electrodes.

With a scanning circuit and a-Si:H, it is possible to provide an a-Si:H monolithic large area solid-state image sensor, although it is necessary to improve high-frequency characteristics.

In the 1960s, research for a solid-state image sensor had progressed to the extent that a video signal could be produced without help from an electron beam. There were two approaches to fabricating the image sensor. In the first, single-crystalline Si MOSFET technology was employed and monolithic sensor readout systems were fabricated in a Si chip (Weckler, 1967; Noble, 1968).

The second approach made use of thin-film technology. Weimer fabricated totally thin-film systems using a thin-film photoconductor combined with a thin-film transistor (TFT) and diode (Weimer *et al.,* 1967).

The Si monolithic image sensor performance has rapidly improved as a result of progress in Si IC technology and/or the advent of charge transfer

ISBN 0-12-752150-X

devices (Boyle and Smith, 1970; Sangster, 1970). As a result, interest in the solid-state image sensor using thin-film technology diminished. The Si monolithic image sensor has been put to practical use in the linear sensor for facsimile and optical character recognition (OCR). Such a sensor was also used in the 1970s as an area image sensor for image pickup in cameras.

Ordinary facsimile or OCR equipment uses the Si monolithic image sensor and a high-magnification lens system, which require considerable space in the equipment to scan the document. In contrast, a contact linear image sensor, whose width is the same size as the document, can be used to construct small machines. The reason for the compactness is that the long linear image sensor does not require such a high-magnification lens system.

There are several problems that are difficult to overcome in the Si monolithic image sensor. The first is the difficulty in fabricating such a long (200–300-mm) linear image sensor. The second is that photosensitivity at short wavelengths is low and the scanning operation is disturbed in strong incident light in the Si monolithic area image sensor because of difficulty in suppressing signal charge leakage to the readout line. Moreover, the effective photosensitive area is small compared to that of vidicons. These limitations are due to the fact that the photosensor and the scanning circuit are built on the same Si chip. Therefore building the photosensor on top of the Si IC scanner makes an attractive device that improves these limitations (Tsukada *et al.*, 1979). In this case, a low-temperature process to fabricate the photosensor array is needed to avoid deteriorating the scanner performance.

Thin-film technology is useful in fabricating these devices. CdS–CdSe photoconductive sensors (Boronkay *et al.*, 1978; Komiya *et al.*, 1981) and As–Se–Te photodiodes (Tsukada *et al.*, 1977) were applied to the long linear-image sensor. As–Se–Te photodiodes (Tsukada *et al.*, 1979) and ZnSe–ZnCdTe photodiodes (Terui *et al.*, 1980) were used in the area image sensor. However, there are some problems with these sensors: long photoresponse time for the CdS–CdSe sensor, thermal instability, and low yield for the As–Se–Te sensor, and a high-temperature process for the ZnSe–ZnCdTe sensor.

Hydrogenated amorphous silicon (a-Si : H) is excellent material from the view point of high photosensitivity in the visible region, short photoresponse time, thermal stability, low-temperature process, and high production yield.

II. Application to a Long Linear Image Sensor

1. Contact Linear Image Sensor Unit

A contact linear image sensor can make the equipment compact, because it does not require a high-magnification lens system. The sensor unit consists of an illuminator, a compact optical guide, and the long linear photosensor array, which is, for example, 210 mm long for the iso-A4 reader, connected to the scanner. There are several structures that can be used for this kind of sensor unit (Tsukada *et al.*, 1979; Borenkay *et al.*, 1978; Komiya *et al.*, 1981; Ozawa *et al.*, 1982; Kaneko *et al.*, 1982). Light-emitting diode (LED) arrays or fluorescent lamps are used for the illuminator. Figure 1 shows an example of the contact linear image sensor (Kaneko *et al.*, 1982). In this figure a pair of LED arrays is used for the illuminator. A rod lens array is used for the optical guide.

Light from the LED arrays illuminates the document and reflected signal light is guided by the rod lens array, which forms an erect real image with unity magnification on the photosensor array. The photosensor array scanned by the scanning circuit converts the reflected light signal to a sequential electric signal. The illuminance for the "white" image is about 35 lux. The optical path from the document to the photosensor array is 17 mm long. Therefore a smaller machine can be built.

2. LINEAR PHOTOSENSOR ARRAY

A photoconductive array or a photodiode array can be used for this kind of sensor.

In the photoconductive sensor, the electrode is ohmic where carrier replenishment occurs. Electron photoconductivity is the dominant mode. In this case, the photocurrent is a secondary current. Photocurrent and photoconductive gain are given by (Bube, 1961)

$$I = qF\mu_n\tau_n V/L^2 \quad \text{and} \quad G = \mu_n\tau_n V/L^2,$$

where F is the total number of free carriers created per second in the sensor, μ_n the electron mobility, τ_n the electron lifetime, V the applied voltage, and L the electrode spacing. The photocurrent increases linearly with applied voltage up to an electric field of about 10^4 V cm^{-1}, where the dark-current and the photocurrent affect the space charges in a-Si : H. Since the photo-

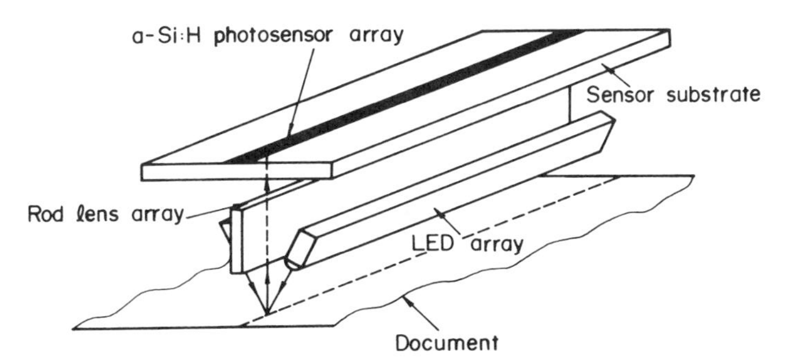

FIG. 1. Exploded view of contact linear image sensor. [From S. Kaneko *et al.*, Amorphous Si : H contact linear image sensor with Si_3N_4 blocking layer. *Technical Digest—International Electron Device Meeting*. Copyright © 1982 IEEE.].

current is proportional to the $\mu_n\tau_n$ product of a-Si:H, an a-Si:H film with uniform $\mu_n\tau_n$ product must be produced to obtain a uniform photocurrent. The $\mu_n\tau_n$ product of a-Si:H produced by the glow-discharge decomposition of silane is about 10^{-4} cm^2 V^{-1} (Zanzucchi; *et al.*, 1978). This gives a photoconductive gain of about 10 at $L = 100$ μm, $V = 10$ V. To increase the photocurrent, Kagawa *et al.* (1982) constructed a linear sensor with interdigital structure. The structure is shown in Fig. 2. One common Al electrode and 32 Al individual electrodes are separated by 10 μm with interdigital geometry. In this case, the photoconductive gain is about 10^3.

The photocurrent for a unit element under 100-lux illumination was 0.1 Å at 10 V applied voltage when undoped a-Si:H was used (Motosugi *et al.*, 1981). This value is large enough for practical use as the linear sensor, while it is one order of magnitude smaller than the CdS–CdSe photoconductive sensor, which has a simple gap geometry. (Komiya *et al.*, 1981).

The photoconductivity increases when the a-Si:H is lightly doped with phosphorus (Anderson and Spear, 1977). However, phosphorus doping causes very slow decay of photoresponse. The photoresponse characteristic for the phototconductive sensor using undoped a-Si:H is shown in Fig. 3. The illumination is the modulated light from a GaP LED. The modulation ratio is defined as $M = (i_1 - i_2)/i_2$, where i_1 is the peak photocurrent and i_2 is the bottom current just prior to the next pulse. Figure 4 shows the modulation ratio of a-Si:H versus the pulse width T, compared to that of the CdS–CdSe photoconductive sensor. The CdS–CdSe sensor modulation ratio decreases as the repetition time becomes shorter. On the other hand, in the a-Si:H photoconductive sensor, the modulation ratio does not decrease

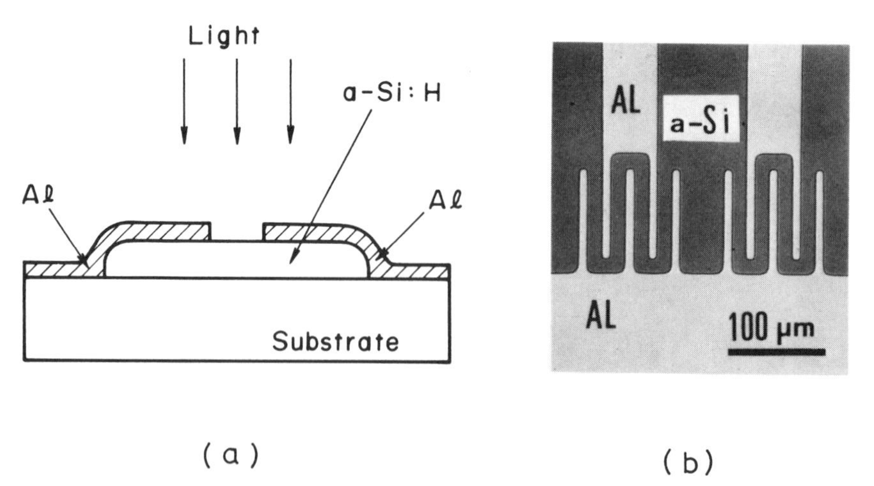

FIG. 2. Photoconductive sensor array. (a) Cross-sectional view. (b) Microphotographic plane view. [From Kagawa *et al.*, (1982).]

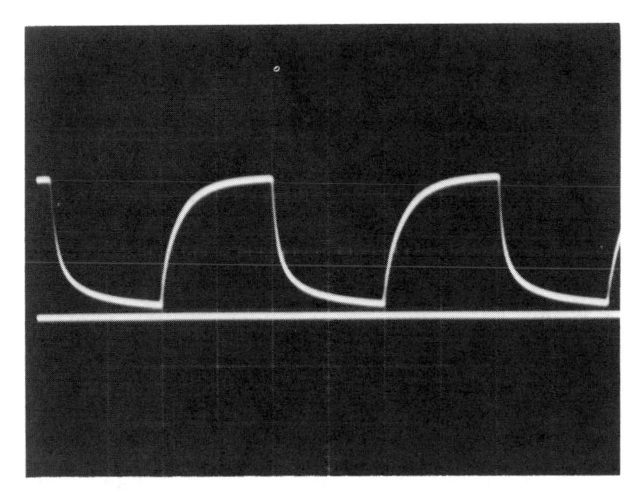

FIG. 3. Photoresponse characteristics for photoconductive sensor. The incident light comes from GaP green LEDs modulated at 500 Hz. Horizontal scale is 0.5 msec/division. [From Motosugi *et al.* (1981).]

rapidly. A modulation ratio of 10 is obtained with a 2 msec square wave. Therefore the a-Si:H photoconductive sensor can be better used for high-speed image processing compared to the CdS–CdSe photoconductive sensor (Kagawa *et al.*, 1982).

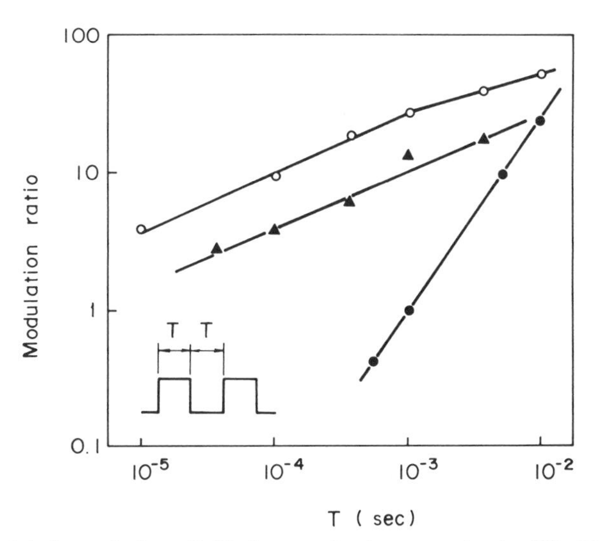

FIG. 4. Modulation ratio for a-Si:H photoconductive sensor [under 200- (○) and 50- (▲) μW cm^{-2} illumination] and CdS photoconductive sensor (●) versus LED driving pulse width. [From Kagawa *et al.* (1982).]

In the photodiode array, photosensitive a-Si : H is sandwiched between a transparent electrode and a metal electrode. Indium tin oxide (ITO) is commonly used for the transparent electrode.

Because both carriers, electrons and holes, can be mobile in a-Si : H, blocking contacts for both electrodes are needed to prevent carrier injection from the electrodes. In this case the photocurrent is a primary current that saturates with unity collection efficiency when $\mu\tau E > d$, where μ is the mobility of the photocarriers that drift in a-Si : H, τ the lifetime of photocarriers, E is the electric field in a-Si : H, and d the thickness of a-Si : H. In the saturation photocurrent region, the photocurrent is not very dependent on a-Si : H film quality and increases linearly with increasing light intensity.

Hamano *et al.* (1982) have fabricated an a-Si : H photodiode array linear image sensor. The sensor structure is shown in Fig. 5. The sensor is constructed by first forming individual electrodes on a glass or a ceramic substrate. Then 1-μm-thick undoped a-Si : H is produced at 230°C by glow-discharge decomposition of silane and finally 1500-Å-thick ITO common electrode, which also acts as an antireflection coating, is deposited by dc sputtering.

The pitch of the individual electrode determines the photosensor resolution (8 elements mm^{-1}). The effective photosensitive area is the overlap portion between the individual electrode and the common electrode, in this case $100 \times 100\ \mu$m.

Figure 6 shows the $I-V$ characteristics both in lighted and dark conditions and their dependence on the individual metal electrode materials for metal/a-Si : H/ITO structures. A fluorescent lamp with peak intensity at 530 nm is

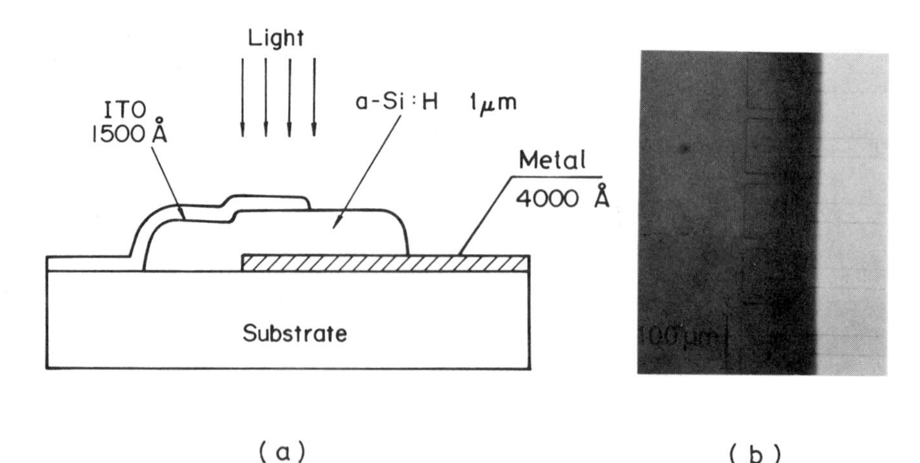

FIG. 5. Photodiode array. (a) Cross-sectional view. (b) Microphotographic plane view. [From Hamano *et al.* (1982).]

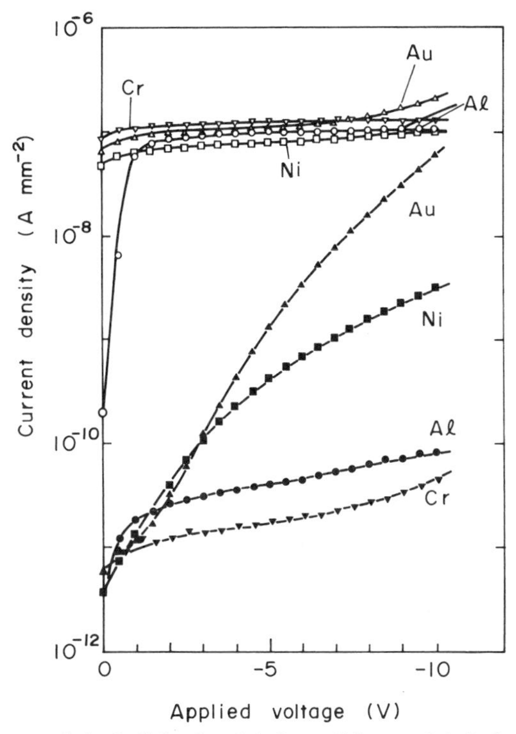

FIG. 6. $I-V$ characteristics in lighted and dark conditions and their dependence on lower metal electrode materials for metal/a-Si : H/ITO structure. Negative bias is applied to an upper ITO electrode. Open symbols: photocurrent under 100 lux; filled symbols: dark current. Circles: Al; triangles: Au; squares: Ni; inverted triangles: Cr. [From Hamano *et al.* (1982).]

used as a light source. Negative bias is applied to an ITO electrode. In this case the photocarriers are electrons. The photocurrent saturates at a very low applied voltage because of the high internal field at the a-Si : H/ITO junction and because of the high electron mobility. A 9×10^{-10} A photocurrent under 100-lux illumination has been obtained. The dark current depends on the lower electrode materials. When Cr or Al is chosen for a metal electrode, the photodiode shows a very low dark current and a high photo- to dark current ratio ($\sim 3 \times 10^3$). Among these materials, Au is a high work function metal, whereas Ni, Cr, and Al are metals with a relatively low work function. The difference in the dark-current levels corresponds to the amount of hole injection from the metal electrode and may be due to the difference in work function for the metal used. In this case, electron injection from the ITO electrode is blocked by the a-Si : H/ITO junction.

Kanoh *et al.* (1981) fabricated almost the same photodiode array structure, in which Pt with high work function is used for the metal electrode. In this photodiode, a very low dark current and saturation characteristics of photocurrent are obtained, regardless of the polarity of bias voltage. However, the blocking of hole injection from the electrode cannot be explained at present.

When a reactively sputtered a-Si : H is used for the photosensitive layer, the voltage at which the photocurrent saturates is about 15 V for undoped a-Si : H. This saturation voltage decreased to 5 V when using nitrogen-doped a-Si : H due to the increased electron mobility (Shimomoto *et al.*, 1982).

To eliminate surface contamination of the photosensor and to protect the surface mechanically, it is possible to construct the protected-surface photosensor on a glass substrate coated with the ITO common electrode. However, a large dark current is observed (Ozawa *et al.*, 1983). This is due to In contamination of the a-Si : H during the deposition of a-Si : H, resulting in ITO/a-Si : H junction deterioration (Kitagawa *et al.*, 1983).

The blocking effect is reinforced by inserting a blocking layer between the a-Si : H and the electrode. Shimizu *et al.* (1981) used thin insulating layers or doped a-Si : H as a blocking layer in electrophotography and vidicon target (see Chapter 4 by Shimizu and 5 by Ishioka). Kaneko *et al.* (1982) fabricated a photodiode array using Si_3N_4 and *p*-a-Si : H blocking layers on both sides of the photosensitive a-Si : H layer as shown in Fig. 7. The photosensitive layer is 3 – 10-ppm boron-doped high-resistivity "intrinsic" a-Si : H, while *p*-a-Si : H is 200 – 500-ppm boron-doped a-Si : H. The photosensor is con-

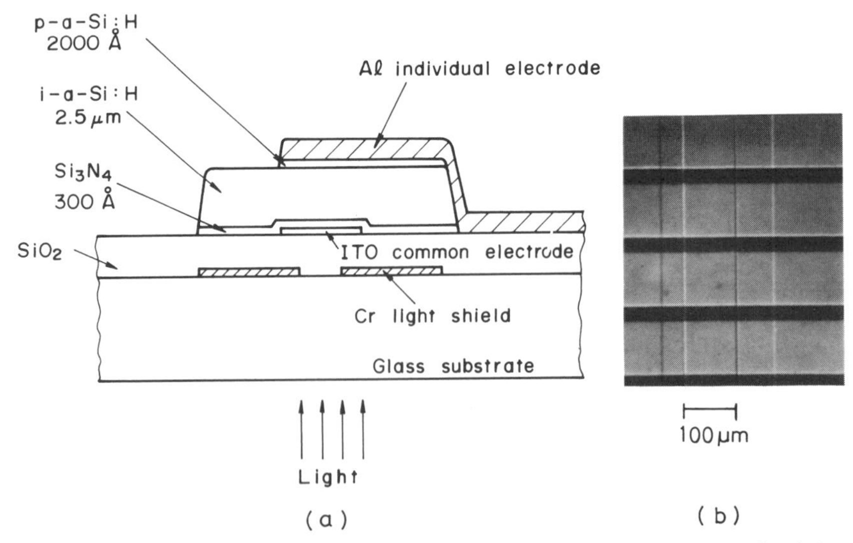

FIG. 7. Photodiode array using Si_3N_4 and *p*-a-Si : H blocking layers. (a) Cross-sectional view. (b) Microphotographic plane view. [From S. Kaneko *et al.*, Amorphous Si : H contact linear image sensor with Si_3N_4 blocking layer. *Technical Digest—International Electron Device Meeting.* Copyright © 1982 IEEE.]

structed on a glass substrate with a Cr light shield layer and SiO_2 insulating layer. The Cr light shield layer has a 100-μm-wide light window. The photosensitive area is the overlap between the light window and an Al individual electrode; its dimension is 100×100 μm. In this sensor, the 300-Å-thick Si_3N_4 and 400-Å-thick ITO act as a good double antireflection coating.

The ITO is positively biased. Electron and hole pairs are generated near the interface between Si_3N_4 and photosensitive a-Si:H layers. The photogenerated electrons tunnel through the Si_3N_4 and photogenerated holes drift toward the Al electrode. There is no hole injection through the Si_3N_4 layer because the ITO electrode is *n*-type and indium contamination from the ITO electrode is suppressed by the Si_3N_4 layer, while the *p*-a-Si:H layer blocks electron injection from the Al electrode (see Fig. 7c). These Si_3N_4, *i*-a-Si:H, and *p*-a-Si:H layers are produced by rf glow-discharge technique. A transition layer, which contains residual nitrogen gas for the Si_3N_4 deposition, is formed unintentionally between the Si_3N_4 and *i*-a-Si:H layers. This nitrogen doping decreases the resistivity (Baixeras *et al.*, 1978; Kurata *et al.*, 1981), resulting in a low electric field in the transition layer and a poor photoelectric conversion characteristic. Therefore this transition layer should be kept thin.

Figure 8 shows the $I-V$ characteristics in both lighted and dark conditions for the photosensors using Si_3N_4 and *p*-a-Si:H blocking layers. In this figure, the $I-V$ characteristics, without one of the two blocking layers, are also shown. These sensors exhibit large dark currents, while the dark current for the photosensor with both blocking layers becomes very small. This confirms that Si_3N_4 and *p*-a-Si:H layers reject the hole and electron injections from each electrode, respectively. The dark current increases with activation energy of 0.8 eV owing to the thermally generated carriers in the intrinsic a-Si:H.

The photocurrent saturates at 4 V (3×10^{-10} A under 35-lux illumination). The photo-to-dark-current ratio obtained is 1×10^3 at 25°C and 1×10^2 at 50°C.

The photocurrent uniformity for these photosensors is mainly dependent on the antireflection layer thickness and the picture element dimensions. A less than $\pm 5\%$ nonuniformity of photocurrent has been obtained.

The spectral response of the photosensor is shown in Fig. 9. The collection efficiency is nearly equal to unity throughout the visible region. Figure 9 also shows the emission intensity of the yellow-green LED used in the contact linear-image sensor. The photosensor has high sensitivity for this illuminator.

Figure 10 shows the photoresponse characteristics, measured with a LED light source (35 lux) modulated at 500 Hz. The upper wave form in Fig. 10 is the photoresponse and the lower wave form is the LED driving pulse. The

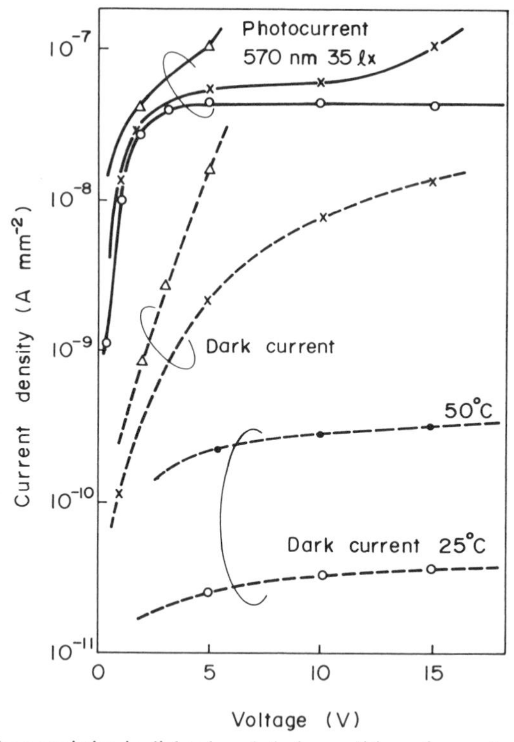

FIG. 8. $I-V$ characteristics in lighted and dark conditions for a photosensor with and without blocking layers. ○, with Si_3N_4 and *p*-a-Si:H blocking layers (25°C); ●, with Si_3N_4 blocking layers (50°C); △, without Si_3N_4 blocking layer; x, without *p*-a-Si:H blocking layer. Positive bias is applied to an ITO electrode. [From S. Kaneko *et al.*, Amorphous Si:H contact linear image sensor with Si_3N_4 blocking layer. *Technical Digest—International Electron Device Meeting.* Copyright © 1982 IEEE.]

photoresponse time is less than 0.1 msec. This value is satisfactory for application to high-speed facsimilie. Almost the same characteristics were obtained in the Cr/a-Si:H/ITO photosensor (Hamano *et al.*, 1982).

3. OPERATION

The video signal is obtained by a set of sequential switches connected to the photosensor array. Two alternative readout modes of operation are possible: dc (or continuous operation) mode and charge-storage operation mode. The dc operation mode has been used in the CdS–CdSe photoconductive linear sensor (Komiya *et al.*, 1981). In this operation mode, the output signal is proportional to the instantaneous photosignal on the photosensor element, at the moment when the photosensor element is addressed by the switch. Because of its high photoconductive gain, the photoconductive sensor is used to obtain a high output signal. The a-Si:H photoconduc-

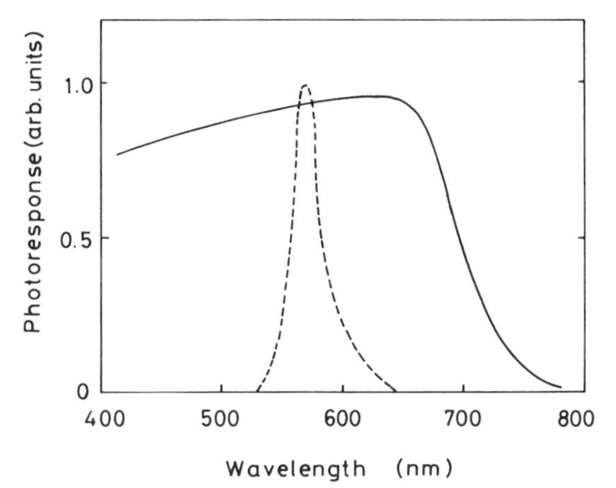

FIG. 9. Spectral response for a-Si:H photodiode using Si_3N_4 and *p*-a-Si:H blocking layers (solid curve). The emission intensity spectrum of the yellow–green LED is also shown (dotted curve). The applied voltage for the photodiode is 5 V. [From S. Kaneko *et al.*, Amorphous Si:H contact linear image sensor with Si_3N_4 blocking layer. *Technical Digest—International Electron Device Meeting.* Copyright © 1982 IEEE.]

tive photosensor has achieved a high photoconductive gain by adopting an interdigital structure (Kagawa *et al.*, 1982). However, the photoresponse time at present is long compared to the photodiode sensor.

The charge-storage mode is a dynamic operation mode in which a transient output pulse is observed. This pulse includes all the charge gen-

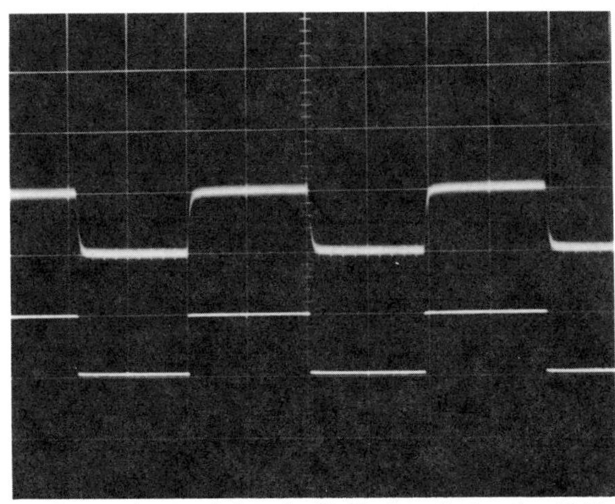

FIG. 10. Photoresponse characteristics of a-Si:H photodiode using Si_3N_4 and *p*-a-Si::H blocking layers. Upper wave form is the photoresponse. Lower wave form is the LED driving pulse. A yellow–green LED (35 lux) modulated by 500 Hz is used. Voltage applied to the photodiode is 5 V. Horizontal scale is 0.5 msec/division and vertical scale is 5 mV/division.

erated by the light incident on the photosensor during the integration time. The charge-storage operation is possible only if the RC time constant of the photosensor exceeds the integration period T_i, where R is the dark resistance of the photosensor and C is the picture element capacitor. The a-Si:H photodiode, having a very low dark current, satisfies this requirement.

Figure 11 shows a scanning circuit for the linear image sensor, which consists of an a-Si:H photodiode array and a scanning circuit, including MOSFET switches and shift resisters. Each photodiode is connected to the MOSFET switch. Silicon MOS ICs are used to operate the photodiode array. When a signal light irradiates the photodiode, a photocurrent is generated. During the integration period T_i, charges in the photodiode are discharged by the photocurrent. When the FET switch is turned on, the photodiode is recharged. The recharge pulse that flows in external load registor R_l contains the total signal charges, which are generated by the signal light during the integration time. Therefore a high output signal is obtained with a gain T_i/T_s, which for continuous sequential scanning is approximately equal to the total number of elements in the array, where T_s is the sampling period.

However, this charge pulse contains switching noise caused by the MOSFETs. Since the linear image sensor have many Si ICs, this switching noise fluctuates because of nonuniformity in MOSFET characteristics. Therefore an efficient noise reduction technique is required for this sensor (Ozawa *et al.*, 1982; Kaneko *et al.*, 1982), as in the case of Si monolithic MOS image sensors (Ashikawa *et al.*, 1973; Aoki *et al.*, 1980).

Figure 12 shows a 210-mm-wide, 1728-element, 8-element-mm^{-1} linear

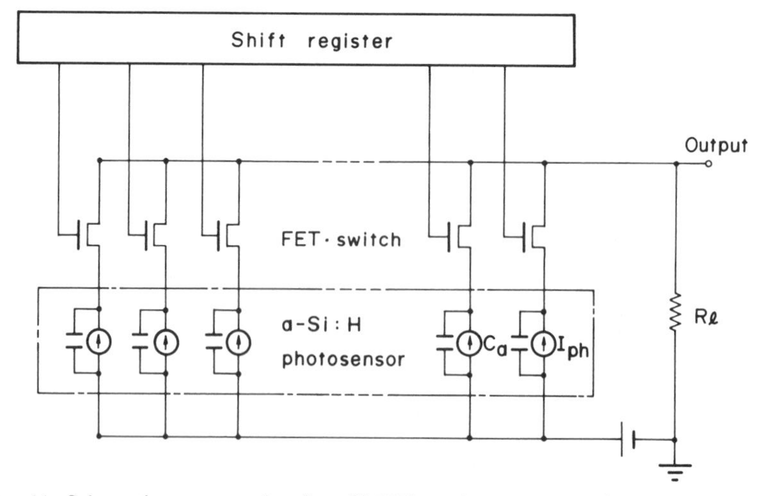

FIG. 11. Schematic representation for a-Si:H linear image sensor using photodiode array. I_{ph} and C_a represent an a-Si:H photodiode.

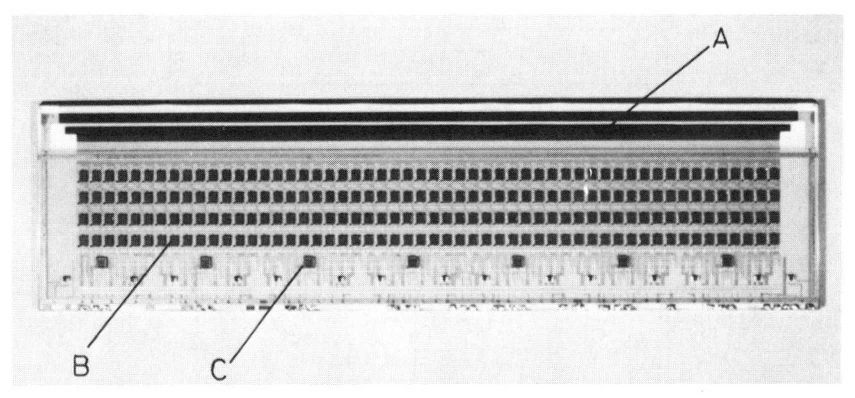

FIG. 12. a-Si : H hybrid linear image sensor. A, a-Si : H photodiode array; B, MOSFET IC; C, shift register. [From Ozawa *et al.* (1982).]

image sensor. The sensor comprises a-Si : H photodiode array, MOSFET switches, and shift registers mounted on a single substrate. Two hundred sixteen MOSFET ICs with 8 channels per chip and 7 chips of shift registers are used in this sensor (Ozawa *et al.*, 1982). The linear image sensor has been operated with 10–20 msec $line^{-1}$ and a clock frequency of about 200–500 kHz. A satisfactorily reproduced image with 8 lines mm^{-1} resolution has been obtained (Ozawa *et al.*, 1982; Kaneko *et al.*, 1982). Since the a-Si : H photodiode has a short photoresponse time, a higher image processing speed, such as 1 msec $line^{-1}$ can be obtained (Suzuki *et al.*, 1984; Sakamoto *et al.*, 1984).

In a long linear image sensor, a large number of MOSFETs must be connected to their associated photodiodes by using wire bonding or elastomer connection (Ozawa *et al.*, 1982; Kanoh *et al.*, 1981). This causes complications in constructing image sensors. Ozawa *et al.* (1983) reduced the number of switching FETs by using a matrix circuit. Figure 13 shows the scanning circuit for a 256-element linear image sensor. Photodiode and blocking diode pairs are connected to the matrix circuit to reduce the number of output terminals. The terminals include 8 common electrodes and 32 individual electrodes. Each terminal is connected to a MOSFET switch, resulting in a decrease in MOSFET switches and associated connections.

The photodiode has an ITO/a-Si : H/Al structure and the blocking diode consists of a Pt/a-Si : H/Al structure, which exhibits a 10^{-12}–10^{-11} A mm^{-1} reverse current. The total leakage current, through the blocking diodes for all the elements, is small enough to allow detecting stored charges in the photodiodes.

Yamamoto *et al.* (1983) also fabricated a linear imge sensor using the

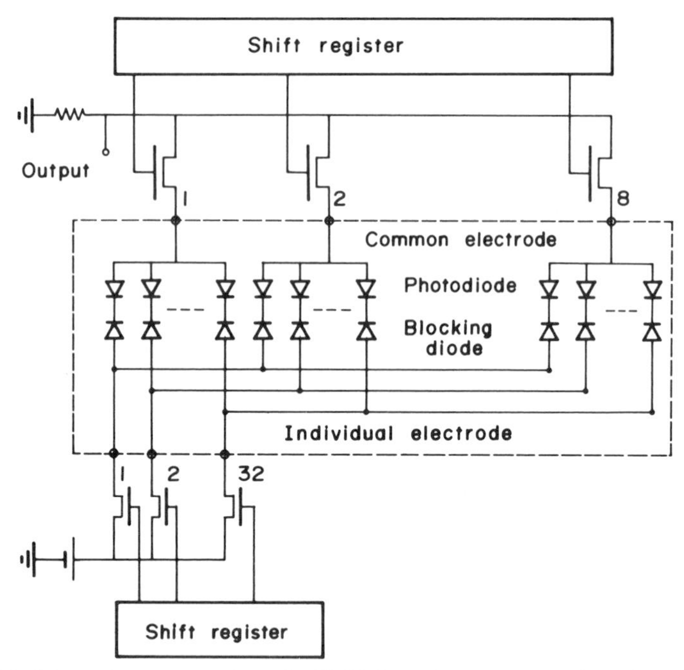

FIG. 13. Schematic representation of a-Si:H linear image sensor using photodiode and blocking diode pair. [From Ozawa *et al.* (1983).]

matrix circuit. In this sensor, a-Si:H $p-i-n$ diodes are used for the photodiodes and the blocking diodes.

III. Application to a Scanning Circuit

A low cost and highly reliable long-image sensor can be obtained by constructing the photosensor array and the scanning circuit on the same substrate. An a-Si:H thin-film transistor (TFT), having a large ON–OFF ratio, can make up the scanning circuit. Matsumura *et al.* (1980) have developed an integrated 8-element linear image sensor using a-Si:H TFTs as switching devices. Figure 14 shows the linear image sensor structure, which consists of interdigital structure photoconductive sensor, a-Si:H TFTs ($L = 10\ \mu$m, $W = 100\ \mu$m) and charge storage capacitors. The photocurrent is integrated in the charge-storage capacitor and transferred to the output, when the a-Si:H TFT gate pulse is applied. Because of high photoconductivity gain and its charge-storage operation, the photosensor has high photosensitivity. The photosensor is driven at 300 Hz and a clear image is obtained under 0.1-lux illumination. However, since the time constant for

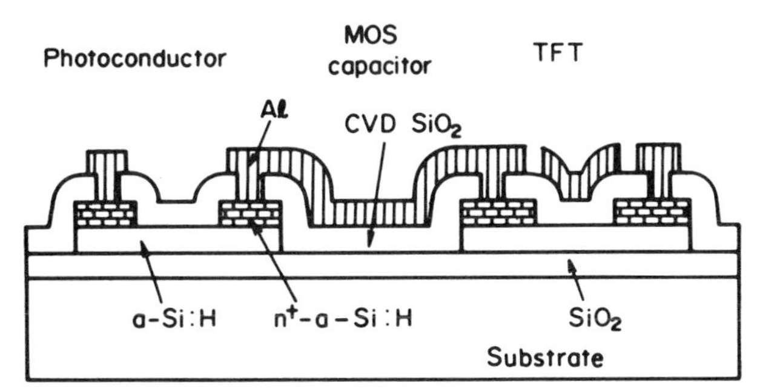

FIG. 14. Unit cell structure for an integrated image sensor. [From M. Matsumura, H. Hayama, Y. Nara, and K. Ishibashi, Amorphous silicon image sensor IC. *IEEE Electron Devices Lett.*,© 1980 IEEE.]

the TFT ON-resistance and storage capacitance is larger than 20 μsec, more than a 100-μsec gate pulse width is required to discharge the storage capacitor, resulting in a 20-kHz maximum operation frequency for this sensor. This long time constant is due to high TFT ON-resistance (80 MΩ) caused by small carrier mobility in the TFT (0.01 cm^2 V^{-1}sec^{-1}). Okumura *et al.* (1983) fabricated a 64-element linear image sensor using a photodiode array and a Si:H TFT with 0.6 cm^2 V^{-1} sec $^{-1}$ carrier mobility. A 5-μsec discharge time is obtained, which has a 100-kHz operation frequency. They also indicate that they can operate the photosensor with higher operation frequency by using a matrix circuit.

Concerning a logic circuit to operate this switching device, Matsumura and Nara (1982) studied inverters that are composed of an enhancement *n*-channel driver TFT and p-channel load TFT. Step-shaped transfer characteristics have been realized. However, the operating frequency is not sufficient at the present time.

The charge-transfer device is one of the important devices used to scan the photosensor array with low noise. Kishida *et al.* (1982) fabricated a-Si:H charge-coupled devices (CCDs), as shown in Fig. 15. A silicon-oxynitride/a-Si:H/silicon-oxynitride structure is sandwiched between staggered transfer electrodes. A four-phase clock pulse is applied to the transfer electrodes. When the clock pulse ϕ_1 is at a high level, signal (excess) electrons gather at the a-Si:H and insulator interface just under electrode A. When the clock pulse ϕ_2 following ϕ_1 rises to a high level, these electrons move to the bulk a-Si:H sandwiched between electrodes A and B, where the potential energy for electrons has the lowest value. Then when ϕ_1 falls to a low level, electrons move to the a-Si:H–insulator interface just above electrode B. The signal electrons are thus transferred along the a-Si:H layer in a serpentine mode. A 0.04% transfer inefficiency has been obtained at 0.5–1-kHz clock frequency. An increase in clock frequency causes an

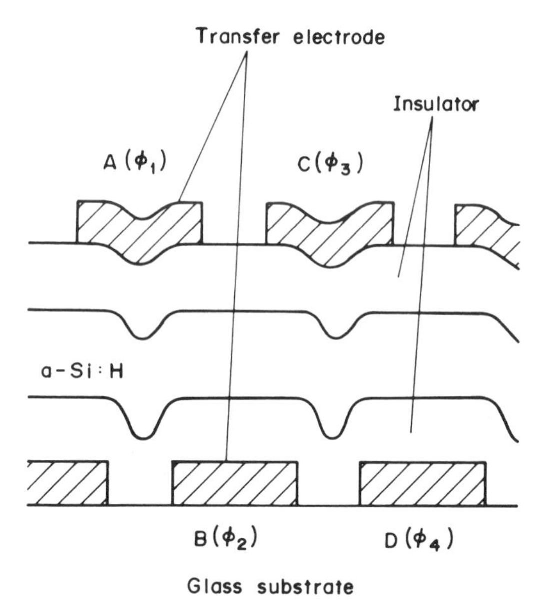

FIG. 15. Cross-sectional view of a-Si : H CCDs. [From Kishida *et al.* (1982).]

increase in transfer inefficiency because of trapping of electrons in deep localized states.

They also fabricated three-phase-clock a-Si : H CCDs with coplanar transfer electrodes. The n^+-a-Si : H resistive layer is set up on and between the electrodes to produce a uniform electric field in the charge-transfer direction. This results in 0.7% transfer inefficiency, at as high as 200-kHz clock frequency (Kishida *et al.,* 1983.)

IV. Application to an Area Image Sensor

The Si monolithic image sensor integrates the photosensor part and the scanner part into a single Si chip. Therefore, the effective photosensitive area is small, compared with vidicons, which have a 100% photosensitive area. Furthermore, the suppression of blooming phenomena, which is caused by strong incident light, is not easy. Placing the photosensor on top of the scanning circuit enlarges the effective photosensitive area and separates the photosensor part from the scanning part, resulting in blooming suppression (Tsukada *et al.,* 1979). Tsukada *et al.* (1981) have developed the image sensor in which the photosensor is built on top of the Si IC scanner. Figure 16 shows the image sensor schematic structure using a-Si : H.

On the Si MOS-switching-FET matrix array, a-Si : H is deposited by rf reactive sputtering in Ar and H_2 atmosphere. This sputtered a-Si : H is also used in vidicons (Imamura *et al.,* 1980). Then the ITO common electrode is

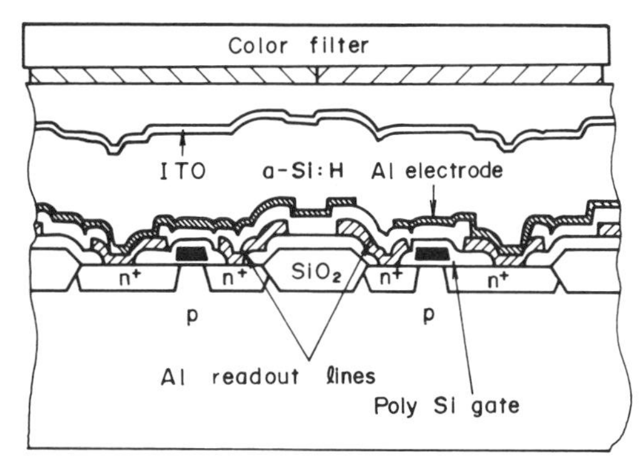

FIG. 16. Cross-sectional view of a picture element. [From T. Tsukada *et al.*, Solid-state color imager using a-Si : H photoconductive film. *Technical Digest—International Electron Device Meeting.* Copyright © 1981 IEEE.]

deposited. The a-Si : H layer is sandwiched between an Al electrode connected to the drain for each MOSFET and the transparent ITO common electrode, resulting in the formation of a blocking diode. The red–green–blue color filter is glued with resin. Thus a single-chip color image sensor, with 485(V)×384(H) picture elements, was fabricated.

Since a-Si : H has a high resistivity, the Al electrode array pitch determines the resolution for this device. One pixel area is $23.5\times 13.5\ \mu$m. The associated Al electrode is $20.5\times 11\ \mu$m. Therefore the ratio between Al electrode area and one pixel area is as high as 0.73.

The ITO electrode is negatively biased relative to the Al electrode. Electron and hole pairs are generated by the incident light and electrons pass through the a-Si : H. Since in undoped a-Si : H the electron mobility is small, nitrogen is added to the sputtering atmosphere to increase the electron mobility. As a result, the photocurrent saturation voltage for the photodiode decreased to 5 V and the photoresponse time decreased to less than 500 μsec (Shimomoto *et al.*, 1982).

Figure 17 shows the schematic representation for this sensor. This device operates in the charge storage mode operation. The photocurrent I_{ph}, generated by the incident light, discharges the capacitance of a-Si : H photodiode (C_a) and p–n junction capacitance (C_d). Vertical MOSFET switches (Q_v) on the same horizontal line are turned on at the same time by the common poly-Si gate line. The signal charges transferred to the vertical readout lines are selected by the horizontal n-MOSFET switches, which are scanned by the horizontal shift register. The sets of vertical MOSFET switches are interlacedly scanned. Thus a sequential video signal is obtained.

At high light intensity, the photocurrent discharges the C_a and C_d rapidly, and the potential difference between the Al electrode and the ITO electrode becomes equal to the open circuit voltage for the a-Si : H photodiode. In this

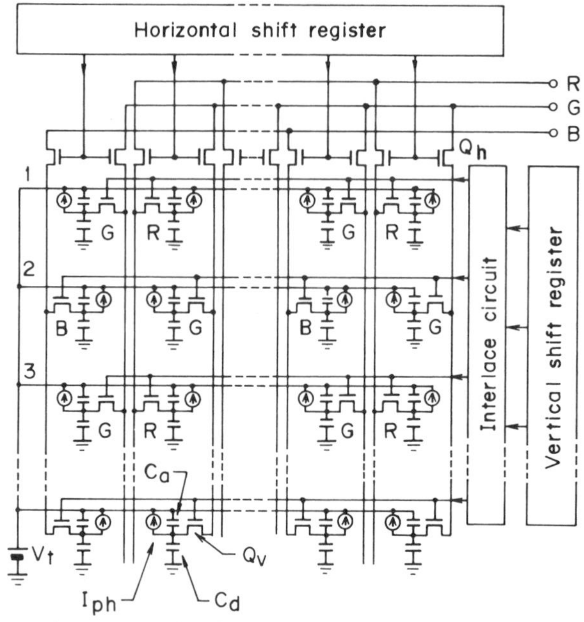

FIG. 17. Schematic representation for an a-Si:H solid-state color image sensor. I_{ph} and C_a represent an a-Si:H photodiode, Q_v and P_h represent n-MOSFET swiches and C_d represents p–n junction capacitance. [From T. Tsukada *et al.*, Solid-state color imager using a-Si:H photoconductive film. *Technical Digest—International Electron Device Meeting.* Copyright © 1981 IEEE.]

FIG. 18. Imagery produced by an a-Si:H solid-state color image sensor. High light image is also shown (reproduced here in black and white). [From T. Tsukada *et al.*, Solid-state color imager using a-Si:H photoconductive film. *Technical Digest—International Electron Device Meeting.* Copyright © 1981 IEEE.]

case, the potential barrier at the ITO/a-Si:H interface disappears and the separation of the photogenerated electron–hole pairs ceases. As a result, the photocurrent I_{ph} becomes zero and the blooming caused by the signal charge leakage to the readout line is suppressed without requiring any additional circuit. Figure 18 shows the image reproduced by the color image sensor. Blooming is suppressed strongly, stopping the photocurrent (Baji *et al.*, 1982). In this blooming suppression mode, a small voltage applied to the photodiode (less than 2.25 V) is required to prevent turning on a parasitic bipolar transistor under the *n*-MOS switching FET. The small applied voltage causes low collection efficiency for this photodiode. Therefore, a photodiode with a lower operating voltage is required (Baji *et al.*, 1982).

In the large negative applied voltage, the collection efficiency is nearly equal to unity over the whole visible light region. The lag for this device is 3% after 50 msec and little burn-in is observed (Tsukuda *et al.*, 1981).

V. Future Expectations

A satisfactory image reproduction has been obtained by using a long a-Si:H photodiode array combined with Si IC scanners. Since a-Si:H photodiode has a short photoresponse time and a high photosensitivity in the visible light region, higher-speed image processing or color image reproduction may be realized. As a result, high-performance and compact equipment may be built in the near future. Much effort on stability and yield will be needed to put this device into practical use.

As mentioned earlier, a long linear image sensor lower in cost and more reliable, may be obtained by fabricating a thin-film scanning circuit on the sensor substrate. The study of scanning circuits using a-Si:H has already begun. However, satisfactory high-frequency characteristics have not yet been achieved. A new device structure suitable to the a-Si:H properties as well as improvement in a-Si:H quality and device fabrication technique should result in good high-frequency performance. Scanning circuits using polycrystalline thin films such as Si, CdS, and Te having higher carrier mobility than that of a-Si:H may be another approach to obtain good high-frequency performance.

In the area-image sensor, it is expected that highly sensitive solid-state image sensors with high resolution, competitive with vidicons, will become available. The solid-state image sensor, which is made of an a-Si:H photodiode array on a Si IC scanner, is the most suitable device for this purpose.

References

Anderson, D. A., and Spear, W. E. (1977). *Philos. Mag.* **36,** 695.

Aoki, M., Ohba, S., Takemoto, I., Nagahara, S., Sasano, A., and Kubo, M. (1980). *Int. Solid-State Circuits Conf., Dig. Tech. Pap.*, p. 26.

Ashikawa, M., Koike, N., Kamiyama, T., and Kubo, S. (1973). *Int. Solid-State Circuits Conf., Dig. Tech. Pap.* p. 128.
Baixeras, J., Mencaraglia, D., and Andro, P. (1978). *Philos. Mag. B* **37,** 403.
Baji, T., Shimomoto, Y., Matsumaru, H., Koike, N., Akiyama, T, Sasano, A., and Tukada, T. (1982). *Jpn. J. Appl. Phys., Suppl.* **21-1,** 269.
Boronkay, S., Gustin, P., and Rossier, D. (1978). *Acta Electron.* **21,** 55.
Boyle, W. S., and Smith, G. E. (1970). *Bell Syst. Tech. J.* **49,** 587.
Bube, R. H. (1961). "Photoconductivity of Solids". Wiley, New York.
Hamano, T., Ito., H., Nakamura, T., Ozawa, T., Fuse, M., and Takenouchi, M. (1982). *Jpn. J. Appl. Phys. Suppl.* **21-1,** 245.
Imamura, Y., Ataka, S., Takasaki, Y., Kusano, C., Ishioka, S., and Hirai, T. (1980). *Jpn. J. Appl. Phys., Suppl.* **19-1,** 573.
Kagawa, T., Matsumoto, N., and Kumabe, K. (1982). *Jpn. J. Appl. Phys., Suppl.* **21-1,** 251.
Kaneko, S., Sakamoto, M. Okumura, F., Itano, T., Kataniwa, H., Kajiwara, Y., Kanamori, M., Yasumoto, M., Saito, T., and Ohkubo, T. (1982). *Tech. Dig.—Int. Electron Devices Meet.* p. 328.
Kanoh, Y., Usui, S., Sawada, A., and Kikuchi, M. (1981). *Tech. Dig.—Int. Electron Devices Meet.* p. 313.
Kishida, S., Nara, Y., Kobayashi, O., and Matsumura, M. (1982). *Appl. Phys. Lett.* **41,** 1154.
Kishida, S., Naruke, Y. Uchida, S., and Matsumura, M. (1983). *J. Non-Cryst. Solids* **59/60,** 1281.
Kitagawa, M., Mori, S., Ishihara, M., Ohno, T., Hirano, Y., and Kohiki, S. (1983). *J. Appl. Phys.* **54,** 3269.
Komiya, K., Kanzaki, M., and Yamashita, T. (1981) *Tech. Dig.—Int. Electron Devices Meet.* p. 309.
Kurata, H., Hirose, M., and Osaka, Y. (1981). *Jpn. J. Appl. Phy.* **20,** L811.
Matsumura, M., and Nara, Y. (1982). *IEEE Trans. Electron Devices* **ED-29** 580.
Matsumura, M., Hayama, H., Nara, Y., and Ishibashi, K. (1980). *IEEE Electron Devices Lett.* **EDL-1,** 182.
Motosugi, G., Kagawa, T., Matsumoto, N., and Kumabe, K. (1981). *Proc. Spring Meet. Jpn. Soc. Appl. Phys.,* p. 382.
Noble, P. J. (1968). *IEEE Trans. Electron Devices* **ED-15,** 202.
Okumura, F., Kaneko, S., and Uchida, H. (1983). *Ext. Abstr. Conf. Solid State Devices Mater., 15th,* p. 201.
Ozawa, T., Takenouchi, M., Hamano, T., Ito, H., Fuse, M., and Nakamura, T. (1982). *Proc. Int. Microelectron. Conf., Dig. Tech. Pap.,* p. 132.
Ozawa, K., Takagi, N., Hiranaka, K., Yanagisawa, S., and Asama, K. (1983). *Jpn. J. Appl. Phys., Suppl.* **22-1,** 457.
Sangster, F. L. J. (1970). *Bell Syst. Tech. J.* **49,** 587.
Sakamoto, M., Okumura, F., Kajiwara, Y., Uchida, H., and Kaneko, S. (1984) *Proc. Electron. Components Conf. 34th,* p. 264.
Shimizu, I., Oda, S., Saito, K., Tomita, H., and Inoue, E. (1981). *J. Phys. Colloq. Orsay, Fr.* **42,** Suppl. 10, C4-1123.
Shimomoto, Y., Tanaka, Y., Yamamoto, H. Takasaki, Y., Sasano, A., and Tsukada, T. (1982). *Jpn. J. Appl. Phys., Suppl.* **21-1,** 263.
Suzuki, K., Saito, T., Suda, Y., Mori, K., Takayama, S., Nakai, T., and Takikawa, O. (1984). *Proc. Electron. Components Conf., 34th,* p. 257.
Terui, Y., Wada, T., Yoshino, M., Kadota, H., Komeda, T., Chikamura, T., Fujiwara, S., Tanaka, H., Ota, Y., Fujiwara, Y., Ogawa, K., Kitashiro, O., and Horiuchi, S. (1980). *Int. Solid-State Circuits Conf., Dig. Tech. Pap.,* p. 24.

Tsukada, T., Yamamoto, H., Matsui, M., Eto, Y., Hirai, T., and Maruyama, E. (1977) *Tech. Dig.—Int. Electron Devices Meet.*, p. 546.
Tsukada, T., Baji, T., Yamamoto, H., Takasaki, Y., Hirai, T., Maruyama, E., Ohba, S., Koike, N., Ando, H., and Akiyama, T. (1979). *Tech. Dig.—Int. Electron Devices Meet.* p. 134.
Tsukada, T., Baji, T., Shimomoto, Y., Sasano, A., Tanaka, Y., Matsumaru, H., Takasaki, Y., Koike, N., and Akiyama, T. (1981) *Tech. Dig.—Int. Electron Devices Meet.* p. 479.
Weckler, O. P. (1967). *IEEE J. Solid-State Circuits* **SC-2,** 65.
Weimer, P. K., Sadasiv, G., Meyer, J. E., Jr., Meray-Horvath, L., and Pike, W. S. (1967). *Proc. IEEE* **55,** 159.
Yamamoto, H., Baji, T., Matsumaru, H., Tanaka, Y., Seki, K., Tanaka, T., Sasano, A., and Tsukada, T. (1983) *Ext. Abstr. Solid State Devices and Mater. 15th,* p. 205
Zanzucchi, P. J., Wronski, C. R., and Carlson, D. E. (1978). *J Appl. Phys.* **48,** 5227.

CHAPTER 9

Charge-Coupled Devices

Masakiyo Matsumura

TOKYO INSTITUTE OF TECHNOLOGY
TOKYO, JAPAN

I. Introduction

This chapter summarizes theoretical and experimental studies of the application of a-Si : H to charge-coupled devices (CCDs) and presents recent results. Amorphous hydrogenated silicon is an interesting material for CCD because it can be deposited on various substrate materials and eventually could be made in the length of a page (e.g., for facsimile transmission). Another advantage is the high electrical resistivity that allows a long charge storage. Finally, a CCD is a simpler structure than a chain of field-effect transistors.

It is well known that single-crystal silicon CCDs are the most important image-sensing devices nowadays. Various theoretical analyses have already been reported (Tompsett, 1973). However, all previous analyses that are based on CCDs in single-crystal semiconductors are not directly applicable to the new a-Si : H CCDs, because a-Si : H has many localized states in the band gap compared to single-crystal semiconductors. When analyzing a-Si : H CCDs, it is necessary to take into account the effects of the localized states. In this chapter we calculate numerically the transfer characteristics of a-Si : H CCDs, assuming that the localized states in a-Si : H have an exponential energy distribution. Then we present an approximate analytical expression, the results of which are compared with numerical and experimental results.

ISBN 0-12-752150-X

II. Theoretical Results on Charge Transfer

1. Numerical Analysis

In the typical CCDs operation in a single-crystal semiconductor [i.e., in the surface channel CCDs (SCCDs)], the signals are minority carriers stored near the interface of a MIS structure operated in the deep depletion mode. Since these carriers are separated from the majority carriers by a space-charge region, this transient state can be kept sufficiently long for the CCD operation. This is because the leakage current or thermally generated dark current in the space-charge and neutral regions is very small. On the other hand, in the accumulation mode of the MIS structure, the signals cannot be stored sufficiently long, because this transient state is as short as the dielectric relaxation time. Thus the accumulation mode cannot be used in the CCD operation.

In a-Si:H, it is difficult to operate CCDs in the same mode as conventional SCCDs. This is because the ideal inversion at the *p*-type a-Si interface has not been observed as yet (Sugiura and Matsumura, 1982), and as for *n*-type a-Si:H, whose interface can be inverted, the drift mobility of the signal charge (i.e., hole) is very low. However, the band gap of a-Si:H is about 1.7 eV, which is much wider than that of a single-crystal silicon. This indicates that the carrier concentration in undoped a-Si:H under thermal equilibrium condition is so low that the accumulation mode can store transient signals sufficiently long for CCD operation, as described later.

Figure 1 shows an equivalent circuit of the a-Si:H CCDs, where we assume that a central electrode is in a "high" state and that electrodes of both sides are in a "low" state. The signal electrons are stored at the a-Si–insulator interface under the central electrode. This storage mechanism can be represented by a capacitor C. Electrons gradually leak to the surrounding area through the adjacent electrodes. The leakage current is

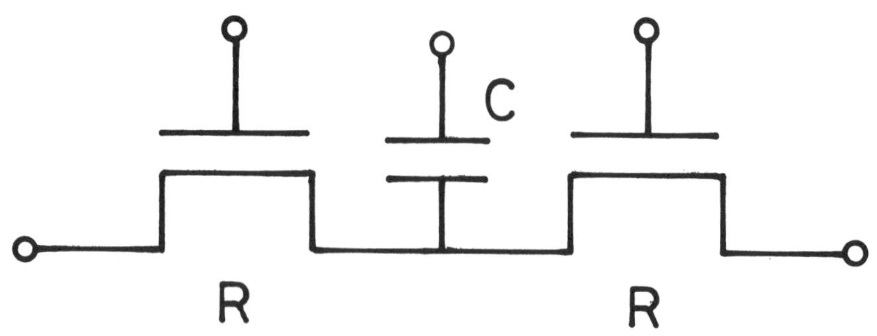

FIG. 1. An equivalent circuit of three transfer gates of a-Si:H charge-coupled devices. A central transfer electrode represented by a capacitor is assumed to be in a "high" state and to store signal electrons. Transfer electrodes of both sides are assumed to be in a "low" state.

controlled by the electrode. Thus the surrounding area can be represented by transistors whose OFF-resistance is R. The time constant τ_r of this relaxation is then given by the RC time constant as follows:

$$\tau_r = C_i L^2 \rho / d, \tag{1}$$

where C_i is the gate capacitance per unit area, L the channel length, ρ the resistivity of a-Si:H, and d the thickness of the a-Si:H layer. When we assume typical values (i.e., $C_i = 4 \times 10^{-8}$ F cm^{-2}, $L = 15$ μm, $\rho = 10^9$ Ω cm, and $d = 0.1$ μm), τ_r is calculated to be 9 sec. Thus a-Si:H CCDs are feasible because the time required for the signal electrons to transfer from one electrode to the next is much shorter than τ_r, as described below.

Since a-Si:H has a large number of localized states in the band gap, most of the signal electrons are trapped by the localized states. Only a small number of free electrons are in the conduction band. These free electrons will drift toward the next electrode while they are experiencing trapping and detrapping by localized states. Since the signal electrons are stored in the thin accumulation layer near the a-Si:H–insulator interface, we assume that they are uniformly distributed in the accumulation layer, and we neglect the trapping and detrapping of electrons by localized states. Then the electrons must satisfy the following equation:

$$\frac{\partial n(E)}{\partial t} = \sigma v_{\text{th}}\{N(E) - n(E)\}n_c - \sigma v_{\text{th}} n(E) n_1, \tag{2}$$

where $N(E)$ is the localized state density at an energy E, $n(E)$ the electron density captured by localized states of energy E, n_c the free electron density, v_{th} the average thermal velocity, σ the capture cross section for electrons, and n_1 is given by

$$n_1 = N_c \exp\{-(E_c - E)/kT\}, \tag{3}$$

where N_c is the effective density of states in the conduction band and E_c is the minimum energy of the conduction band. When we neglect diffusion of electrons, the motion of the signal electrons under the storage electrode is characterized by the transit time τ,

$$\tau = L/\mu F, \tag{4}$$

where F is the fringing field strength along the channel, μ is the electron mobility in the conduction band, and L is the channel length. Charges must satisfy the conservation law as

$$\frac{\partial n_c}{\partial t} = -\int \frac{\partial n(E)}{\partial t}\, dE - \frac{n_c}{\tau}. \tag{5}$$

When we assume typical values, that is, $\mu = 10$ cm^2 V^{-1} sec^{-1} (LeComber and Spear, 1979), $L = 15$ μm, and $F = 150$ V cm^{-1}, τ will be 10^{-6} sec.

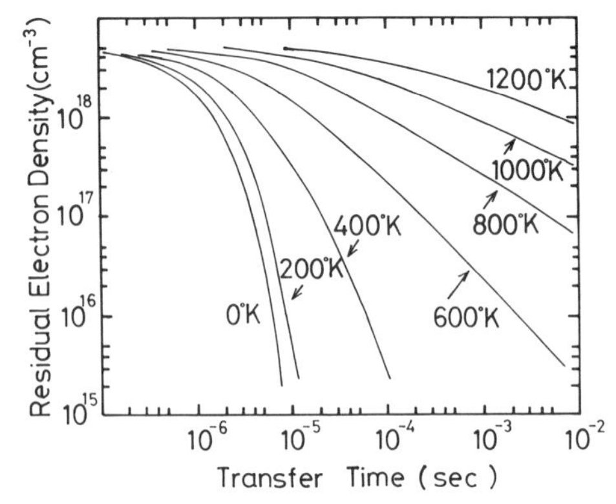

FIG. 2. Numerical results of residual electron density as a function of time for various values of characteristic temperature T_G. The values of other parameters used in the calculation are tabulated in Table I.

Next, we assume that the localized state density distribution is approximated by

$$N(E) = N_G \exp\left(-\frac{E_c - E}{kT_G}\right), \tag{6}$$

where N_G is the localized state density just below the conduction-band edge and T_G is the effective characteristic temperature, a parameter that describes the exponential density distribution of localized states. By using Eqs. (2)–(6), and assuming an initial electron density N_0, we can evaluate the residual electron density under the electrode, when the transfer begins at the time $t = 0$.

Figure 2 shows the numerical results of the residual electron density for various values of T_G. The values of other parameters used in the calculation are tabulated in Table I. The values of T_G reported to date range from 312

TABLE I

VALUES OF PARAMETERS USED IN THE CALCULATION

σ	2×10^{-16} cm^2
v_{th}	10^7 cm sec^{-1}
N_C	1.3×10^{19} cm^{-3}
N_G	2×10^{20} cm^{-3} eV^{-1}
μ	10 cm^2 V^{-1} sec^{-1}
T_G	600°K
T	300°K
τ	10^{-6} sec
N_0	5×10^{18} cm^{-3}

(Tiedje *et al.*, 1981) to ~600°K. If we assume a typical (though poor) value of $T_G = 600°K$, more than 95% of initially stored signal electrons can be transferred under the next electrode within 250 μsec. And when T_G decreases further, we can obtain a lower residual electron density.

2. APPROXIMATE ANALYSIS

For the purpose of grasping intuitively the physical phenomena, the simple approximate analysis is more suitable than the cumbersome numerical calculation. Therefore in this section we deal with approximate analysis.

Before the transfer starts, the energy distribution of electrons takes the form of a Fermi–Dirac distribution function. While the number of electrons is decreasing steadily with time, the distribution of electrons keep the form of a Fermi–Dirac distribution function. This constancy of the distribution is due to the fact that the capture rate of free electrons by the localized states is much faster than the loss of free electrons caused by the transfer when the occupation probability of localized states is not approximately one. Therefore, electrons are considered to be in their quasi-thermal equilibrium condition; i.e., the energy distribution of electrons is described by quasi-Fermi energy E_F. Then the total density n_t of electrons captured by the localized states per unit volume can be written as

$$n_t = \int_{E_V}^{E_c} N(E) f_{FD}(E)\, dE, \tag{7}$$

where $f_{FD}(E)$ is the Fermi–Dirac distribution function. The function $f_{FD}(E)$ can be approximated (Kishida *et al.*, 1983) by

$$f_{FD}(E) \simeq \begin{cases} 1 - \exp\left(\dfrac{E - E_F}{kT}\right) + \dfrac{1}{2}\exp\left(\dfrac{3}{2}\dfrac{E - E_F}{kT}\right), & E \leq E_F, \\ \exp\left(\dfrac{E_F - E}{kT}\right) - \dfrac{1}{2}\exp\left(\dfrac{3}{2}\dfrac{E_F - E}{kT}\right), & E \geq E_F. \end{cases} \tag{8}$$

Then, in the case $T_G > T$, we obtain

$$n_t = N'_G\, kT_G \exp\left(-\frac{E_c - E_F}{kT_G}\right), \tag{9}$$

where N_G is given by

$$N'_G = N_G \frac{T_G^2(T^2 + 9T_G^2)}{(T_G^2 - T^2)(9T_G^2 - 4T^2)}. \tag{10}$$

A detailed derivation was described elsewhere (Kishida *et al.*, 1983). On the other hand, the free electron density n_c in the conduction band is also given by E_F as

$$n_c = N_c \exp\left(-\frac{E_c - E_F}{kT}\right). \tag{11}$$

And between the n_t and n_c the following conservation equation must be satisfied:

$$\partial n_c/\partial t + \partial n_t/\partial t = -(n_c/\tau). \tag{12}$$

From (9) and (11), we obtain

$$n_c = N_c(n_t/N'_G kT_G)^{T_G/T}. \tag{13}$$

Inserting (13) into (12), we get

$$\frac{d}{dt}\left\{N_c\left(\frac{n_t}{A}\right)^\alpha + A\left(\frac{n_t}{A}\right)\right\} = -N_c\frac{(n_t/A)^\alpha}{\tau}. \tag{14}$$

Then the density of residual electrons ($=n_t$) at time t is obtained as

$$\alpha \ln(n_t/A) - B(n_t/A)^{-(\alpha-1)} = -(t/\tau) + C_0. \tag{15}$$

where A, B, α, and C_0 are given by

$$A = kT_G N'_G, \tag{16}$$

$$B = (N'_G/N_c)(kTT_G/(T_G - T)), \tag{17}$$

$$\alpha = T_G/T, \tag{18}$$

$$C_0 = \alpha \ln(N_0/A) - B(N_0/A)^{-(\alpha-1)}. \tag{19}$$

When $t \gg \tau C_1$, (15) can be simplified to

$$n_t = A(\tau B/t)^{1/(\alpha-1)}. \tag{20}$$

Figure 3 shows the analytical results obtained from (15) and (20). For reference purposes, numerical results are shown by dotted curves. It can be seen that even the coarsest approximation [i.e., (20)] agrees with the numerical results in the region $n_t(t) \ll N_0$, which is a necessary condition for CCD operation. The important feature of (20) is that $\ln(n_t)$ forms a straight line. The slope of the line is determined by a function of α (i.e., T_G/T) and is independent of other material constants or of the initial density N_0 of signal electrons. The transfer inefficiency ϵ of the CCDs is given by

$$\epsilon = n_t/N_0 = A/N_0(\tau B/t)^{1/(\alpha-1)}, \tag{21}$$

where in case of four-phase clocking, t is given by clock frequency f as

$$t = 1/4f. \tag{22}$$

Therefore $\ln(\epsilon)$ versus $\ln(f)$ relation becomes a linear form and the steep-

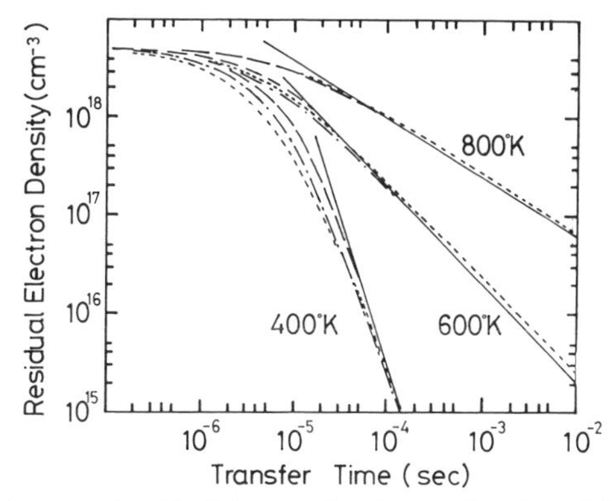

FIG. 3. Analytical results of residual electron density as a function of time for various values of characteristic temperature T_G. The values of other parameters used are tabulated in Table I. The dashed curves are the analytical results obtained from (15), and the solid lines are those obtained from (20). For reference purposes numerical results are shown by dotted curves.

ness of the line yields T_G as

$$T_G = T\left\{\frac{\partial(\ln f)}{\partial(\ln \epsilon)} + 1\right\}. \tag{23}$$

III. CCD Structure for High-Frequency Operation

There are three methods for shortening the transit time τ. They are (1) to increase μ, (2) to decrease the cell size L, and (3) to increase F. However, μ is a material constant and L is determined by the application of the device. Thus to shorten the transit time, we must increase the drift field F parallel to the electrode.

The first a-Si : H CCD operation has been verified experimentally by an insulator/a-Si : H/insulator structure sandwiched between staggered transfer electrodes.(Kishida *et al.*, 1982).

Figure 4 is a partial cross section of our original a-Si : H CCD. When both

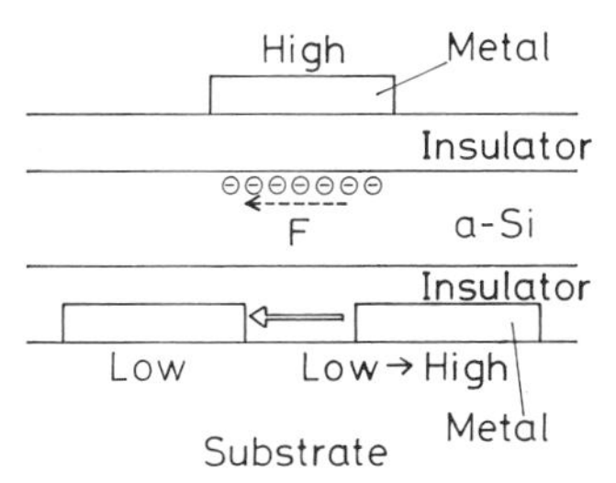

FIG. 4. Partial cross section of the first a-Si : H CCDs.

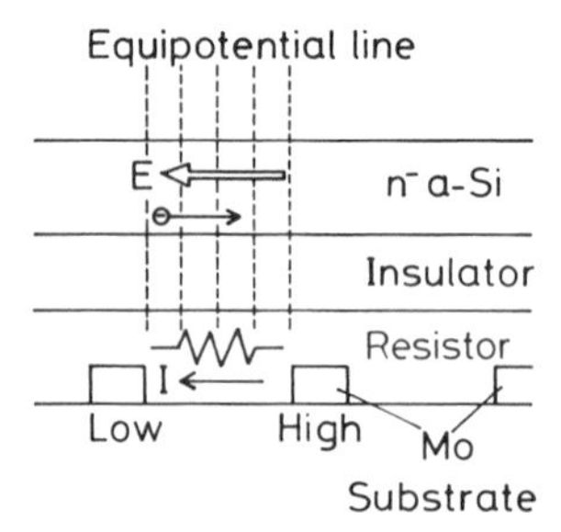

FIG. 5. Partial cross section of the new a-Si:H CCDs for high-speed operation.

of the lower electrodes are at a low level and the upper electrode is at a high level, signal electrons are gathered just under the upper electrode. When the right-hand-side electrode on the lower surface changes to a high level, a strong electric field is generated between the lower electrodes. Thus electrons begin to drift toward the right. However, the total thickness of the a-Si:H layer and insulating layers is less than 1 μm. On the other hand, the electrode is longer than 10 μm. The parallel electric field generated between the lower electrodes is shorted by the upper electrode. Thus the drift field for electrons near the upper electrode becomes weak. This was the reason the earliest device could not operate at more than 10 kHz. Thus for high-frequency operation, we must eliminate the shorting effect of large metallic electrodes while keeping a strong coupling between the electrodes.

Figure 5 shows a new structure for high-speed operation. The electrodes with narrow stripe and wide gap are on the same surface, and the gap between the electrodes is buried in a resistive layer. This structure is the same as that of the resistively connected CCDs (Kim and Snow, 1972). In the present device, however, the transfer electrode must be designed to be as narrow as possible and the gap must be designed to be as wide as necessary. When clock pulses are applied to the electrodes, the difference in potential makes the current flow through the resistive layer, resulting in a linear potential gradient in the gap. Since the insulator and the a-Si:H layers are thin compared to the length of the gap, a similar potential gradient is generated in the active a-Si:H region. This results in a strong drift field for electrons. Since the strong field continues from one electrode to the next, we can obtain a strong coupling between the electrodes.

IV. Experimental Results

Figure 6 shows a cross-sectional view of the latest device. We used a Corning 7059 glass substrate with molybdenum as the transfer electrodes and input gate and output gate electrodes. After patterning molybdenum,

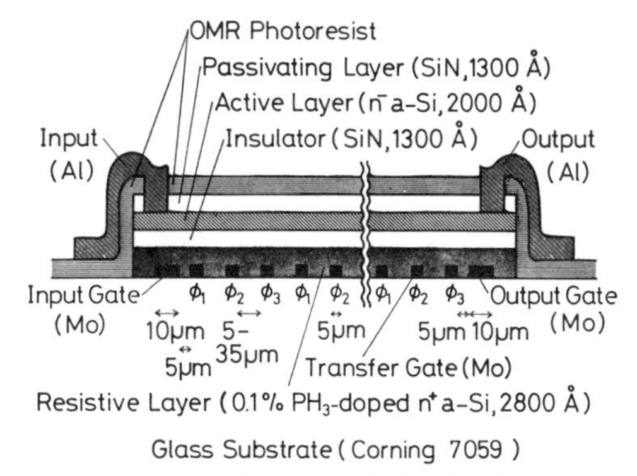

FIG. 6. Cross-sectional view of the new a-Si:H CCDs for high-speed operation.

we successively deposited phosphorus-doped a-Si:H as a resistor, silicon nitride as an insulator, undoped a-Si:H as an active layer, and silicon nitride as a passivating layer. These layers were deposited within one vacuum pumpdown by hot-cathode arc-discharge deposition method (Uchida *et al.*, 1982). All deposited layers except at the active area were etched off by CF_4 plasma. Next, the source and drain contact holes were opened. The photoresist layer was not removed but was used as an isolation layer. Aluminum was evaporated and patterned for the source and drain electrode and interconnection lines. Thicknesses of the resistive layer, gate insulator, and active a-Si:H layer were 0.28, 0.13, and 0.2 μm, respectively. The resistivity of the n^+ a-Si:H resistive layer was 200 Ω cm. The drift mobility of electrons in the active region evaluated by field-effect transistors was 0.11 $cm^2 V^{-1} sec^{-1}$. The length and width of transfer electrodes were 5 μm and 1 mm, respectively. The gap between the transfer electrodes was 5 μm or 35 μm. The device had 24 transfer electrodes and was driven by a three-phase clock pulse. Constant high voltages were applied to the drain and output gate electrodes. The source electrode was connected to the ground. When the input gate electrode becomes high, electrons are injected from the source to the first transfer electrode. Then they are transferred toward the right synchronized with the clock pulse. As soon as they arrive at the last transfer electrode, they cross the output gate electrode to the drain. The output current was detected by an emitter follower.

Figure 7 is a photograph of the new device. Narrow vertical lines are the 5-μm-wide transfer electrodes. The three wide horizontal lines are clock lines. Square in the clock lines are through-holes between the clock lines and transfer electrodes.

Figure 8 shows the input and output waveforms for 100-kHz operation of a 10-μm-pitch device. Clock pulses were 20–30 V. Seven continuous pulses

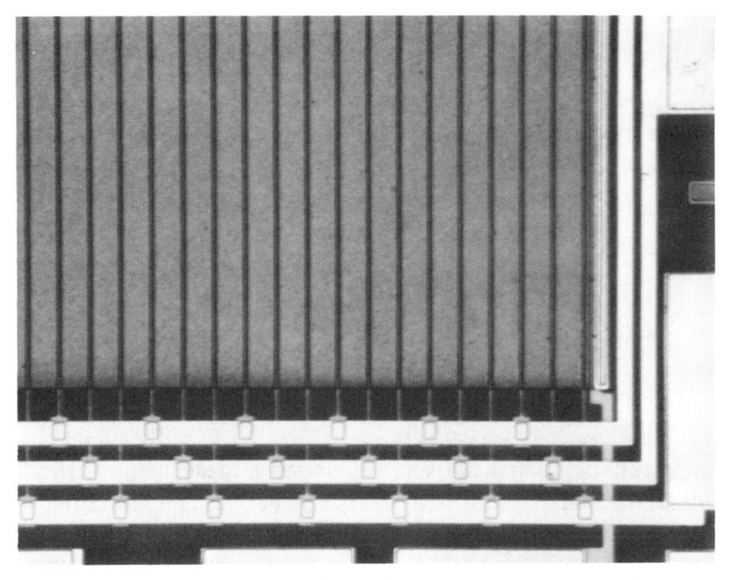

FIG. 7. Microscopic photograph of a-Si:H CCD with 24 transfer gates with 40-μm pitch. Narrow vertical lines are 5-μm Mo gates.

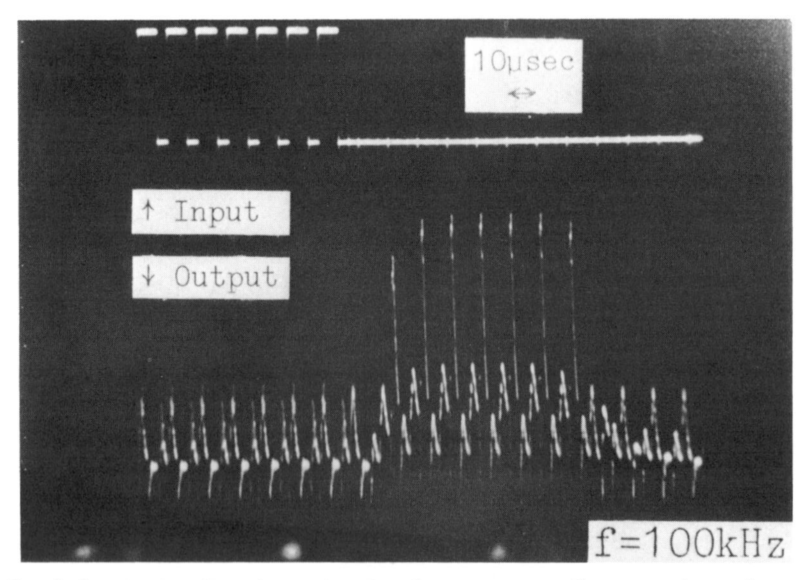

FIG. 8. Input gate voltage (upper trace) and output current (lower trace) waveforms of a-Si:H CCD with 24 transfer gates with 10-μm pitch. Clock frequency was 100 kHz; transfer inefficiency was less than 0.7%.

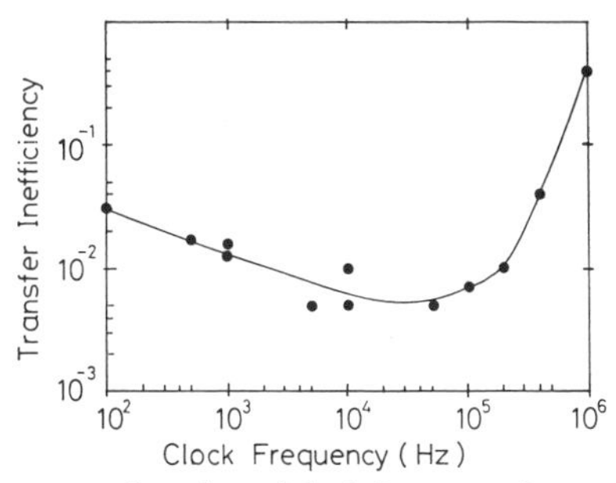

FIG. 9. Transfer inefficiency as a function of clock frequency for the improved a-Si : H CCD.

were applied to the input gate. We can see seven output pulses that are delayed behind the input pulses by just eight cycles. Thus the device operates at high efficiency. The height of the first output pulse was lower than that of the following pulses. From this difference we can evaluate the transfer inefficiency.

Figure 9 shows the transfer inefficiency per transfer as a function of clock frequencies. From 1 to 200 kHz, a transfer inefficiency of less than 1% was observed. The increase of inefficiency at higher frequencies appears to result from the fact that the period of the clock pulse approaches either the transit time of the conduction electrons or the *CR* time constant of the gate capacitance and the resistance of the resistive layer. Thus it seems possible to achieve higher-frequency operation by improving a-Si : H and a-Si : H–insulator interface properties and by crystallizing the resistive layer. The increase of inefficiency at lower frequencies will be caused by the leakage current through the a-Si : H or the gate leakage current.

A device with a 35-μm gap was also examined and a 20-kHz operation was observed.

V. Conclusion

When analyzing the operation of a-Si : H devices, the localized states must be taken into account. In this chapter the transfer inefficiency ϵ of a-Si : H CCDs is evaluated numerically and analytically on the basis of the assumption that the localized states in a-Si are distributed exponentially in energy. We have clarified that $\ln(\epsilon)$ versus $\ln(f)$ is linear and that its coefficient is determined by T_G and T. This feature agreed with our experimental results. And the rate equation and the conservation equation used in this analysis can be applied to a-Si : H FETs.

The resistively connected gate a-Si : H CCD described in this chapter operates at high clock frequencies. A 10-μm-pitch device operated at 200 kHz and a 40-μm-pitch device operated at 20 kHz. Thus the device is sufficiently fast for contact-type image sensors. The most important problem remaining is to increase the transfer efficiency.

Acknowledgment

I would like to thank Mssrs. S. Kishida and Y. Uchida for their collaboration and Profs. S. Furukawa and K. Takahashi of Tokyo Institute of Technology for their encouragement. A part of this work was supported by the Ministry of Education, Science, and Culture of Japan with a Grant-in-aid for Special Project and Scientific Research by Hoso-Bunka Foundation, and by the Research and Development Corporation of Japan.

References

Kim, C., and Snow, E. H. (1972). *Appl. Phys. Lett.* **20,** 514.

Kishida, S., Nara, Y., Kobayashi, O., and Matsumura, M. (1982). *Appl. Phys. Lett.* **41,** 1154.

Kishida, S., Naruke, Y., Uchida, Y., and Matsumura, M. (1983). *Jpn. J. Appl. Phys.* **22,** 551.

LeComber, P. G., and Spear, W. E. (1979). *In* "Amorphous Semiconductors" (M. H. Brodsky, ed.), p. 265. Springer-Verlag, Berlin and New York.

Sugiura, O., and Matsumura, M. (1982) *Trans. Inst. Electron. Commun. Eng. Jpn.* **J65-C,** 914.

Tiedje, T., Cebulka, J. M., Morel, D. L., and Abeles, B. (1981). *Phys. Rev. Lett.* **46,** 1425.

Tompsett, M. F. (1973). *IEEE Trans. Electron Devices* **ED-20,** 45.

Uchida, Y., Kobayashi, O., and Matsumura, M. (1982). *Jpn. J. Appl. Phys., Suppl.* **21-2,** 225.

SEMICONDUCTORS AND SEMIMETALS, VOL. 21, PART D

CHAPTER 10

Optical Recording

M. A. Bosch

AT&T BELL TELEPHONE LABORATORIES
HOLMDEL, NEW JERSEY

I. Summary

An optical recording medium should satisfy simultaneously the requirements of high-sensitivity and long-term stability (> 10 years). In most cases a low melting temperature material—though chemically less stable—is selected. Intricate geometrical arrangements, like the multilayer structures, or enhancements using chemical reactions, led to high-sensitivity media.

In this chapter we describe modifications of chemically stable materials to achieve sensitive archival storage media. The following sensitizing approaches will be discussed: incorporation of a gas component into the atomic network in bound form, texturing of the media surface, and electrically amplified writing.

ISBN 0-12-752150-X

II. Introduction and Scope

The optical recording of submicrometer "marks"† provides a technology to store information at ultrahigh bit densities in digital, analog (video), or any other format. The inherent advantages of a laser-driven *direct read after write* (DRAW) system are very high data rate, rapid access time, and contactless no-wear readout. An optical recording system presently can store information in excess of 10^{10} bits per disk. Erasability seems thus of secondary importance. The development of practical and archival optical recording media is most important. Major research efforts on optical recording media are concentrated on ablative, magnetooptical, and phase change methods. Optical data storage became possible with advances in optical media and laser and microcomputer technologies.

Optical recording (OR) is a universal technology for mass information storage and can be realized in various ways and formats. Major research and development efforts started around 1968. It was recognized then that a serial real-time recording technique (as compared to a "parallel" holographic one) would span a huge array of applications. Additionally, the optical readout technique is inherently robust and reliable. This eventually materialized with the introduction of the read-only systems: the video laser disk (VLD) in the late 1970s and the digital audio disk (or compact disk) in 1982. The potential benefits of the VLD proved most beneficial in education and training and eventually in interactive intelligent systems. Consumer acceptance of VLD has been slow despite the fact that it is the highest-quality playback medium (cf. videotape and other disk formats). Large research activities in media came to fruition with the current introduction of optical disk recorders. We anticipate that OR could drive a revolution in information storage.

Tellurium-based recording media are widely used in commercially available products because of the unique optical and thermal properties of tellurium (Bulthuis *et al.*, 1979). Silicon media have promising features but lack high sensitivity; thus their modification through material engineering is desirable. One approach is gas sensitizing that can be achieved either intrinsically or extrinsically. Intrinsic gas incorporation leads to archival storage properties if (i) the gas component is chemically bound within the host network, (ii) its release has a threshold energy, and (iii) the resulting compound maintains or improves the initial corrosion resistance of the recording medium. Amorphous hydrogenated silicon is well suited for demonstration of retention, release, and effusion of hydrogen gas. Optical writing can be demonstrated in several ways resulting from gas evolution

† Mark (or imprint) is used in a generic sense to describe a small area where the reflectivity changes locally due to previous laser interaction (writing).

within the network. Sensitization by extrinsic gas generation from an adjacent layer is briefly outlined.

Electrically amplified optical recording is realized by using heat generation from a photocurrent in a suitable sandwich structure. We distinguish between two variants. In the first case the photoconductor is concurrently a triggering and writing layer. In a more advanced variant, the optical recording layer is divided into two active parts: a photoconduction layer for carrier and heat generation and a writing layer utilizing the locally transferred heat to induce an optical change.

III. Optical Recording in a-Si

Early investigations of optical recording included a-Si as a favorite archival medium. Although it was known that the writing energy requirement was high, the mechanical and chemical stability of silicon was viewed as an asset. The amorphous-to-crystalline phase transition was utilized to record optical information by Brodsky (1973). Simultaneous investigations of reversible phase changes in chalcogenide glasses by Ovshinsky and Fritzsche (1973) revealed erasable optical recording in chalcogenide alloys, dependent on composition and sufficient cross-linking within the network. The laser-induced crystalline-to-amorphous phase transition in silicon films was not accomplished because of the very high quench rate required for silicon. It is now known that c-Si can be amorphized only with very short light pulses, most favorably with photon energies close to or above the direct band gap (Liu *et al.*, 1979). Hence a-Si is not a likely candidate for erasable optical recording.

Renewed interest in the optical properties of silicon films prepared by various new methods, like chemical vapor deposition (CVD), glow-discharge deposition, and sputter deposition, also stimulated investigations of the characteristics of these films for optical recording.

1. Deposited a-Si Films

Thin films of a-Si seem to satisfy some of the requirements of high-density optical storage media: negligible degradation on repeated readout, low cost of fabrication, and long-term stability. As already mentioned, Liu *et al.* (1979) proposed the crystalline-to-amorphous phase transition of silicon for erasable optical memory systems. The reflectivity change is less than 20% of the initial value, thus limiting the available readout signal-to-noise ratio. Recently, Janai and Moser (1982) have demonstrated that high-contrast, high-resolution images can be obtained in a-Si in the transmission mode. They discuss the use of such recordings in video disk systems and the application of this process in integrated circuit microfabrication. Their studies were made in a laser annealing mode that will be compared with data from similar work by Bosch and Lemons (1982).

Janai and Moser (1982) have used chemical-vapor-deposited amorphous silicon films that were deposited at 600°C on silica (fused quartz) substrates. Information was recorded in films with thickness d between 2500 and 5000 Å by irradiation with a ruby laser pulse of 50 nsec duration and an energy density ranging from 0.4 to 1.5 J cm^{-2}. The upper energy limit is known to be above the threshold for laser melting in a-Si (Baeri *et al.*, 1980). To determine the optical transmission density difference

$$\Delta D \approx 0.43(\alpha_a - \alpha_{pc})d \tag{1}$$

between the amorphous (a) and the polycrystalline (pc) phase, Janai and Moser measured the optical absorption coefficient α of large-area-laser-annealed films as a function of wavelength. The reflectance change is neglected in Eq. (1). However, the correction for reflectance changes is similar in the amorphous and polycrystalline phases and contributes little to ΔD if $\Delta\alpha d \geq 1$ ($\Delta\alpha = \alpha_a - \alpha_{pc}$). Figure 1 shows the optical absorption coefficient versus wavelength of CVD a-Si and the pulsed laser annealed film (PLA). For wavelengths shorter than 5500 Å ($h\nu \geq 2.2$ eV) the absorption difference $\Delta\alpha$ is at least 10^{-5} cm^{-1}. In the thickness range of samples used, $d > 2000$ Å, the lateral heat diffusion in the a-Si film is comparable to the heat loss into the substrate; thus a high threshold results for optical recording. The optical density difference for the preceding conditions is $\Delta D > 1$, which gives a much larger contrast ratio than that for reflective readout.

Investigations of the amorphous-to-crystalline phase transition in electron-beam-evaporated silicon films using continuous wave (cw) laser heating have been reported by Bosch and Lemons (1982). Two facts are noteworthy: (i) The amorphous-to-crystalline phase transition in a-Si starts at 600°C as determined by heating experiments and (ii) the crystallization process passes through two distinct states, possibly involving "nano-crystallization."

Silicon films that were electron beam evaporated at a rate of 5 nm sec^{-1} on silica substrates at 440°C were subsequently irradiated with an Ar$^+$ laser. The rapidly scanned Gaussian beam formed a smooth lateral temperature gradient in the film; hence it provided a simple means to study the crystallization mechanism. The laser-heated track reveals two easily discernible areas. A 1-μm-thick film showed color changes from black to deep red at the margins of the track to light yellow in the middle of the track. Despite the smooth fall of the laser intensity, the different boundaries are abrupt. Optical absorption measurements of the respective areas are also displayed in Fig. 1. The curve E440 represents the as grown evaporated film and is in

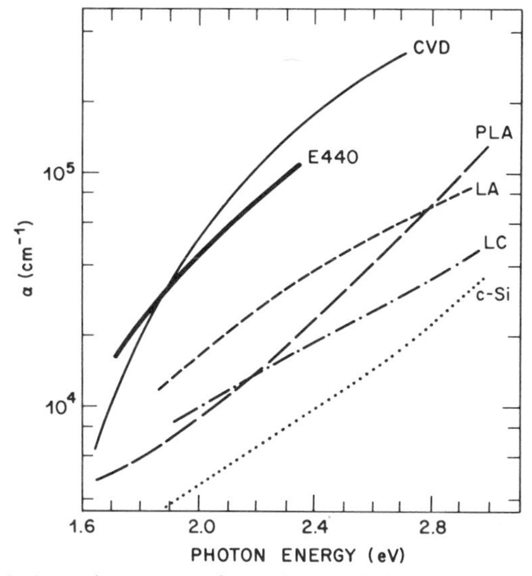

FIG. 1. Optical absorption spectra for various a-Si films and laser treatments. E440 represents the absorption coefficient for electron-beam-deposited a-Si at 440°C substrate temperature, whereas LA and LC represent the laser-annealed and laser-crystallized spectra, respectively. The absorptivity curve of the CVD film is from Janai and Moser (1982); after pulse crystallization the curve PLA was measured.

close agreement with the data from Janai and Moser (1982) of their CVD a-Si films. The curve labeled LA represents the red area, whereas curve LC is characteristic for the yellow stripe.

The microstructure of the mentioned areas has been studied in some detail. The central yellow region contains large grains up to 10 μm wide. In the middle portion of the red area, crystals with submicron dimensions have formed. No grain structure in the vicinity of the track boundaries is detectable with a scanning electron microscope (SEM) under 10-nm resolution. The first crystallization process can be understood by solid phase nano-crystallization of the amorphous material. The latent enthalpy of the a-Si is released at a temperature of 600°C, forcing the temperature to increase locally, but this increase is sufficient only for a solid phase transformation leading to a fine crystalline–amorphous mixture. Transient heating effects are not considered yet but should influence the optical marking mechanism. Explosive crystallization was found by scanned electron beam irradiation of a-Si films (Lemons and Bosch, 1980). Figure 2 reveals periodic explosive crystallization patterns in a-Si induced by the rapidly scanned electron beam. The periodic crescent-shaped topography can be explained by the release of latent heat stored in the random network. This energy is liberated if the electron beam heats the material above the critical temperature

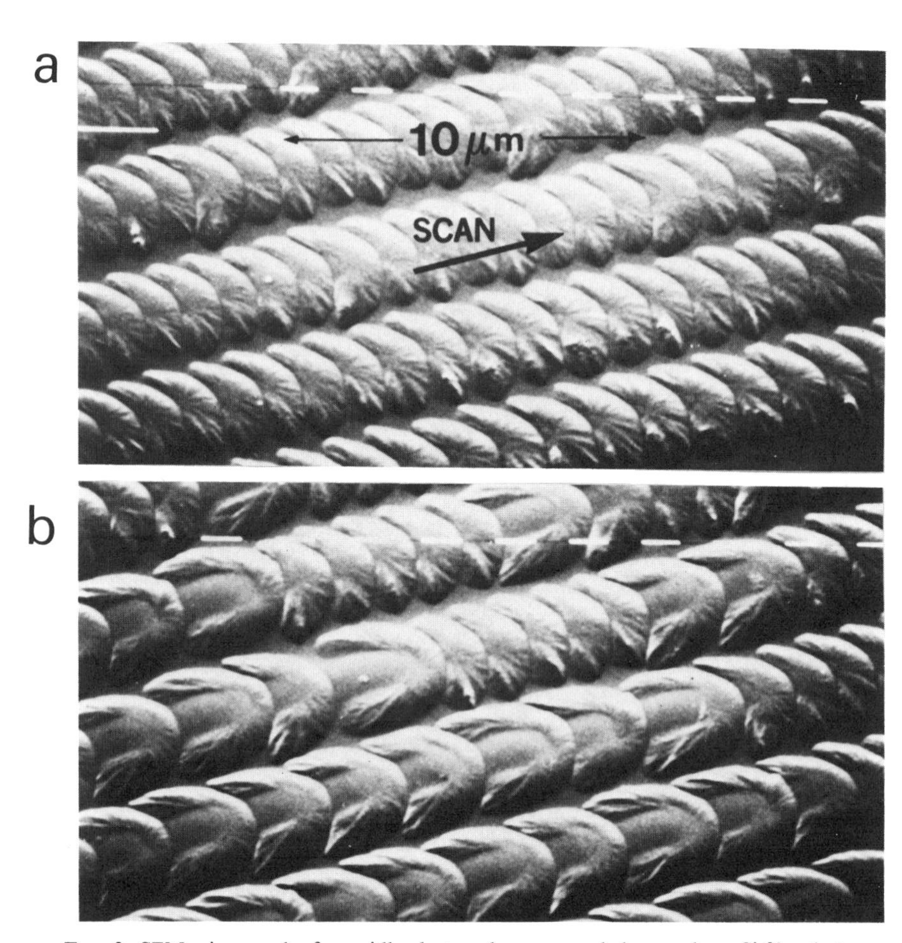

FIG. 2. SEM micrograph of a rapidly electron-beam-annealed amorphous Si film that was thermally evaporated on a silica substrate. The 20-keV, 1-μA electron beam was focused to 2 μm and raster scanned at a rate of 2.5 cm sec^{-1}. (a) Shows a periodic crescent-shaped explosive crystallization pattern. (b) Two different types of explosive crystallization features appear at slightly higher scan speed.

($T_C \approx 600$°C), at which crystallization occurs. The released latent heat raises the local temperature further, and thermal conduction maintains self-sustaining explosive transformation until the crystallization front has propagated down the thermal gradient to a point below T_C. Pulsed electron beam heating experiments indicate that a-Si melts at 1170°K (Baeri *et al.*, 1980).

Rapid scanned laser annealing of a-Si films revealed similar explosive crystallization mechanisms (Bensahel and Auvert, 1983a,b). Experiments involving various scan speeds and ion-implanted impurities (Bensahel and Auvert, 1983a) revealed a wealth of crystallization behaviors. Figure 3 shows selected explosive crystallization patterns. At a moderate scan speed of 85 cm sec^{-1} both solid phase (SP-XCR) and liquid phase (LP-XCR) explosive crystallization is observed, as shown in Fig. 3a. Liquid phase explosive crystallization produces micrometer-sized grains oriented perpendicular to the temperature gradient. At higher scan speed and moderate laser intensity, multiple solid phase explosive crystallization is observable [see Fig. 3b]. At even higher speed, periodic growth is maintained laterally only as shown in Fig. 3c. Above 11 m sec^{-1} periodic motion of both solid phase

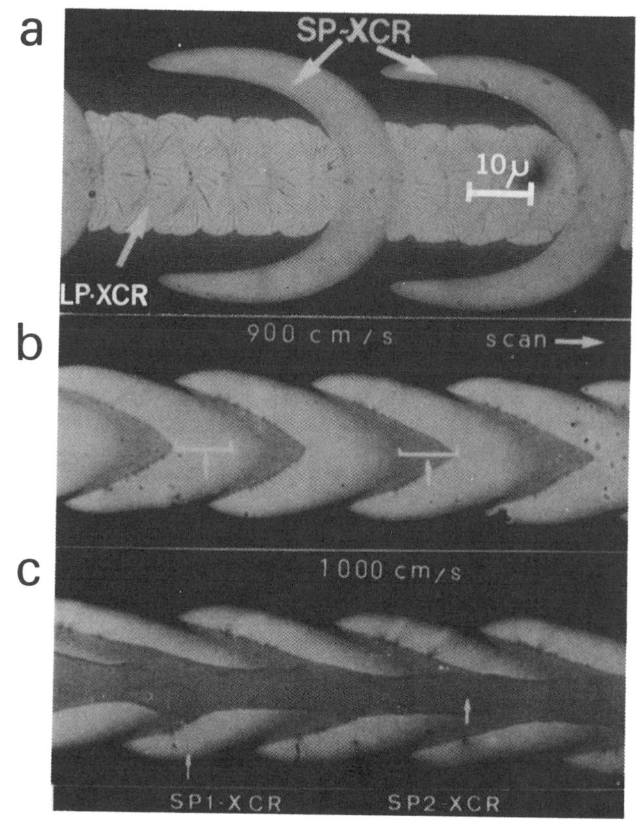

FIG. 3. Explosive crystallization in a-Si by rapid argon-laser scanning. (a) At a scan speed of 85 cm sec^{-1} both explosive solid-phase crystallization and liquid-phase crystallization are exhibited. (b) Two different types of solid phase crystallization are observed at a scan speed of 900 cm sec^{-1}. (c) The situation is the same as that in (b) but an even higher scan speed of 1000 cm sec^{-1} is used. [From Bensahel and Auvert (1983b).]

explosive crystallization mechanisms is absent. We notice that linear velocities of about 10 m sec^{-1} are practical in optical recording.

From the laser annealing experiments mentioned earlier it is not clear whether a-Si would provide sufficient resolution for optical recording purposes. Nevertheless, Janai and Moser (1982) have estimated, from their photographic printing experiments, line densities of at least 800 lines mm^{-1}.

2. Textured a-Si Films

Surface texture for optical recording purposes was proposed by Stephens (1981). Two modes of operation were described, namely, wetting and dewetting of the textured surface, depending on the average depth of the surface modulation. An average cavity spacing of $\lambda/20$ for the texture must be considered. A further development of a dewetting textured medium with ablation of the flowing layer (Te) has been outlined by Kivits *et al.* (1981b). In this approach the surface roughness of the substrate material increases the sensitivity of the high-surface-tension overlayer (e.g., Te alloy) through efficient coupling of the laser light and improved dewetting property.

Texturing of crystalline silicon by reactive sputter etching was reported by Gittelman *et al.* (1979). They textured c-Si in a chlorine plasma and proposed the use of microtexture to eliminate reflection losses of solar absorbers. If the microcolumnar structures are small compared to the wavelength of the light, the radiation encounters a transition layer in which the optical constants vary smoothly over the interface, and reflection is practically eliminated. Craighead *et al.* (1980) have textured a-Si films by reactive ion etching (RIE) with chlorine- and/or fluorine-containing plasmas. Diode RIE was used with cathode self-bias of 500 V at a power density of 0.55 W cm^{-2}. The feed gas mixture consisted of equal parts of CCl_2F_2, Ar, and O_2 at a pressure of 20 mTorr. Aluminum or iron contamination self-masking was found to generate the microtexture in the amorphous silicon films. Figure 4a shows a SEM micrograph of an edge-on view of the textured a-Si (Craighead *et al.*, 1980). The cross-sectional dimension of the columns ranges between 400 and 1000 Å.

Textured a-Si was proposed as an archival optical storage medium by Craighead and Howard (1981). Thermally isolated amorphous silicon films are textured with the random seeding and masking RIE technique described earlier. The textured a-Si surface is practically black (Gittelman *et al.*, 1979; Craighead *et al.*, 1980), the reflectance R below a wavelength of about 1 μm is less than 1% for a texture depth of about a wavelength. Optical writing employs the local meltdown of the silicon columns as a consequence of the laser heating [see Fig. 4b], and the recorded data may be retrieved by detecting the increased reflectance of the pit. The polycrystalline writing mark could approach a reflectance of about 40%. An almost ideal contrast

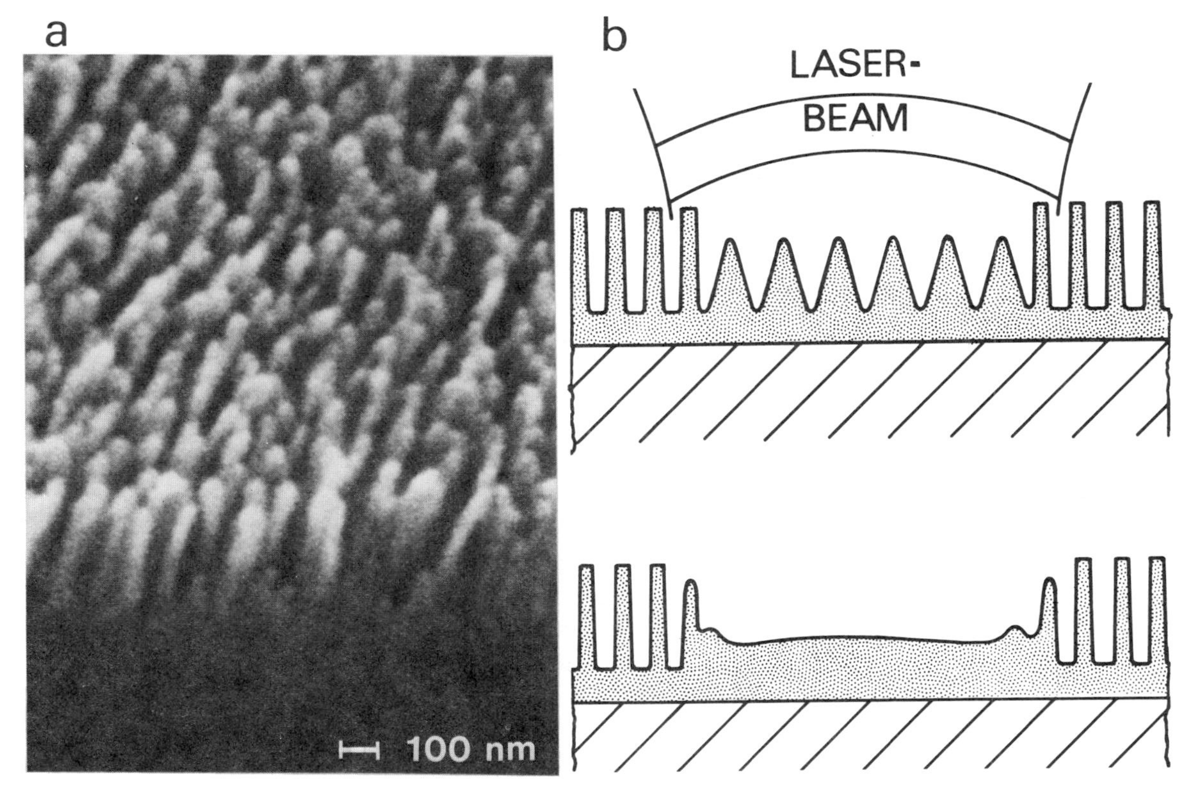

FIG. 4. (a) SEM micrograph of a cleaved edge of a textured a-Si film, with a viewing angle of 45°. [From Craighead, Howard, and Tennart (1980).] (b) Meltdown process of a textured surface is schematically indicated. The formed pit provides higher reflectivity for readout.

ratio $C = (R_c - R_{tex})/(R_c + R_{tex}) \approx 1$ thus should result. Craighead *et al.* (1982) have shown writing characteristics of a textured Si surface; the reflection ratio R_c/R_{tex} as a function of pulse energy is shown in Fig. 5. The written spot should have a high reflection ratio because the Gaussian tail of the readout light beam is fully absorbed by the textured surrounding. Scattering from the melted area as a consequence of the "debris" remaining from the SiO_2 cap that was formed as a self-passivation layer lowers the expected spot reflectivity. Furthermore, the SiO_2 capping layer could also prevent complete meltdown of the columns at the edge of the melt; some evidence supporting this conclusion is given in the micrographs of pits reported by Craighead and Howard (1981).

A theoretical analysis of the optical writing process in textured media (Suh, 1983) predicts the absence of an energy barrier for pit formation. The writing threshold is given by the energy to melt the columns. Since the columns are small and thermally decoupled, the free energy of the surface is the driving force for mass flow. Furthermore, the high aspect ratio of the columnar structure favors high surface tension; hence melting of the columns provides the initialization of the pit-forming process. Under assumption of low mass flow and small velocity gradients in the mass redistribution, the pressure ΔP_{tex} for surface initialization is (Suh, 1983)

$$\Delta P_{tex} > (2\eta l^2/\tau h^2) + (\rho l^2/\tau^2), \tag{2}$$

where η is the viscosity, ρ the density, and l the average distance for the mass redistribution. The average columnar height is h and τ the pit-forming time. We compare this pressure with the surface pressure required to form a hole in a continuous film. This surface pressure is calculated from the Laplace–

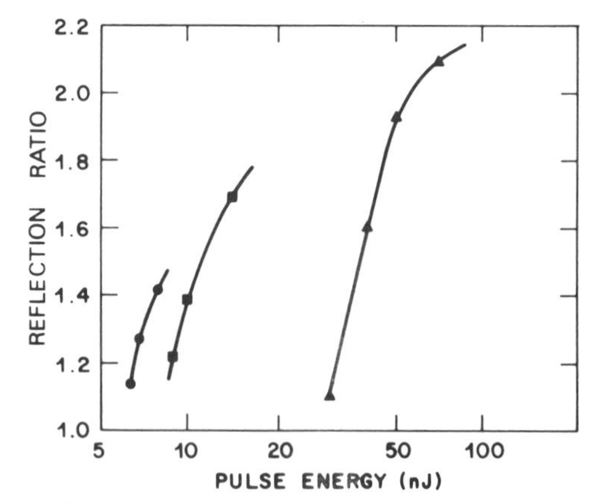

FIG. 5. Reflection ratio of textured a-Si versus laser pulse energy. Data for three different pulse widths are given. ●, 100 nsec; ■, 200 nsec; ▲ 1 μsec. [After Craighead *et al.* (1982).]

Young equation (Isenberg, 1978):

$$\Delta P_{\mathrm{h}} = \nu(r_1^{-1} + r_2^{-1}). \tag{3}$$

Here, ν represents the surface tension of a smooth film, whereas r_1 and r_2 are the principal radii of the formed hole. Suh (1983) has shown that for the case of a tellurium film, $\Delta P_{\mathrm{tex}} \ll \Delta P_{\mathrm{h}}$. Thus the energy barrier to restore a planar surface is small and pit formation ensues for a melted textured surface.

We have seen that texturing offers a very high conrast ratio and increased sensitivity. The disadvantages are related to noise, surface detection capability, and stability. Bell (1983) has estimated that for an adequate playback signal-to-noise ratio (SNR) the column (needle) diameter must not be more than 200 Å and it must be uniform. It is not obvious whether preformatting of a textured medium can be accomplished; furthermore, the medium must have reasonable characteristics to provide servo information. Since a textured surface is optically black, the reflected laser beam energy is very small and optical servo information is difficult to retrieve, in analogy to detecting a black hole. Therefore this medium encounters difficulties for application in an all optical recorder. The texturing process increases the surface area enormously; henceforth only very stable materials can withstand chemical attack.

IV. Optical Recording in a-Si:H

Numerous ways to hydrogenate silicon have been devised. We mention and describe here only relevant techniques and refer the reader to earlier chapters for detailed information on growth and fabrication methods of hydrogenated silicon films.

3. Microcolumnar Films

Several processes for optical recording are available on microtextured surfaces. We have discussed thus far the surface-tension-driven meltdown of a columnar a-Si film. Two other processes based on capillary forces are wetting or dewetting of the columnar structure by an overcoat layer with suitable characteristics. Silicide formation might alleviate one of the major drawbacks of plain textured films: the optical blackness. If a microcolumnar silicon surface is coated with a properly selected metal layer, the reflectivity of the medium can be tailored for satisfactory optical servo-focusing. Optical writing can be accomplished by surface tension meltdown and alloying. Naturally, the light is less efficiently coupled into the sensitive layer. Amorphous hydrogenated silicon can be grown or fabricated in microtextured forms. Three different texturing processes are described below.

a. Grown Columnar a-Si : H *Films*

Structural studies on glow-discharge-deposited a-Si : H films revealed nano-columnar growth under appropriate plasma conditions (Knights and Lujan, 1979). The relevant growth parameters to influence texture for rf-diode deposition are the silane–argon concentration, applied rf power, and substrate temperature. A transmission electron microscope (TEM) micrograph of a representative sample, grown at 25°C in a mixture of 5% silane in argon, is shown in Fig. 6a. The average lateral dimension of the columns is 100 Å (10 nm). Columnar morphology propagates through a-Si : H films up to a thickness of 10 μm. Knights and Lujan (1979) give evidence that the columnar growth habit originates from island nucleation and that the high density columns are surrounded by lower-density regions comprised of polysilane chains $(SiH_2)_n$. Nevertheless, monohydride saturation of dangling bonds remains the primary hydrogen incorporation mechanism. Disturbed rapid growth was found to propagate from microscopic

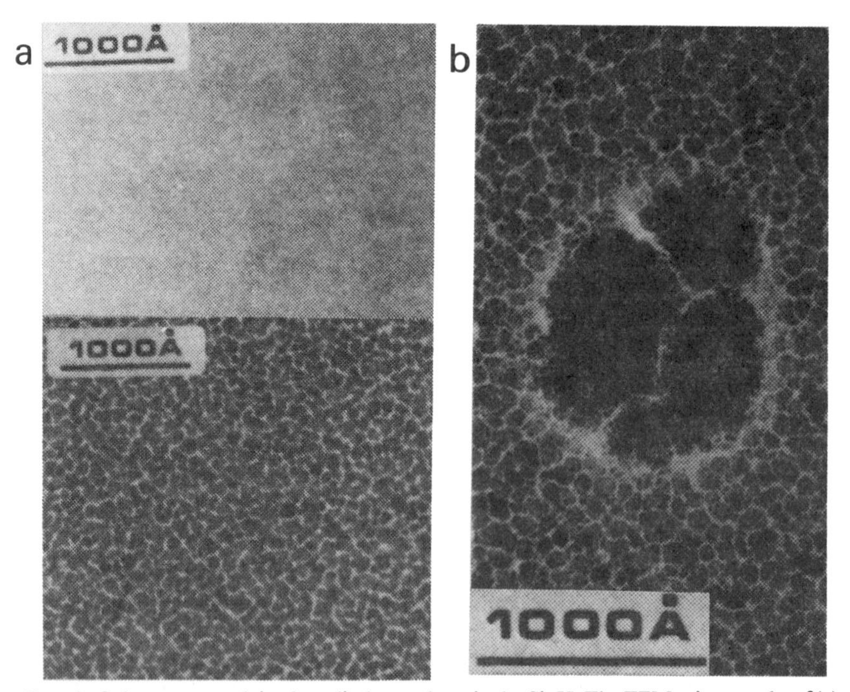

FIG. 6. Columnar growth in glow-discharge-deposited a-Si : H. The TEM micrographs of (a) reveal the film structure from a silane glow discharge on the top and from a 5% SiH_4–Ar mixture on the bottom. (b) shows an area with film growth distortion probably caused by surface defect. [From Knights and Lujan (1979).]

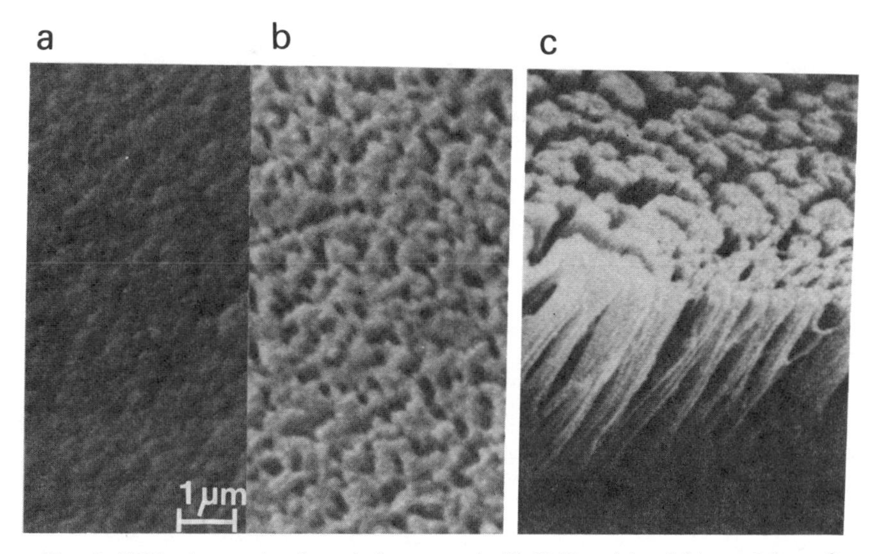

FIG. 7. SEM micrographs of reactively sputtered a-Si : H. Parts (a) and (b) reveal the surface morphology before and after chemical etch. Part (c) shows clearly the columnar structure of these films. [After Ross and Messier (1981).]

imperfections or possibly debris on the substrate. Such a growth disturbance region is depicted in Fig. 6b.

Surprisingly, microstructure has also been discovered in sputter-deposited a-Si : H films (Ross and Messier, 1981). The film preparation is briefly outlined below. Samples were grown by rf reactive sputtering of a Si target in an argon–hydrogen plasma. The morphology of the a-Si : H film is strongly dependent on the total gas pressure. Subsequent chemical etching of the layer with a $HF : HNO_3 : CH_3COOH$ (1 : 1 : 1) etchant enhances the texture, as seen in the SEM micrograph of Figs. 7a and b. Films on smooth substrates exhibit a dense texture of fine microcolumns (≤ 1000 Å). At higher sputter gas pressure coalescence of the microcolumns into clusters is observed, as shown in Fig. 7c. Rough substrate surfaces affect the morphology of the texture; this fact might be utilized to fabricate a predetermined pattern (e.g., preformatting).

b. Reactive-Ion-Etched Texture

Moustakas (1982) has recently reported texturing of sputter-deposited silicon films by reactive ion etching. Very fine texture was achieved in a hydrogen plasma. Figure 8a shows a SEM micrograph of the surface. Hydrogen reacting with the silicon may form volatile SiH_x radicals, thus providing a mechanism for selective etching. Studies of reactive ion etching

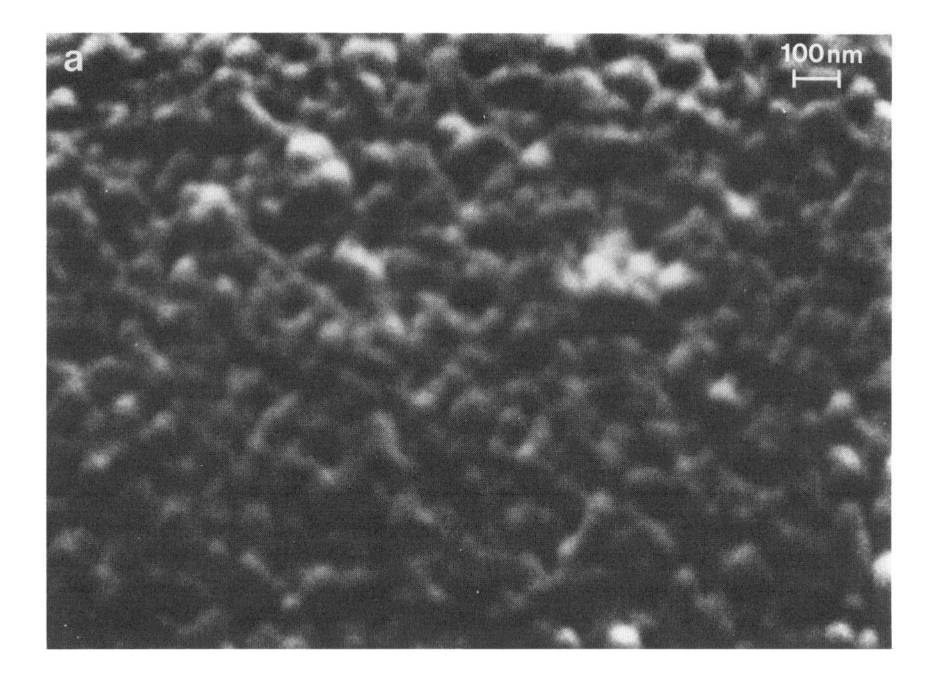

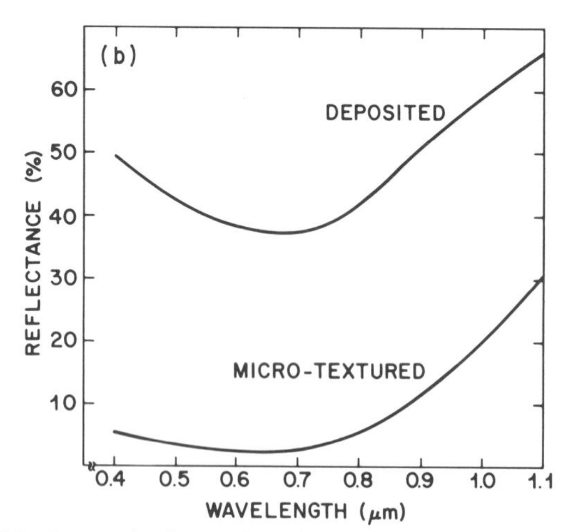

FIG. 8. (a) SEM micrograph of a reactively ion-etched a-Si surface. (b) Reflectance versus wavelength of RIE-treated film. [From Moustakas (1982). Copyright 1982, North-Holland Publ. Co.]

on films with intrinsic columnar structure, as mentioned earlier, are not available. It should be noted at this point that RIE in a hydrogen plasma is a slow process.

Reflectance measurements on films prior to and after plasma treatment are shown in Fig. 8b. These films would allow a maximum contrast ratio of 0.87 at a wavelength of 633 nm based on the data of Fig. 8b. This demonstrates again that a very high contrast ratio is an inherent property of textured media.

c. *Plasma Hydrogenated* a-InAs : H

Optical writing in a-InAs : H films has been investigated by B. Tell, M. A. Bosch, and T. Y. Cheng (unpublished, 1982). Films of InAs grown by molecular beam deposition are hydrogenated in a hydrogen discharge. Depending on the excitation method and plasma condition present, either hydrogenated plain films or hydrogenated textured films are produced. The hydrogenated films are remarkably stable against corrosion. Simultaneous hydrogenation and texturing are demonstrated by the following writing experiment. Instead of impinging the laser beam onto the textured surface, we have written the optical marks with a beam incident on the substrate film interface, as depicted in Fig. 9a. The optical writing mechanism is indicated in this figure, too. The SEM micrograph (Fig. 9b) demonstrates (i) fine texture of the film and (ii) hydrogenation of the bulk film. Writing has generated tiny bubbles. Since the film is deposited on a glass substrate, simultaneous incorporation of hydrogen from the plasma is evident from the writing mechanism. The laser energy (1 mW, 1 μsec) is absorbed in the unetched portion of the film, heating the film mostly at the glass interface, where dehydrogenation builds up and internal gas pressure eventually deforms the film into a bubble. This "porcupine"-like structure can easily be recognized in Fig. 9b.

4. Laser–Solid Interactions

Attempts to understand the hydrogen incorporation and chemical bonding of laser-annealed hydrogenated silicon were undertaken on H-ion-implanted amorphized (self-implanted) crystalline silicon (Peercy and Stein, 1978). Hydrogen concentration profiles before and after laser interactions were studied using the $^{1}H(^{19}F,\alpha\gamma)^{16}O$ resonant nuclear reaction analysis. The results demonstrate that hydrogen inhibits pulsed-laser-induced crystallization. Moderate-quality crystallization could be achieved at low concentration and also with hydrogen "drive-out."

Continuous wave laser annealing studies of rf-sputtered a-Si : H films on crystalline silicon substrates by Thomas *et al.* (1981) revealed microcrystallization of the deposited films. Dark field TEM micrographs show crystal-

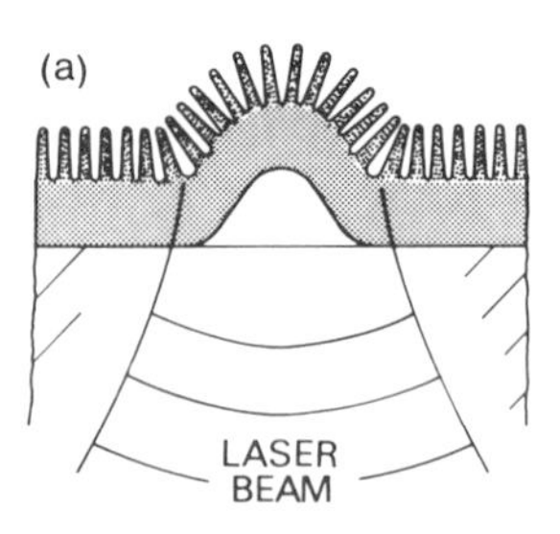

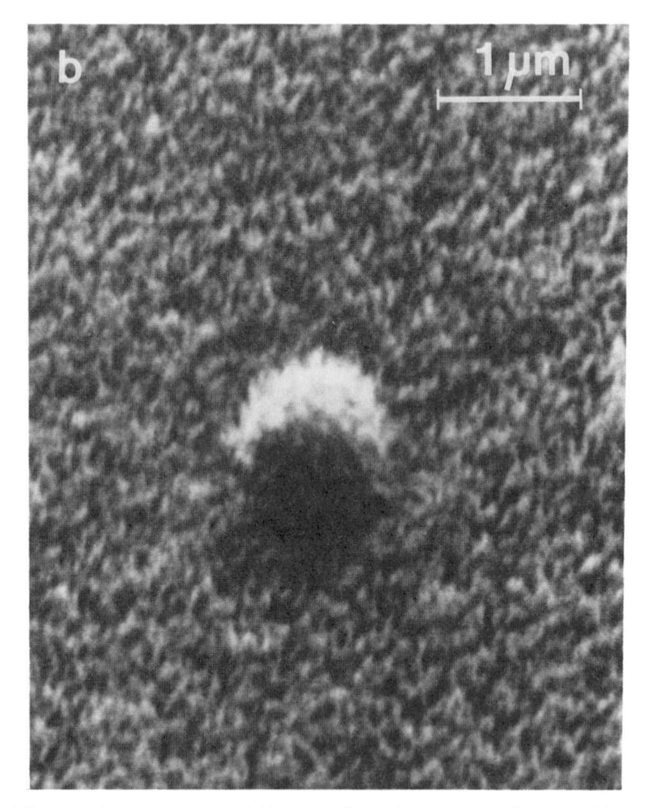

FIG. 9. (a) Schematic cross-sectional diagram of writing geometry for recording experiment in a-InAs: H. (b) The "porcupine"-like written structure is shown in a SEM micrograph.

lites of about 500-Å size produced at 1.2 kJ cm^{-2}. The grain size increases at higher irradiation energy. The hydrogen and argon depth profile measurements proved that crystallization is strongly dependent on the drive-out of either species. Figure 10 represents the hydrogen depth profile from the $^{1}H(^{15}N,\alpha\gamma)^{12}C$ resonant nuclear reaction analysis of the as-grown sample

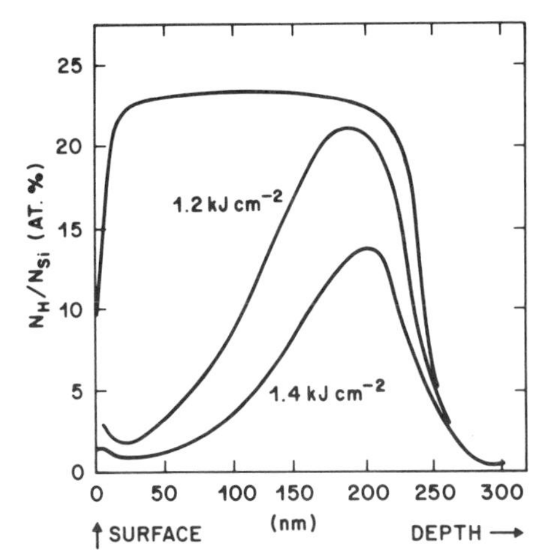

FIG. 10. Hydrogen depth profiles obtained from the resonant nuclear reaction of an as-deposited a-Si : H film and after laser annealing at different intensities. [After Thomas *et al.* (1981).]

and after laser annealing. The original hydrogen content of 22% was nearly depleted at the surface (≈ 1 at. %) because of the laser heating. In contrast to the preceding findings, Pankove *et al.* (1980) reported partial crystallization of glow-discharge-deposited a-Si : H films on single-crystal silicon without significant loss of hydrogen for pulsed laser annealing.

5. OUT-DIFFUSION OF HYDROGEN

Laser annealing experiments on glow-discharge-deposited a-Si : H films have been reported by Staebler (1979); see also Carlson (1980). His experimental findings demonstrate pronounced optical changes in undoped and doped a-Si : H films on cw laser annealing. Doping was achieved by adding B_2H_6 or PH_3 to the silane in the dc discharge deposition system, where the substrate was placed on the cathode. Hydrogen incorporation is dependent on the substrate temperature. Deposition at room temprature is known to yield amorphous polysilane films (Brodsky *et al.,* 1977) rather than a Si network in which the Si dangling bonds are saturated by hydrogen. Substrate temperatures of 200°C or more yield good-quality hydrogenated amorphous silicon films with a minimal density of gap states (Madan, *et al.,* 1976).

Optical absorption spectra of as-grown and laser-annealed areas are shown in Fig. 11 for an undoped and a B-doped a-Si : H film (Staebler, 1979). Phosphorus-doped films have an absorption characteristic straddling between the undoped and the boron-doped films but showed additional illumination effects. These curves revealed that all films darken on exposure

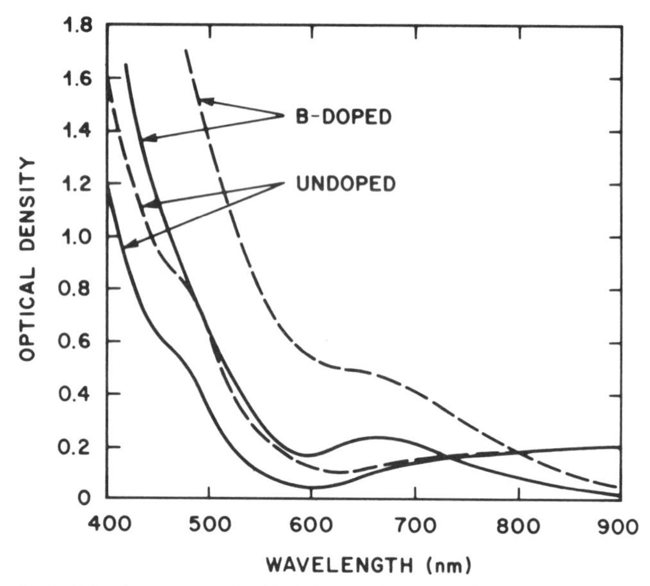

FIG. 11. Optical density spectra of a-Si : H films before and after exposure to raster scans of a 90-μm-diameter beam of 488-nm wavelength at the indicated power level. The film thicknesses are 200 nm for the boron-doped and 85 nm for the undoped samples. — —, 140 mW exposure at 90 μm; ———, unexposed. [After Staebler (1979).]

to laser irradiation. No microcrystallization was detected by electron diffraction examination; thus the films remained in the amorphous state, but the changes were irreversible. It is assumed that the dwell time of the cw laser annealing allowed some of the hydrogen to out-diffuse at relatively low temperature.

The darkening process includes several stages. Staebler and Wronski (1977) observed an interesting photoelectronic effect, which might contribute to the darkening process. Initial exposure of the films below a threshold energy results in reversible bleaching. The origin of this effect was not clearly established but was tentatively attributed to photostructural changes of gap states in a metastable fashion, which are maintained as long as no hydrogen is desorbed from the network (see Chapter 11 by Schade in Volume 21B). During the exposure but at an intensity below the irreversibility threshold, the samples darken because of the thermal narrowing of the band gap. The temperature rise on irradiation with a Gaussian beam profile with $1/e$ radius r_0 is given approximately (Weiser *et al.*, 1973) by

$$\Delta T = P/2\sqrt{\pi}\kappa r_0. \tag{4}$$

P is the absorbed power and κ represents the thermal conductivity of the substrate. If this tempeature rise is substantially above 100°C, nonreversible optical darkening occurs. Thermal biasing of the film indicates that the coloration process is of a thermal nature. Figure 12 shows the coloration time as a function of the calculated temperature using Eq. (4) and the data of Fig. 11. All coloration curves follow the functional dependence

$$t_c = t_0 \exp(E_a/kT), \tag{5}$$

where E_a is the activation energy and k the Boltzmann constant. The quantity t_0 ranges from 10^{-8} sec to $\sim 3 \times 10^{-10}$ sec. This Arrhenius behavior gives strong evidence that the darkening process involves smooth out-diffusion of hydrogen. Figure 13 shows laser recorded spots in a phosphorus-doped film. Spot pairs are recorded with increasing light intensity from an argon ion laser. Figure 13a depicts several darkened spot pairs, two of them permanently written. Compositional changes in the film are made evident by wet etching of the amorphous silicon in a basic etch solution, as shown in Fig. 13b.

Since the time required for hydrogen out-diffusion depends on the temperature rise at the laser-heated spot, estimates of temperature and time can be extrapolated. For instance, real-time video recording would require a temperature that would melt the film. It is then not surprising that preliminary tests (Staebler, 1979) showed that recording marks with rough edges resulted and thus degraded considerably the signal-to-noise ratio. Staebler (1979) proposed the use of the out-diffusion process in glow-discharge-deposited a-Si : H films for dry recording of updatable microform systems. Resolution of the process and media lifetime consideration were assumed to be sufficient.

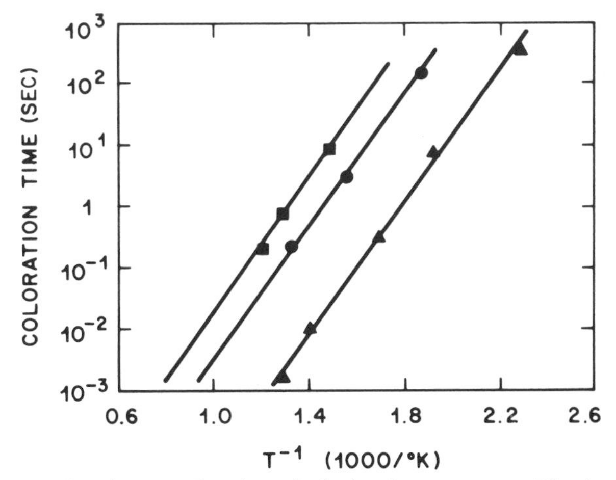

FIG. 12. Coloration time as a function of calculated temperatures. The data are taken as indicated in the text. ●, undoped; ■, P doped; ▲, B doped. [After Staebler (1979).]

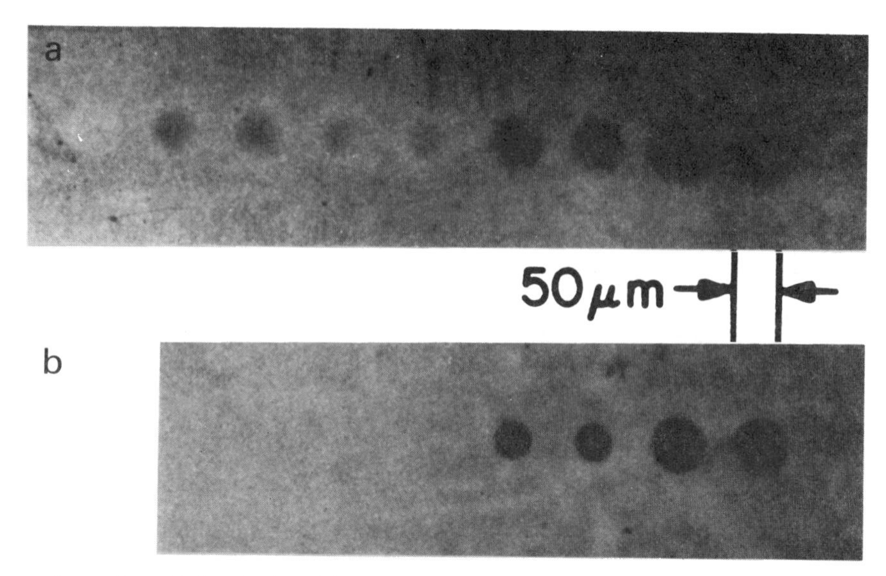

FIG. 13. A P-doped film with pairs of spots recorded at different argon laser beam intensities. The lower micrograph depicts the spots after chemical etching. [After Staebler (1979).]

6. Electron-Beam Writing and Reading

A unique application of irradiation effects in thin hydrogenated amorphous silicon films has been developed by Schade and Hockings (1983). Low-energy-electron exposure (1 keV) of glow-discharge-deposited a-Si:H films results in a resistivity increase of the irradiated volume (Schade and Pankove, 1981). In turn, these resistivity increases can be detected with Auger electron spectroscopy. The higher resistance of the electron-exposed film area leads to charging by the scanning Auger beam. The amount and sign of charging depend on the characteristics of the secondary electron emission from the a-Si:H. For sufficient charging, up or down energy shifts of the Auger electrons can be analyzed and then displayed. A scanning Auger microscope (SAM) micrograph of an electron-beam-written pattern is presented in Fig. 14. The resolution of the pattern depends mostly on the write beam, since the read beam provides the energy analysis. The electron-beam-induced resistivity changes are reversible: Patterns in a-Si:H films can be removed by thermal annealing at about 150°C.

7. Optical Recording by Rapid Hydrogen Effusion

Hydrogenated amorphous semiconductors may represent a new class of recording materials. The recording mechanism is based on rapid hydrogen evolution or effusion within the recording layer. The recording properties of

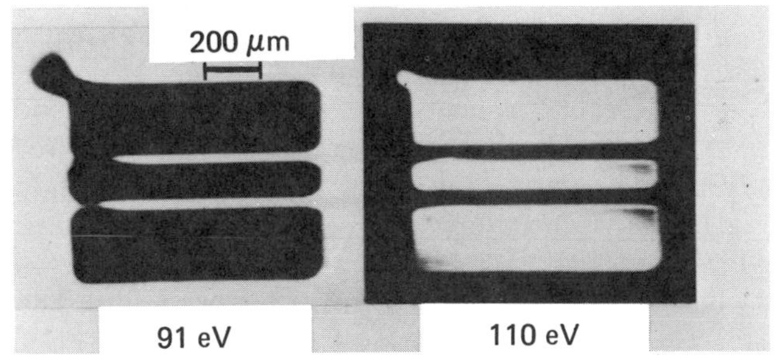

FIG. 14. Scanning Auger electron micrograph of a pattern in a-Si:H previously irradiated by an interrupted scanning electron beam. [After Schade and Hockings (1983).]

reactively sputtered silicon and germanium films reveal that the laser heating involves three different imprint processes (Bosch, 1982a,b): (i) Hydrogen evolution in an amorphous hydrogenated silicon film deforms the film into microbubbles, (ii) ablation of thick a-Si:H films shows holes with sharp edges, and (iii) the effusion of hydrogen in a-Ge:H can be described as microswelling; here the laser imprint resembles a sponge with cavities that are a few hundred angstroms in diameter. None of the observed processes involves melting of the film. The impressions of cases (i) and (iii) diffract or scatter the radiation away from the collection optics, providing changes in the intensity of the readout light.

a. Fabrication of the Medium

The hydrogenated semiconductor films have been produced by reactive sputtering (Paul *et al.*, 1976). The sputtering gas mixture consisted of 25–35% hydrogen in argon at 10 mTorr. The films were sputtered onto microscope slides held at a bias temperature of 250°C. Deposition rates of 3–5 Å sec^{-1} were achieved at 1 W cm^{-2} rf power. The recording films contained less than 10 at. % hydrogen. A higher hydrogen concentration is attainable with other preparation techniques. Homogeneous chemical vapor decomposition (Scott *et al.*, 1981) of silane (SiH_4) can be achieved at low substrate temperature; the films acquire up to 30-at. % hydrogen. Pyrolytic decomposition of higher silanes can lead to an appreciable alloying of hydrogen in the deposit (Gau *et al.*, 1981). Hydrogen concentrations up to 50 at. % in a-Si:H films have been reported from plasma decomposition of silane (Brodsky *et al.*, 1977; Pankove and Carlson, 1977).

Thin a-Si:H films are mechanically tough, scratch resistant, and chemically stable, hence this material might be a potential archival storage medium. The optical writing was performed under quasi-static conditions. The sample was scanned on a precision microscope stage moving through a mechanically chopped (3-μsec) argon ion laser beam (5145 Å). The laser light was focused by a 20× long-working distance microscope objective (beam waist ~2 μm).

The absorbed laser energy heats the recording film locally and induces the hydrogen to evolve. The hydrogen release from the network occurs at moderately low temperature and is strongly dependent on the semiconductor material (~150°C in a-GaAs:H, ~600°C in a-Si:H (Paul *et al.*, 1980)). Breaking the hydrogen bond is accompanied by a temporary softening of the network, which allows mechanical deformation of the film. Thermodynamical considerations of the metastable hydrogen binding configuration in amorphous silicon have been given by Oguz and Paesler (1980). The various hydrogen release mechanisms are discussed in detail below.

b. Recording in a-Si:H *Films*

On laser heating, hydrogen evolves within the hydrogenated amorphous silicon layer; the gas pressure deforms the film into bubbles (or bulges), as illustrated in Fig. 15a. Figure 15b shows an optical Nomarski micrograph of the microbubbles in an a-Si:H film on glass. A thin a-Si:H film supports both bubble formation and bulging at the substrate–semiconductor interface, whereas in thick films only bulges are formed. The reflectivity of the irradiated area changes from 40% for the as-deposited film to a few percentage points reflected from the bubble, which provides a good contrast ratio. The morphology of the microbulges is more apparent in the scanning electron microscope picture of Fig. 15c. The film remains continuous and no cracks or ripples are observable with a resolution of 200 Å. This recording mechanism produces geometrically smooth and symmetric bubbles of ≤1-μm diameter. The advantage of a bubble-forming medium lays in the signal-to-noise ratio, since no rims or debris are encountered. Furthermore, the writing of small bubbles is not limited by the smooth amorphous material but by the optics. Scrubbing with organic solvents and the cellophane tape test did not alter or obscure the recordings in this durable and scratch-resistant film.

Ablative hole formation without melting the a-Si:H layer occurs at higher energy density (about twice the energy threshold for bulge formation). Figure 16a illustrates the ablation of the a-Si:H film, which resembles a lift-off process. Although this process is perhaps less attractive for applications, the appearance of holes with sharp rims is remarkable. Figure 16b demonstrates that lift-off is achieved for a diameter-to-film-thickness ratio of 2:1, which is quite surprising but characteristic for hydrogen effusion. This ablative type of hole formation is probably connected to explosive release of the hydrogen. A double pulse ablation recording is shown in Fig. 16c.

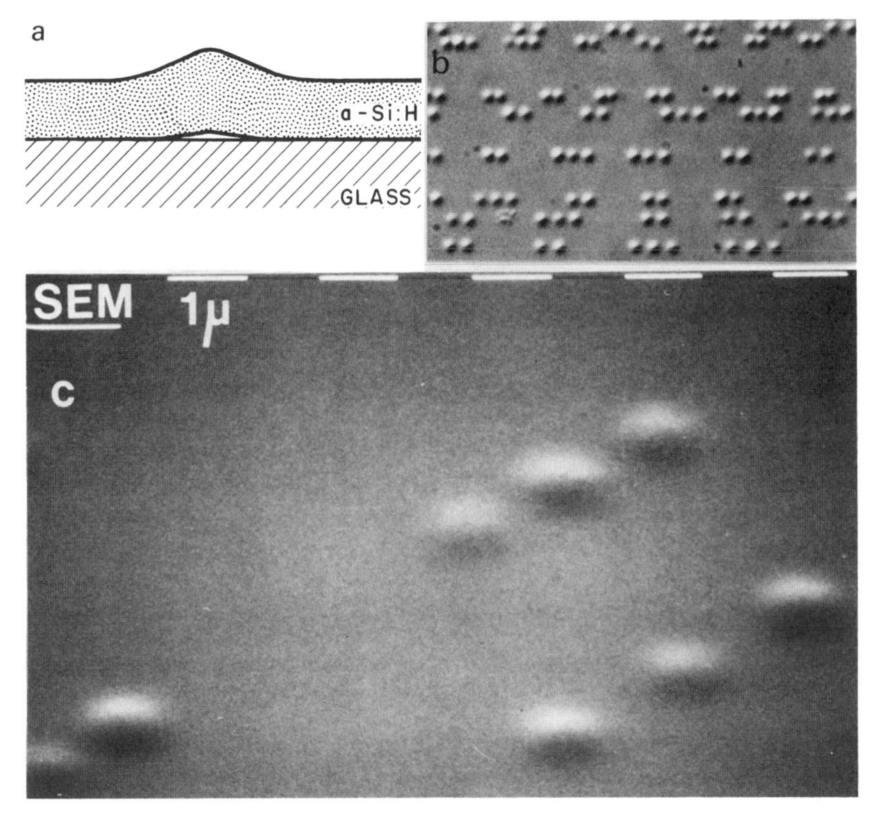

FIG. 15. (a) Schematic illustration of optical writing process based on hydrogen evolution in a-Si:H. (b) Optical differential interference contrast micrograph of bubbles in a 0.5-μm-thick film written with a laser pulse of 3 μsec and 28 mW power measured before the optics. (c) SEM micrograph of microbubbles.

c. Recording in a-Ge:H *Films*

Optical recording tests on a pure germanium film of 900 Å revealed large irregular multiple warping of the surface on laser irradiation. In contrast, writing in a-Ge:H films of various thicknesses exhibited small symmetrical recording marks at lower threshold energy. It is thus assumed that the writing mechanism in a-Ge:H is caused by the gas evolution. The recording marks appear as a microswelling of the recording layer depicted in Fig. 17a. Microscopic examination revealed a spongelike structure, which is reproduced in Fig. 17b.

Laser irradiation induces hydrogen dissociation, which results in the expansion of the network. Hydrogen is then partially released, leaving deep

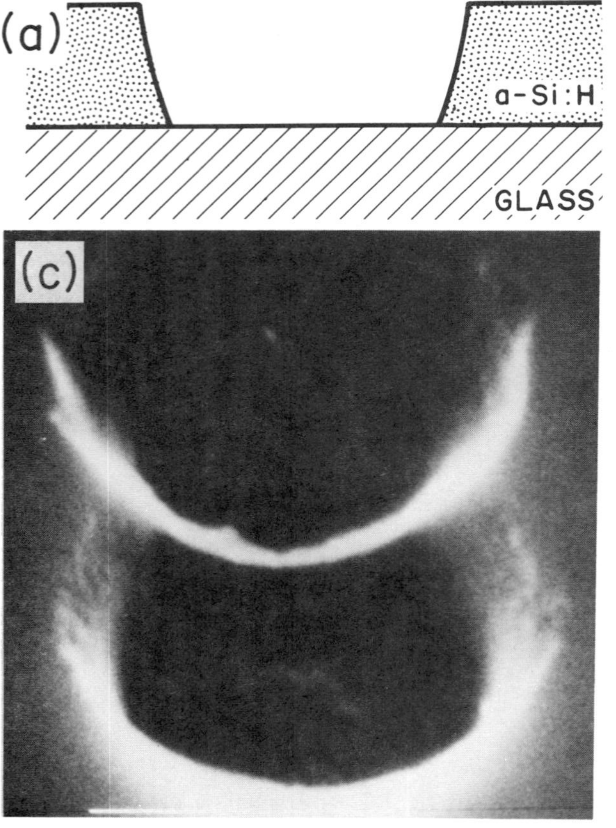

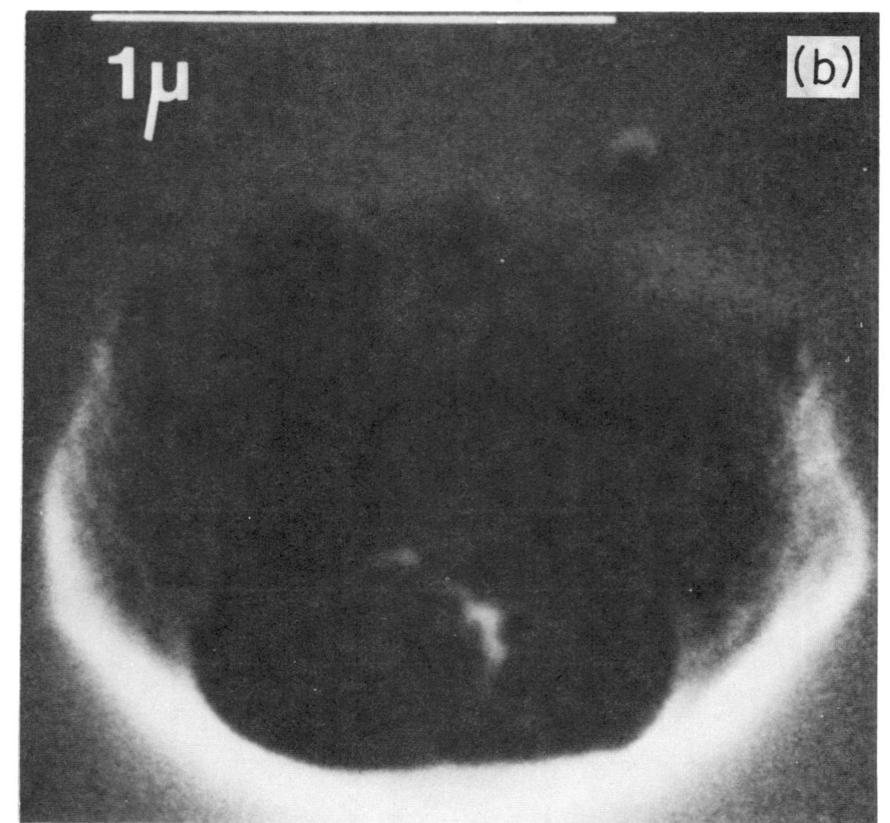

FIG. 16. (a) Illustration of the ablation process in a-Si : H by lift-off. (b) SEM micrograph of a hole recorded in a 0.5-μm layer of a-Si : H on glass. (c) SEM micrograph of closely spaced holes.

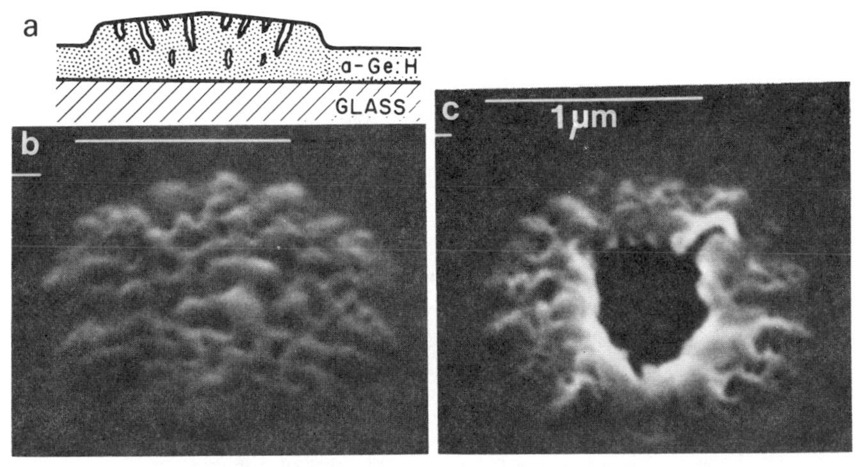

FIG. 17. (a) Microswelling mechanism occurring in a-Ge : H. (b) SEM micrograph of a mark in 0.1-μm-thick a-Ge : H film on glass written with a power of 14 mW. (c) SEM micrograph of a partially ablated mark written at 20 mW.

holes 300 Å in diameter in the reaction zone. The dehydrogenated spot, though slightly larger than in the case of a-Si : H, is still considerably smaller than the beam waist and has rather sharp edge definition. The minimum center separation for writable, successive marks was slightly larger than one diameter. Microswelling is highly reproducible; a standard deviation of 4% in the diameter of the recording marks was determined. Ablation in a-Ge : H films shows a less pronounced threshold, as was the case for a-Si : H. Hole opening is only partial. Figure 17c shows a partially ablated mark (bullhead-shaped hole) that was preceded by microswelling and partial melting. It would be desirable to have a higher hydrogen concentration in the recording medium. Glow-discharge decomposition of germane might provide films with higher hydrogen content and probably improved writing characteristics.

d. Analysis of the Writing Mechanism

Hydrogenated semiconductors have exhibited several different optical-mark-producing mechanisms. The bubble-forming mode needs the least amount of laser energy. Data on composite layer bubble-forming media have been reported by Cornet *et al.* (1981) and Robbins *et al.* (1981). We omit discussion of these extrinsic gas generation media. The energy requirements for laser heating of thin films can be calculated for special cases (Ghez and Laff, 1975; Lemons and Bosch, 1981b). Here we estimate roughly the caloric energy required to dehydrogenate a film and compare it to ablating a Te film. With the assumption of constant specific heat c_p and no heat diffusion, the energy to melt a film is $c_p(T - 300°\text{K}) + (H_f)$. The enthalpy of melting H_f dominates the energy requirement. We neglect the fact that superheating might occur for ablative optical recording in thin films (Kivits *et al.*, 1981a). Melting of Te requires an energy of 1.2 nJ μm^{-3}, whereas five times more energy is needed to melt a-Si. Since our observations confirm that no melting takes place for dehydrogenation, the corresponding energy values for hydrogenated semiconductors can be less than 1 nJ μm^{-3}. Losses contribute up to a factor of 10 in the energy balance and hence dictate the use of films less than 1000 Å thick to achieve an optimal energy sensitivity of ≤ 1 nJ μm^{-2}.

Tellurium, which has low thermal diffusivity, can be used in very thin films (~300 Å), which accounts for its excellent writing sensitivity. Hydrogenated amorphous silicon is a wide band gap semiconductor; reducing the film thickness is impractical, since the film has a lower absorption coefficient in the red. It has been shown that ablation of a very thin Ge film in a trilayer structure requires three times more energy than Te (Bell and

FIG. 18. Optical differential interference contrast micrograph of recorded a-Te : H surface at original 2000×. Bubble forming from standard "Laser Vision" video recording, $f_m = 8.1$ MHz. A broadcast quality picture was achieved at a recording radius of 85 mm.

Bartolini, 1979). Dehydrogenation of a-Ge:H could be a competitive recording process. The writing power for a-Ge:H was measured before the optics and is therefore an upper limit. A realistic evaluation of the writing energy sensitivity requires video rate writing, where the mechanisms are energy limited. Although the energy sensitivity of the material is not known, hydrogenated semiconductors could be used for low cost archival disks, which would also be utilized as a master for replication purpose. We have characterized three different recording processes, which are based on hydrogen evolution of hydrogenated semiconductor films.

e. Other Hydrogenated Semiconductors

Until now we encountered two bubble-forming hydrogenated semiconductors, a-Si:H, a-Ge:H and a-InAs:H. There is no reason that other materials should not be useful for this optical writing process. In fact, we have hydrogenated Te and have found that a-Te:H is a good medium for analog video recording. In the optical micrograph of Fig. 18, smooth bubble formation in the a-Te:H layer is demonstrated. The hydrogen passivates the Te, making the medium more stable than a Te–Se alloy according to rapid corrosion tests.

V. Electrically Amplified Writing

8. Ablative Optical Writing

The thrust of research on optical recording media is to achieve high sensitivity and long-term stability simultaneously. Since these are conflicting properties, a compromise between sensitivity and stability is usually made. In the future, the availability of visible semiconductor lasers with high output power will relax the sensitivity requirement. Finally, the recording media should require no further processing steps (Jamberdino, 1981).

In this section we describe a recording medium based on a thin-film geometry that shifts the burden of writing power from the laser to a voltage source. A photocurrent is triggered by the laser pulse, and the dissipated heat provides the writing energy. This geometry is shown schematically in Fig. 19a. The photoconductor, in our case hydrogenated amorphous silicon, with a low dark conductivity is deposited between conductive thin electrodes so that a voltage can be applied on the photoconductor. If one of the electrodes is semitransparent, a focused laser beam is absorbed in the photoconductor and will switch it to a high conductivity state. At the end of the laser pulse the photocurrent starts to decay. Therefore the current pulse greatly increases the electrical power dissipated in the illuminated area. This can result in the localized ablation of the entire sandwich structure, leaving a pit with high optical contrast.

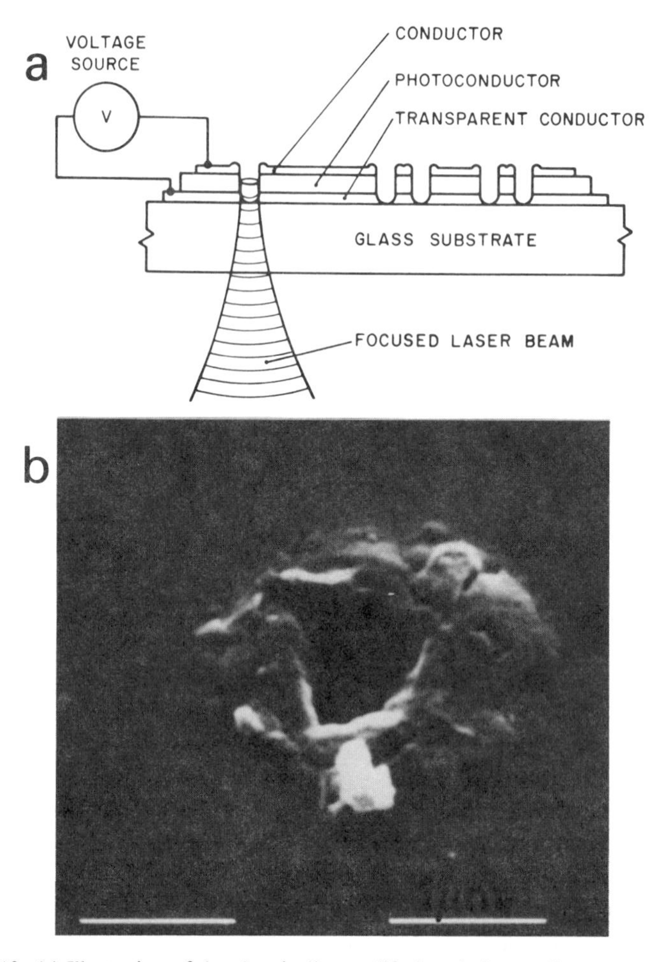

FIG. 19. (a) Illustration of the electrically amplified optical recording geometry. (b) SEM micrograph of the pits produced by electrical heating in the illuminated area.

Photoconductors have been used in other optical recording geometries. For example, in the Ruticon the photoconductor is used to switch an electric field to the elastomer recording medium (Sheridon and Berkovitz, 1976). Terao *et al.* (1973) investigated photoconductor sandwich structures, in which the optically sensitive film is part of a composite photoconductor layer. The optical writing occurred at laser-illuminated spots when voltage pulses were applied to the electrodes. In the magnetooptic photoconductor sandwich geometry, the photoconductor is used to facilitate the switching of a magnetooptic material (Krumme *et al.,* 1975, 1977). However, in the present experiment the photoconductor sandwich itself is the recording medium.

The electrical power P_{el} dissipated in the illuminated region is simply

$$P_{el} = \sigma(I)V^2A/d, \tag{6}$$

where V is the applied voltage, $\sigma(I)$ the intensity-dependent conductivity, A the illuminated area, and d the thickness of the photoconductive layer.

A reactively sputtered film of a-Si:H with a thickness of $d \approx 2\ \mu$m (Lemons and Bosch, 1982) was found to provide the properties suitable for this experiment. This film was deposited by rf sputtering a Si target with a 1:3 mixture of hydrogen in argon. The rf power density was about 2 W cm^{-2}, giving a deposition rate of $\sim$4 Å sec^{-1}. During deposition the substrate was held at 250°C. This film had a dark resistivity of $\sim 10^9\ \Omega$ cm.

Several transparent electrode materials were tried, including indium–tin oxide, or 5-nm-thick layers of Cr or Nb. Although the thin-metal films have an optical density of 0.2–0.3, they produce less debris on the surface during laser writing. The opposite electrode was typically a 20–30-nm-thick film of Cr or Nb deposited by evaporation or ion-beam deposition, respectively. For test purposes, this layer was patterned with a shadow mask in 3.5-mm-diameter circles that could be connected independently to the voltage source.

Applying 40 V across this sandwich structure was sufficient to achieve writing with 0.5 mW from a He–Ne laser (6328 Å). Because of the low dark conductivity of the a-Si:H photoconductor, the power dissipation in the dark is only about 0.01 W cm^{-2} at this voltage. The laser light was focused to a spot of about 2 μm with a 20× 0.4-NA microscope objective and was mechanically chopped to yield 100-μsec pulses. By increasing the applied voltage to $\sim$50 V, the writing threshold was reduced to $\leq$0.1 mW of laser power. However, at this higher voltage, defects in the film sometimes produced spontaneous shorting.

In the absence of an applied electrical field, the optical power threshold for writing in this hydrogenated amorphous silicon photoconductor structure was $>$20 mW. Thus if the voltage is turned off, there is no danger of accidental writing during laser beam incidence for the readout process. This external control over the writing threshold is an important advantage in this optical recording geometry. The light-induced recording marks show a high optical contrast and a relatively constant diameter. The structure of the pit can be seen in detail in the SEM micrograph of Fig. 19b. The material in the center of the illuminated spot appears to have ablated, leaving a submicrometer hole through the thickness of the sandwich. This hole is surrounded by an elevated ridge that extends to $\sim$2 μm in diameter. The irregularity of the pit is a consequence of ablating a very thick film. This writing mechanism would not provide a high signal-to-noise ratio. Thin, pinhole-free a-Si:H films are difficult to produce. In the next section we discuss a different improved electrically amplified recording medium.

9. Transferred Bubble Writing

In the preceding section we saw that amplification could make a medium more sensitive without suffering the disadvantage of reducing its lifetime. When the photoconductor sandwich itself acts as the recording layer, however, the recording marks are often irregular in shape, with debris along the rim. For high signal-to-noise optical recording it is desirable to have marks that are symmetric and very uniform.

Here we describe an improved version of the electrically amplified medium in which the optical recording occurs in a polymer layer on top of the photoconductor sandwich. The marks formed in the polymer are highly symmetric bumps that should exhibit a large signal-to-noise ratio.

This four-layer optical recording geometry is similar to the magnetooptic photoconductor sandwich (MOPS) reported by Krumme *et al.* (1975, 1977). In the MOPS geometry the photoconductor sandwich was deposited on a magnetooptic crystal film that was grown epitaxially on a garnet substrate. Heating inside the photoconductor reduces the magnetic anisotropy field and thus facilitates the reversible switching of magnetization. This change can be detected by the magnetooptic Faraday rotation. The advantage of recording in a polymer film is that the contrast and readout efficiency are large, the resolution is high, and the polymer can be deposited on arbitrarily large substrates.

The geometry we use is shown in Fig. 20a. A semitransparent conductor, a photoconductor, a second semitransparent conductor, and a polymer film are successively deposited on a glass substrate. The semitransparent conductors are about 10-nm-thick layers of Nb deposited by ion-beam sputtering. Although the films have an optical density of ~ 0.25, they are very stable and easy to contact. More transparent conductors, such as indium–tin–oxide, can be used if the subsequent processing temperature remains below 250°C.

The photoconductor is a 1.8-μm-thick film of reactively sputtered a-Si:H. This film was deposited under the same condition as described in the preceding section. The second conductor layer was deposited through a shadow mask to produce 3.5-mm-diameter circles that could be connected independently to the voltage source. Finally, a 1.5-μm-thick layer of positive photoresist (Shipley 1350J) was applied by spinning at 6000 rpm for 30 sec. This polymer film was baked for 30 min at 70°C. Small contact areas were opened by conventional exposure and development process, and the sample was again backed for 30 min at 70°C.

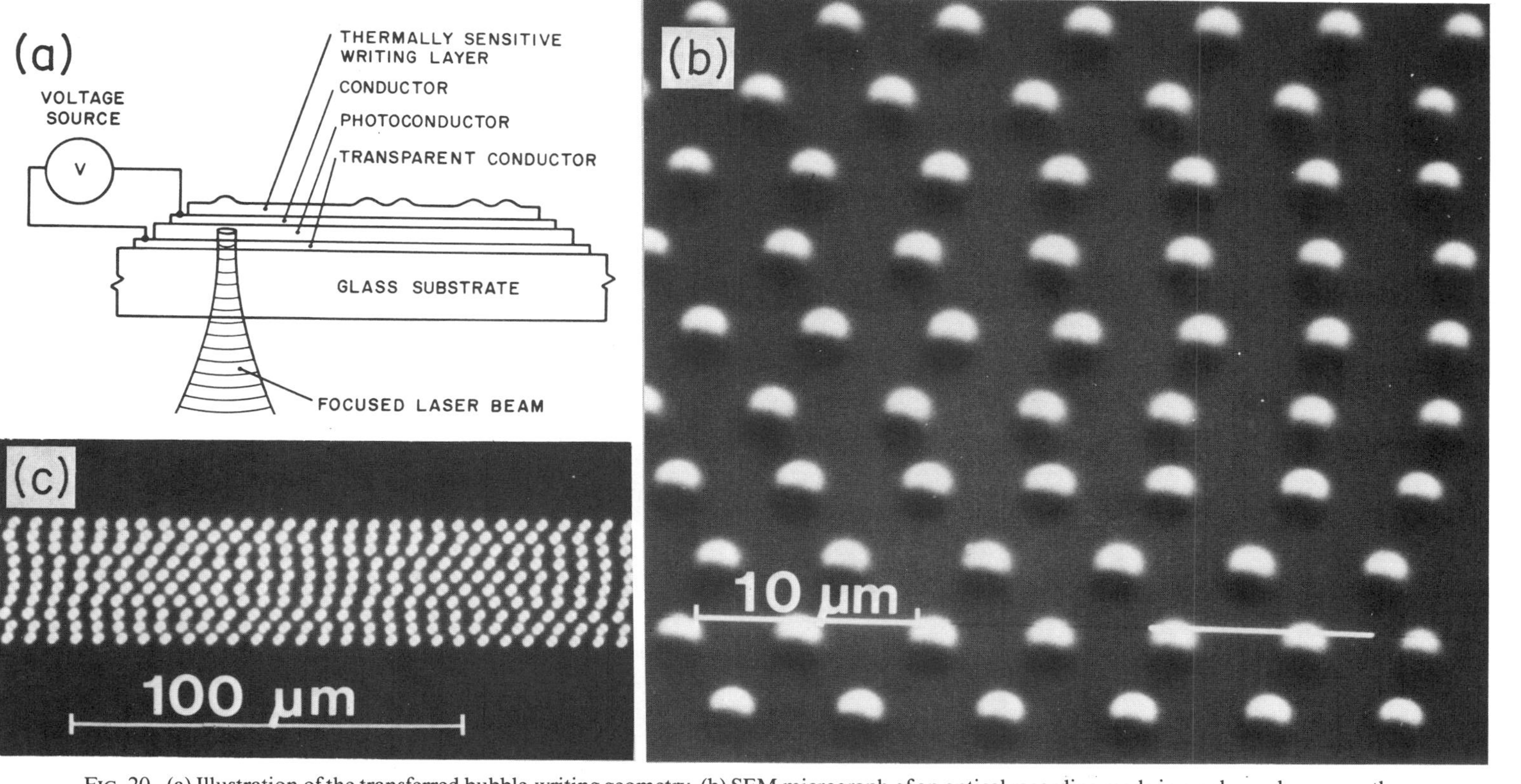

FIG. 20. (a) Illustration of the transferred bubble-writing geometry. (b) SEM micrograph of an optical recording made in a polymer layer over the photoconductive sandwich at 0.5 mW power and 25 V across the photoconductor. (c) Optical differential interference contrast micrograph shows high contrast of the bubble-forming process.

For testing convenience the device was mounted on a microscope stage in which the recording laser beam is focused from below through the glass substrate, as shown in Fig. 20a. This allows the recording surface to be directly observed from above. Since both the polymer and the top conductor are semitransparent, the performance is comparable to that obtained if the laser beam illuminates the photoconductor from the top surface.

Micrographs of the optical recording marks in the polymer film are shown in Fig. 20. The bubbles were written with 100-μsec laser pulses at a peak power of 0.5 mW and with an applied voltage of 25 V. The size and uniformity of the bubbles are clearly exhibited in the SEM micrograph of Fig. 20b. Each dot is circular in cross section with a 1.8-μm diameter. The resolution is partially limited by thermal diffusion over a distance comparable to the photoconductor thickness. Higher resolution can be achieved with a higher-numerical-aperture lens and with a thinner photoconductor layer. The optical differential-interference contrast micrograph of Fig. 20c shows the high contrast of the bubble-forming process.

Identical optical marks have been written with laser powers from 0.1 to 10 mW by proper adjustment of the voltage. The voltage therefore provides an independent control of the sensitivity of the medium. If the voltage is turned off, there is no danger of accidental writing during readout. Indeed, the same laser power used for writing can be applied for readout to improve the signal-to-noise ratio. The sensitivity can be adjusted to compensate for variation from one recording machine to another, for aging of the writing laser, or for aging of the medium itself if it should occur.

The writing threshold voltage as a function of laser power was measured by varying the attenuation of the He–Ne laser beam. This beam with $\lambda = 0.633$ μm was chopped with a rotating disk to produce 100-μsec pulses and then was focused on the photoconductor with a $20 \times 0.4 =$ NA microscope objective. The laser power (P_1) was measured immediately after the attenuator rather than at the sample plan.

At successively lower laser powers the voltage was increased until consistent optical recording was achieved. These data are plotted in Fig. 21. By applying 28 V across the photoconductor the threshold laser power can be reduced nearly two orders of magnitude to 0.2 mW. The background heating is negligible because of the high dark resistivity of the hydrogenated amorphous silicon film.

In the simplest model the threshold power for writing (P_{th}) is

$$P_{\text{th}} = \sigma(I, V)V^2A/d + P_{\text{o}} \tag{7}$$

where V is the applied voltage, $\sigma(I, V)$ the intensity and voltage-dependent conductivity, A the illuminated area, d the photoconductor thickness, and P_{o} the absorbed optical power. At voltages below threshold ($V \leq 10$ V) we

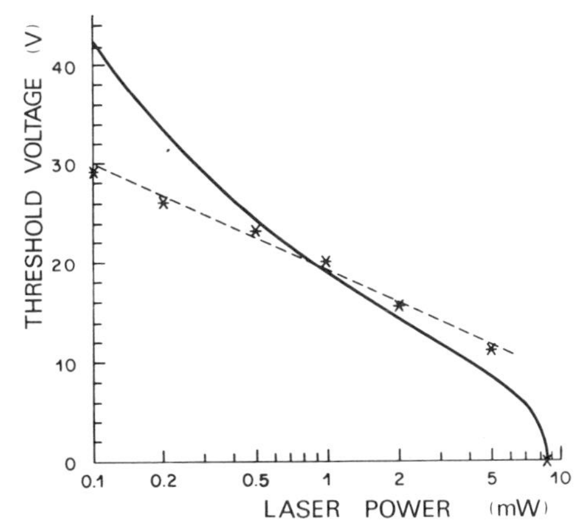

FIG. 21. Threshold voltage versus laser power for optical writing in the electrically amplified polymer medium. The solid line is the theoretical curve predicted by Eq. (9).

find empirically that

$$\sigma \approx \beta V P_o / A \tag{8}$$

is a good approximation with $\beta = 1 \times 10^{-7}$ cm V^{-3}. Because of losses in the optics and attenuation by the lower electrode, half of the laser power is absorbed in the photoconductor ($P_o = P_1/2$). The observed threshold for $V = 0$ is $P_1 = 8.5$ mW. Thus we assume $P_{th} = 4.3$ mW is a constant. Substitution of these results into Eq. (7) yields

$$V = [(P_{th} - P_o)d/\beta P_o]^{1/3}. \tag{9}$$

This expression is plotted as a solid curve in Fig. 21. At low laser power the equation predicts higher threshold voltages than observed. This might be due to the fact that the conductivity has a superlinear dependence on high voltage. In this range, the data are better approximated by a straight line on the semilogarithmic plot. The threshold power P_{th} increases with decreasing pulse width.

VI. Outlook

Modification of stable materials to improve the sensitivity for optical recording has been described. The data and models are certainly not complete and need more sophistication to characterize the media fully. Dynamical testing and accelerated life tests should be undertaken to give relevant data for recording. Optical recording is growing into a very important storage technology and is now in a vigorous developmental stage. Silicon-based media might capture a niche in this field for archival storage.

REFERENCES

Baeri, P., Foti, G., Poate, J. M., and Cullis, A. G. (1980). *Phys. Rev. Lett.* **45,** 2036.
Bartolini, R. A. (1981). *J. Vac. Sci. Technol.* **18,** 70.
Bell, A. E. (1983). *Proc. Soc. Inf. Display* **24,** 17.
Bell, A. E., and Bartolini, R. A. (1979). *Appl. Phys. Lett.* **34,** 275.
Bensahel, D., and Auvert, G. (1983a). *J. Appl. Phys.* **54,** 392.
Bensahel, D., and Auvert, G. (1983b). *In* "Laser-Solid Interactions and Transient Thermal Processing of Materials" (J. Narayan, W. L. Brown, and R. A. Lemons, eds.), p. 165. North-Holland, Publ., Amsterdam.
Bosch, M. A. (1982a). *Appl. Phys. Lett.* **40,** 8.
Bosch, M. A. (1982b). *Proc. Soc. Photo-Opt. Instrum. Eng.* **329,** 181.
Bosch, M. A., and Lemons, R. A. (1982). *Appl. Phys. Lett.* **40,** 167.
Brodsky, M. H. (1973). U.S. Patent 3,716,844.
Brodsky, M. H., Cardona, M., and Cuomo, J. J. (1977). *Phys. Rev. B***16,** 3556.
Bulthuis, K., Carasso, M. G., Heemskerk, J. P. J., Kivits, P. J., Kleuters, W. J., and Zalm, P. (1979). *IEEE Spectrum,* **16,** 26.
Carlson, D. E. (1980). *In* "Polycrystalline and Amorphous Thin Films and Devices" (L. L. Kazmerski, ed.), p. 201. Academic Press, New York.
Cornet, J., Lehureau, J. C., and LeCarvennec, F. (1981). *Tech. Dig.—Conf. Laser Electro-Opt., 1981, Washington, D.C.,* Optical Society of America.
Craighead, H. G., and Howard, R. E. (1981). *Appl. Phys. Lett.* **39,** 532.
Craighead, H. G., Howard, R. E., and Tennant, D. M. (1980). *Appl. Phys. Lett.* **37,** 653.
Craighead, H. C., Howard, R. E., Smith, R. W., and Snyder, D. A. (1982). *Proc. Soc. Photo-Opt. Instrum. Eng.* **329,** 202.
Gau, S. C., Weinberger, B. R., Akhtar, M., Kiss, Z., and MacDiarmid, A. G. (1981). *Appl. Phys. Lett.* **39,** 436.
Ghez, R. A., and Laff, R. A. (1975). *J. Appl. Phys.* **46,** 2103.
Gittlemen, J. I., Sichel, E. K., Lehmann, H. W., and Widmer, R. (1979). *Appl. Phys. Lett.* **35,** 742.
Isenberg, C. (1978). "The Science of Soap Films and Soap Bubbles," Wood Spring Press, Somerset, England.
Jamberdino, A. A. (1981). *J. Vac. Sci. Technol.* **18,** 65.
Janai, M., and Moser, F. (1982). *J. Appl. Phys.* **53,** 1385.
Kivits, P., de Bont, R., and Zalm, P. (1981a). *Appl. Phys.* **24,** 273.
Kivits, P., Jacobs, B., and Zalm, P. (1981b). *J. Vac. Sci. Technol.* **18,** 68.
Knights, J. C., and Lujan, R. A. (1979). *Appl. Phys. Lett.* **35,** 244.
Krumme, J. P., Hill, B., Kruger, J., and Witten, K. (1975). *J. Appl. Phys.* **46,** 2733.
Krumme, J. P., Heitmann, H., Mateika, D., and Witten, K. (1977). *J. Appl. Phys.* **48,** 366.
Lemons, R. A., and Bosch, M. A. (1981a). *Appl. Phys. Lett.* **39,** 344.
Lemons, R. A., and Bosch, M. A. (1981b). *Appl. Phys. Lett.* **39,** 962.
Lemons, R. A., and Bosch, M. A. (1982). *IEEE Electron Device Lett.* **EDL-3,** 254.
Liu, P. L., Yen, R., Bloembergen, N., and Hodgson, R. T. (1979). *Appl. Phys. Lett.* **34,** 864.
Madan, A., LeComber, P. G., and Spear, W. E. (1976). *J. Non-Cryst. Solids* **20,** 239.
Moustakas, T. D. (1982). *Sol. Energy Mater.* **8,** 187.
Oguz, S., and Paesler, M. A. (1980). *Phys. Rev. B* **22,** 6213.
Ovshinsky, S. R., and Fritzsche, H. (1973). *IEEE Trans. Electron Devices* **ED-20,** 91.
Pankove, J. I., and Carlson, D. E. (1977). *Appl. Phys. Lett.* **31,** 450.
Pankove, J. I., Wu, C. P., Magee, C. W., and McGinn, J. T. (1980). *J. Electron. Mater.* **9,** 905.
Paul, D. K., Blake, J., Oguz, S., and Paul, W. (1980). *J. Non-Cryst. Solids* **35/36,** 501.

Paul, W., Lewis, A. J., Connell, G. A. N., and Moustakas, T. D. (1976). *Solid State Commun.* **20,** 969.
Peercy, P. S., and Stein, H. J. (1978). *AIP Conf. Proc.* **50,** 331.
Robbins, W., Freese, R., Smith, T., and Willson, R. (1981). *Tech Dig. Conf. Laser Electro-Opt. 1981, Washington, D.C.,* Optical Society of America.
Ross, R. C., and Messier, R. (1981). *J. Appl. Phys.* **52,** 5332.
Schade, H., and Hockings, E. F. (1983). *J. Vac. Sci. Technol., A* **1,** 592.
Schade, H., and Pankove, J. I. (1981). *J. Phys. Colloq. Orsay, Fr.* **42,** C4-327.
Scott, B. A., Plecenik, R. M., and Simonyi, E. E. (1981). *Appl. Phys. Lett.* **39,** 73.
Sheridon, N. K., and Berkovitz, M. A. (1976). *Proc. Soc. Photo-Opt. Instrum. Eng.* **83,** 69.
Staebler, D. L. (1979). *J. Appl. Phys.* **50,** 3648.
Staebler, D. L., and Wronski, C. R. (1977). *Appl. Phys. Lett.* **31,** 292.
Stephens, R. B. (1981). U.S. Patent 4,245,229.
Suh, S. Y. (1983). *Proc. Soc. Photo-Opt. Instrum. Eng.* **382,** 196.
Tell, B., Bosch, M. A., and Cheng, T. Y. (1982). (unpublished).
Terao, M., Yamamoto, H., and Maruyama, E. (1973). *J. Jpn. Soc. Appl. Phys.* **42,** 233.
Thomas, J. P., Fallavier, M., Affolter, K., Lüthy, W., and Dupuy, M. (1981). *J. Appl. Phys.* **52,** 476.
Weiser, K., Gambino, R. J., and Reinhold, J. A. (1973). *Appl. Phys. Lett.* **22,** 48.

CHAPTER 11

Ambient Sensors

A. D'Amico and G. Fortunato

ISTITUTO DI ELETTRONICA DELLO STATO SOLIDO
CONSIGLIO NAZIONALE DELLE RICERCHE
ROME, ITALY

I. Introduction

In the past few years there has been a growing interest in particular solid-state devices capable of interfacing the ambient with analog and digital microelectronics. By the word *ambient* we shall intend its broadest meaning, i.e., either the surrounding atmosphere or any closed or open environmental space where organic and/or inorganic chemical species are present in either or both the gas or liquid phase under equilibrium or nonequilibrium conditions.

Examples of the ambient can be the living atmosphere, a drop of blood, a cooled cryostatic chamber in which interacting or noninteracting gases are present, etc. Thus the ambient can be characterized by physical parameters such as temperature, pressure, density, and/or chemical parameters such as concentration, chemical potential, and activity.

Measurement of the previously mentioned parameters has proved to be extremely important in the framework of the many industrial and scientific problems at the instrumentation and control levels. In the context of this work, solid-state device that can sense chemical species or physical quantities and give an electric signal as output will be called sensor.

Furthermore, since we shall be dealing mainly with chemical quantities associated with the ambient, such a device will be concisely called an ambient chemical sensor (ACS).

The stimulating and increasing activity in the development of research on

ISBN 0-12-752150-X

ACSs has its roots in the continuous, unremitting progress made in the field of microelectronic technologies. In fact, the wish to utilize as a reference point for their development the spread of solid-state electronic devices such as Schottky diodes, MIS structures, FETs, and MOSFETs has been in researchers' minds for many years.

Nevertheless, only recently have we witnessed a ferment of research activity in universities, public and private research institutes, and now, more consciously, industry, which sees in ACSs the possibility for new and promising markets. Merely to give an idea of the importance of the development of the ACSs let us list a few areas, stressing those dealing with the detection of chemical species and the measurement of their concentration.

Biomedics: (a) Cardiopulmonary bypass (N^+, Na^+, Ca^{2+}, P_H, P_{O_2}, P_{CO_2}, CN^-, $C1^-$, F), (b) blood analysis (P_{O_2}, P_{CO_2}, N^+, K^+, Ca^{2+}, $C1^-$, H^+), (c) renal dialysis and renal hepatic disfunction (uric acid, creatinine, phosphates, P_H, acetates K^+, bicarbonates, bilirubin, etc.).

Microbiology (enzymes, markers, P_H, etc.).

Neuropharmocology: (P_{O_2}, K^+, etc.).

Industrial control: (a) Combustion (CO, CO_2, SO_2, H_2, etc.), (b) central biogas sytems (CH_4, CO_2, H_2, etc.), (c) wastes (CO, CO_2, O_2, NH, etc.), (d) pollution in operating rooms (CO_2, N_2O, etc.).

Atmospheric control: (a) Primary atmospheric pollution sources (Hg, Pb, Cr, Zn, SO_2, NO, CO, etc.), (b) secondary atmospheric pollution sources (NO_2, CO_2, O_3, aldehydes, H_2SO_4, etc.).

Security systems: For large concentrations of H_2 in energy storage.

An ideal ACS should be (1) chemically selective, (2) highly sensitive, (3) reversible, (4) reliable, (5) small, (6) low in noise, (7) noncontaminating, (8) nonpoisonous, (9) insensitive to radiation, (10) insensitive to temperature variation, (11) simple to prepare, (12) rugged and solid in construction, (13) simple to calibrate, (14) compatible with local preprocessing, (15) sufficiently fast, and (16) low in cost. In practice, among the many applications the ACSs could certainly not satisfy any of the preceding constraints; in fact, at the design level, according to the particular need, a compromise between cost and performance will dictate the most convenient choice in selecting the ultimate characteristics.

Up to now, the single-crystal silicon (c-Si) structures that have been intensively employed as ACSs are the following: (1) the ion-selective field-effect transistors (FETs)(Bergueld, 1971, 1972; Matsuo and Wise, 1974; Bergueld and de Rooij, 1981; Ohta *et al.*, 1982; Akiyama *et al.*, 1982) and the ion-controlled diode (ICD) (Wen *et al.*, 1978, 1979) for measuring the H^+, CA^{2+}, Na^+, K^+ concentrations and (2) the Pd gate MOSFET (Lundström, 1981; Lundström and Soderberg, 1981), and MIS (Fonash *et al.*,

1982; Keramati and Zemel, 1982) for both hydrogen fluxes and concentration measurements. At this point it must be recalled that other semiconductors in their thin-film form (e.g., ZnO, CdS, SnO_2-doped Sb) have also been tested for sensing gases, but their selectivity has not yet proved to be adequate.

The last decade has revealed the importance of hydrogenated amorphous silicon (a-Si : H) in a wide range of low-cost thin-film applications such as solar cells, photoconductors, memory cells, FETs, etc. Recently, the a-Si : H has been examined as a new semiconductor for ACS applications by the authors of this chapter. The important role that we envision for this material in ACS devices has already been proved.

In fact, obstacles related to the time stability of this material that were encountered in early years have been almost entirely overcome, as has been widely demonstrated by the time stability of some a-Si : H-based devices such as MOSFETs (LeComber *et al.,* 1979). Furthermore, the relatively limited speed response of such MOSFETs is not a problem for the ACSs, considering the rather long time constants (seconds or minutes) needed to reach the equilibrium condition between the ACS and the liquid or gaseous ambient for chemical measurements.

This work is divided into four main sections. First, in Part II, we discuss the influence of adsorbate effects on the conductivity of a-Si : H material. In Part III the possibility of using a-Si : H for thermistor applications is considered and justified. Part IV covers in some detail the studies of a-Si : H-based MIS structures for hydrogen detection, while Part V deals with ACS devices based on a-Si : H FETs. We will consider the specific example of a Pd MOSFET H_2 sensor.

We will only consider applications involving inorganic membranes. The use of organic membranes will require further development of ion-selective compounds. The proven practicability of making MIS and MOSFET structures and integrated readout systems such as CCDs (Kishida *et al.,* 1982; Matsumura *et al.,* 1984), shows that there already is sufficient background to start combining ACSs with simple local preprocessing systems integrated in the same substrate to make "intelligent sensors."

II. Adsorbate Effects

In this section we deal briefly with adsorbate effects on the electrical conductance of a-Si : H, which would represent the easiest application of this material in ACSs.

It is well known that adsorbates can alter the surface potential of crystalline semiconductors (Brattain and Bardeen, 1953), changing the conductance of the sample in the space-charge region. In the past few years, it has been shown by work mostly developed at the University of Chicago that the electrical conductance of a-Si : H films can be strongly affected by adsorbates (Tanielian *et al.,* 1978, 1980, 1981; Fritzsche and Tsai, 1979; Tanielian 1982; Fritzsche and Tanielian, 1981; Suzuki *et al.,* 1980) and that the largest changes have been observed on thin films (0.5-um thickness) with a low density of gap states (Tanielian, 1982). Typical variations of the electrical conductance are shown in Fig. 1 for a *p*-type sample and for three different adsorbates (Tanielian *et al.,* 1978). The conductance changes are due to the creation at the surface of accumulation or depletion layers for electrons, depending on adsorbate behavior as electron donor or acceptor, respectively. The relative conductance variations in the planar configuration depend on the parallel combination of the resistance of the space-charge region with the bulk resistance. Consequently, the conductance depends on both the type of adsorbate and the Fermi-level position.

In particular, evidence of the role of adsorbed water as an electron donor has been presented in which the position of the Fermi level was varied (see Fig. 2) (Fritzsche and Tanielian, 1981). It has also been demonstrated that NH_3, CO, and $(CH_3)_2O$ (dimethyl ether) act as electron donors, while O_2 acts as an electron acceptor. It is interesting to note that in order to restore

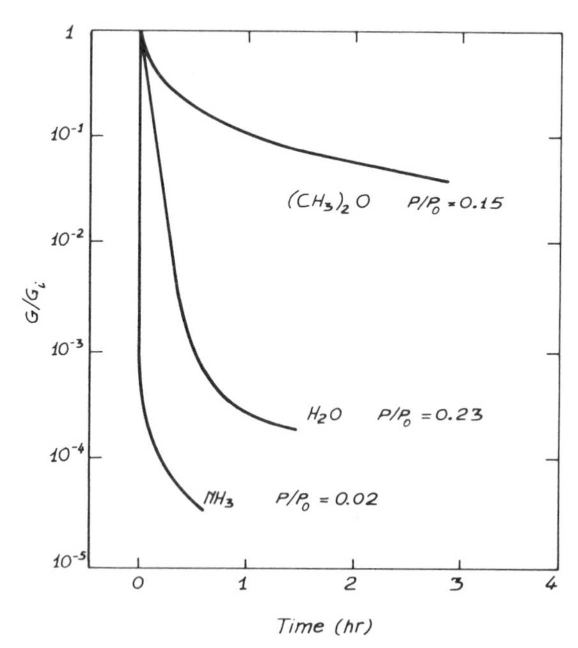

FIG. 1. Relative change of conductance G of a 0.36-μm-thick a-Si : H film due to exposure to three different adsorbates at room temperature. The p/p_0 is the ratio of the partial pressure of the gas admitted to its equilibrium vapor pressure at 250°C. [From Tanielian *et al.* (1978).]

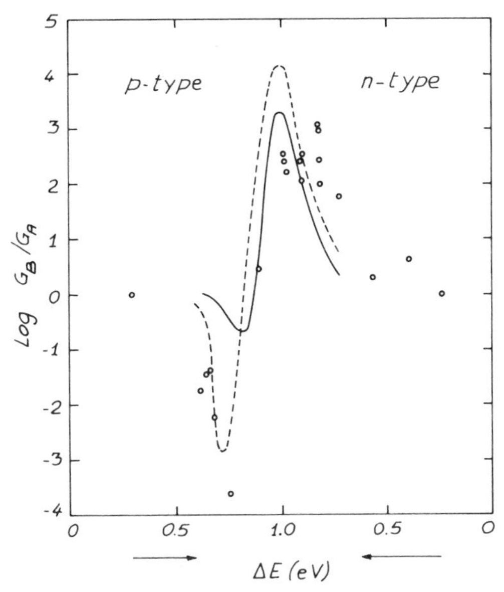

FIG. 2. Effect of adsorbed water on conductance as a function of doping. The curves marked 0.3 (dashed) and 1.0 μm (solid) are the results of model calculations for two film thicknesses in which water donates a charge Q to the bulk creating a space-charge layer. [From Fritzsche and Tanielian (1981).]

the initial surface sensitivity of *p*-type samples to adsorbates after NH_3 or H_2O exposure, extended pumping (Tanielian, 1982) or annealing in vacuum at 160°C (Tanielian *et al.*, 1978) was necessary.

Long-term changes in the electrical properties of a-Si : H have been noted. For example, variations of about 0.2 eV in the activation energy of nearly intrinsic materials over a period of several months have been reported, suggesting that the spontaneous SiO_x formation can create donorlike states in the gap (Tanielian, 1982). The difficulty of operating under atmospheric conditions, i.e., the instability of the initial condition, the irreversible component of the surface contamination, and the need for specific filtering membranes, does not suggest a practical future for a-Si : H sensor based on adsorbate-induced conductance changes. Nevertheless, it is worth pointing out that because of the very large relative conductance changes observed for several gases, the utilization of a-Si : H could remain attractive.

III. Thermistors

The importance of thermistors is universally known. They have found widespread acceptance and use not only in almost every branch of electronics and industrial control, but also in research and development in such fields as medicine, biology, and meteorology. Their success is mainly due to their simplicity and versatility, especially when used as calibrated thermometers, flowmeters, and so on.

Thermistors are thermally sensitive resistors whose resistance value varies, usually decreasing with increasing temperature. Generally speaking, all kinds of semiconductors are candidates for thermistor application, and because of the particular temperature-dependence behavior of their conductance, they are frequently called negative temperature coefficient (NTC) resistors. Because of the technological possibility of obtaining rapidly varying resistivity with temperature, a-Si:H also appears to be a suitable semiconductor material for making thermistors. In fact, for this material it is commonly verified for the temperature range (0–250)°C that the conductivity follows the form

$$\sigma = \sigma_0 \exp(-\Delta E/kT),$$

which is generally attributed to extended state transport above the mobility edge. For any thermistor made by any kind of semiconductor, the resistivity ρ_T at absolute temperature T can be expressed by the following general relationship over a wide range of temperatures:

$$\rho_T = \exp \sum_{i=1}^{n} \frac{C_i}{T^i},$$

where the C_i are constants, and the average thermistor temperature can be obtained by

$$T = \left[\sum_{i=1}^{n} a_i (\ln \rho_{\mathrm{T}})^i \right]^{-1},$$

where the a_i are constants.

The preceding two equations imply that, for high accuracy, high n values must be considered. In practice $n = 3$ gives, for the known thermistors, a good fit and an accuracy of about 10^{-3}°C. For a suitable narrow range of temperatures the following simplified equation may be conveniently used to represent the ρ_T characteristics of an NTC resistor:

$$\rho_T = \rho_{T_{\mathrm{ref}}} \exp\left[\beta\left(\frac{1}{T} - \frac{1}{T_{\mathrm{ref}}}\right)\right],$$

but the conductivity versus temperature characteristic can be considered as well:

$$\sigma_T = \sigma_{T_{\mathrm{ref}}} \exp\left[-\beta\left(\frac{1}{T} - \frac{1}{T_{\mathrm{ref}}}\right)\right],$$

where β is a constant dependent on the thermistor material and $\sigma_{T_{ref}}$ is the conductivity at reference temperature T_{ref}. In this case, if both the reference temperature (normally the room temperature) and β value are known, it is possible to determine the average temperature of the thermistor, which is given by

$$T = \left[\frac{1}{T_{ref}} + \frac{1}{\beta} \ln \frac{\rho_T}{\rho_{T_{ref}}} \right]^{-1}.$$

In dealing with thermistors the following important relationships have to be taken into consideration: the zero-power temperature coefficient of resistance α_T (as defined in thermistor terminology MIL-T-23648A)†

$$\alpha_T = (1/\rho_T)(d\rho_T/dT)$$

and the β value given by

$$\beta = \alpha_T T^2 = \Delta E/k,$$

where ΔE is the activation energy of the semiconductor and k the Boltzmann constant. It is worthwhile to point out that the α_T and β parameters are connected to the activation energy of the material.

In practice, it is possible to get doped or undoped a-Si:H thin films by several techniques: glow discharge (GD), thermal evaporation, rf and dc reactive sputtering, rf magnetron sputtering, ion-beam sputtering, and dual ion-beam sputtering. These techniques give the opportunity to make thin-film thermistors characterized by activation energy values ranging from 0.2 up to 0.9 eV, as shown, for instance, in Figs. 3 and 4 for the GD- and rf-sputtered material. This means that β values from 2320 to 10440°K or α_T values from 2.6 to 11.7% °C^{-1} can be obtained.

The ability to get such broad range of α_T values, which is not common to other semiconductors, indicates the extreme versatility of the a-Si:H material. Furthermore, according to the particular needs, and taking into account the high flexibility of the deposition techniques, thermistors made with this material can also be tailored as far as both ρ_T and α_T values are concerned.

The stability of the thermistor, that is, "its ability to retain specified characteristics after being subjected to designated environmental or electrical test condition" (MIL-T-23648A),† is the fundamental problem that requires the most attention. Although in some cases stable materials have

† Thermometrics—The Source of Thermistor Expertise, Cat. No. 181-B, U.S. Government Printing Office, Washington, D.C.

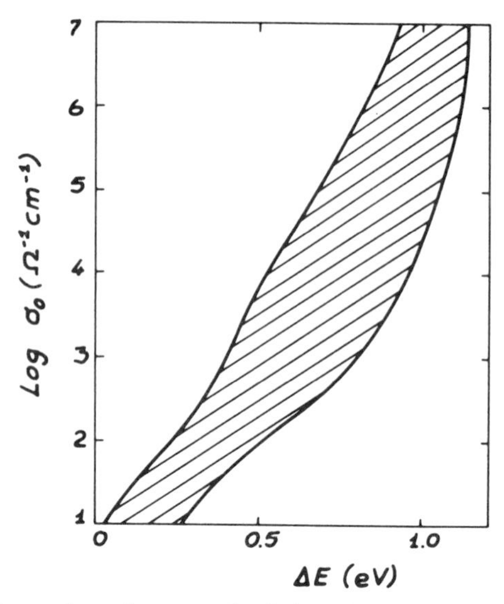

FIG. 3. The dashed area shows the range of variation of σ_0 with the activation energy ΔE for the glow-discharge material produced at the Universities of Chicago and Dundee and at RCA. [From Fritzsche and Tanielian (1981).]

been obtained, aging and light-induced instability (Staebler–Wronski effect) as well as chemisorption effects can significantly alter the performance of the a-Si:H material, especially at intermediate temperatures (100–200°C). Furthermore, applications at temperatures higher than 250°C should be avoided because of the possible thermal evolution of hydrogen out of the film. In thermistor applications, photoconductive effects should also be suppressed by operating in the dark or under a suitable protective coating.

Furthermore, for thermistor applications it is very important that noise analysis should be done, not only to have more useful information on the material itself but to quantify the certainly present $1/f$ noise.

This low-frequency noise must be considered superimposed to the thermal noise whose rms voltage is given by: $v_{rms} = (4kTR\,\Delta f)^{1/2}$. A straightforward expression for the thermal noise is the following: $v_{rms} = 4 \times 10^{-9}$ $(R \times 10^{-3})^{1/2}$ for $T = 298$°K and $\Delta f = 1$ Hz.

IV. MIS Diodes for Hydrogen Detection

In this section we will deal with the basic principles of H_2 sensors based on the combination of a catalytic metal (Pd, for instance) with an a-Si:H thin film. As for the sensitivity of these devices, we will explore the variation of

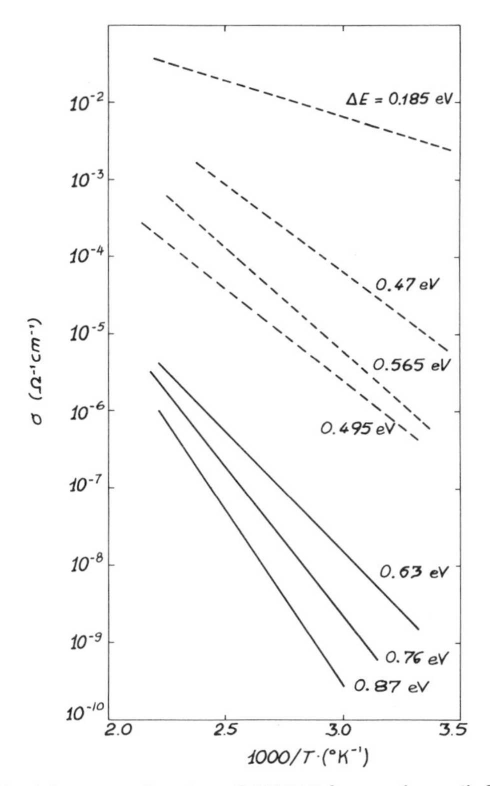

FIG. 4. The conductivity σ as a function of $1000/T$ for reactive radiofrequency sputtered undoped a-Si : H (———) and Sb-doped by cosputtering (---------) grown at different hydrogen partial pressures. [From Fortunato *et al.* (1983a).]

the transport properties, i.e., the dependence of the current–voltage characteristics at a given pressure and temperature and as a function of H_2 concentration.

The basic principle of H_2 sensitivity is connected with the dissociation of H_2 molecules on the catalytic metal surface and on the subsequent adsorption and diffusion of H atoms throughout the metal itself. This process is reversible in the presence of O_2. In fact, at the metal surface, chemical reactions between O_2 and H_2 take place with H_2O, OH, H_2O_2 formation until all the H present in the Pd film is completely removed (Lundström, 1981).

Two main hypotheses about the catalytic behavior of Pd have emerged from studies of Pd/SiO_x/Si structures. The first one is concerned with a change in the contact potential due to several possible mechanisms: (1) formation of a dipole layer by H atoms at the metal–insulator interface

(Lundström, 1981; Lundström and Soderberg, 1981; Lundström *et al.*, 1975; Shivaraman *et al.*, 1976), (2) absorbed H reaction with Pd to form a hydride with lower work function (Steele *et al.*, 1976), and (3) formation of a Pd–H system without chemical bonding of H atoms, with a lower work function (Lewis, 1967). The second one concerns the penetration of H atoms through the oxide layer and the formation at the SiO_x/Si interface of trapping states that determine tunneling of electrons between the metal and the semiconductor (Keramati and Zemel, 1982; Zemel *et al.*, 1980; Keramati, 1980).

Ultraviolet photoemission spectroscopy(UPS) measurements of a Pd/SiO_x/a-Si : H structure indicate (Fortunato *et al.*, 1984) that the mechanism responsible for the transport property variations is a change in the contact potential. In these experiments a few angstroms of a-Si have been deposited by an *in situ* evaporation onto a Pd/SiO_x substrate. Figure 5 shows the photoemission spectra obtained by synchrotron radiation at a photon energy of 30 eV and for three different conditions: (1) after the a-Si deposition, (2) after the H_2 exposure at 10^{-2} Torr for 2 min, and (3) after O_2 exposure at 5×10^{-5} Torr and 110°C for 7 min.

Exposure to H_2 gives rise to two main effects. The first is the hydrogenation of the a-Si with a relative increase of the band gap (Fortunato *et al.*, 1984), which causes a change in the valence-band discontinuity between the SiO_x and a-Si : H. The second deals with the H_2 adsorption in the thin film of Pd. The presence of this second effect can be put in evidence in an O_2 atmosphere in which the reversible desorption process can be triggered. In fact, curve c in Fig. 5 shows a rigid shift of the top of the valence band of both SiO_x and a-Si : H and also shows the recovery of the initial position of the top of the SiO_x valence band.

To discriminate between the two mechanisms related to the two starting hypothesis, it is worth noting that a rigid shift as that observed in the EDC (curves b, c) implies a change in the contact potential (Fortunato *et al.*, 1984). The hypothesis of the existence of interface states, which determine the transport properties by tunneling without an effective change in the built-in potential can be excluded, in first approximation, because such states would imply no variation of the top of the valence band of both the semiconductor and the oxide. It is therefore possible to argue that the reason for the H_2 sensitivity of the Pd/SiO_x/a-Si : H structures is a change in the work function of the metal. As a consequence any structure sensitive to the metal work function is a candidate for hydrogen detection.

Among the devices with characteristics sensitive to the metal work function, the most simple structure is the Schottky barrier. Nevertheless, the Pd/a-Si : H structure does not show any sensitivity to H_2. This behavior is analogous to that described by other authors (Fonash *et al.*, 1982; Shivaraman *et al.*, 1976; Keramati and Zemel, 1982), for Schottky barriers on

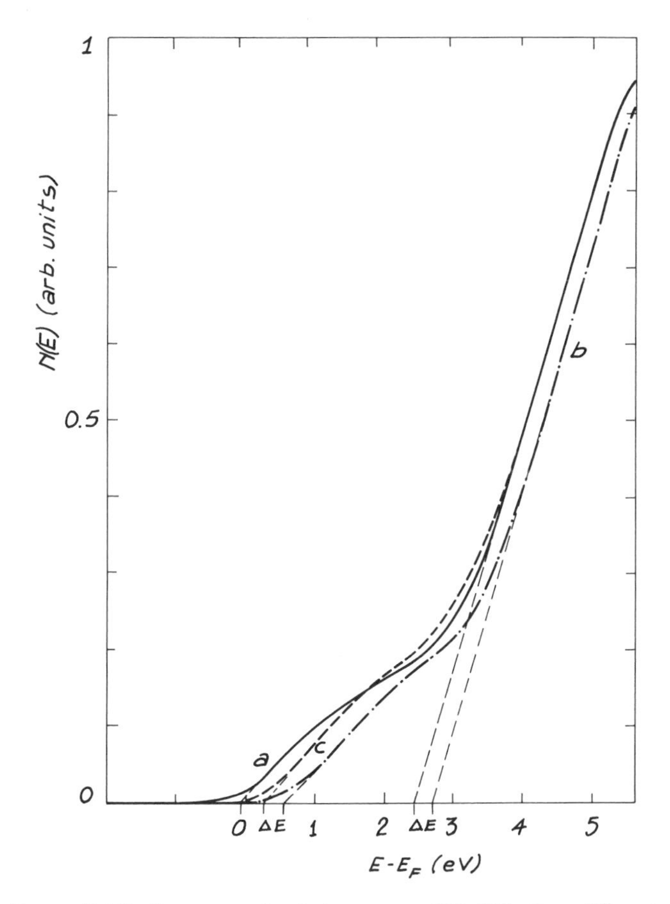

FIG. 5. Energy distribution curves at a photon energy of 30 eV for three different conditions: (a) after a-Si deposition; (b) after H_2 exposure at 10^{-2} Torr for 2 min; (c) after O_2 exposure at 5×10^{-5} Torr for 7 min at 110°C. The tops of valence band of SiO_x and a-Si:H are also extrapolated (dashed lines). The top of SiO_x VB shifts from 2.7 eV under the Fermi level to 2.45 eV ($\Delta E = 0.25$ eV), while the top of a-Si:H shifts from 0.6 to 0.35 eV, going from condition (b) to condition (c).

crystalline silicon, namely, the formation of Pd silicide at the interface. Such a compound, which can grow also at room temperature, can control the barrier height in a Pd/a-Si:H structure (Thompson *et al.*, 1981).

The formation of Pd silicides can be avoided by modifying the structure and interposing an insulating layer between the metal and the semiconductor, thus obtaining an MIS structure. The insulating layer must be thin enough (10–30 Å) to allow carriers to communicate via tunneling between the metal and the semiconductor. Hydrogen-sensitive MIS structures can be obtained (D'Amico *et al.*, 1983a) by depositing onto a Cr film 300 Å of *n*-type a-Si:H by glow discharge using a SiH_4 + 1% PH_3 mixture to provide an ohmic contact; subsequently, an undoped layer of a-Si:H about 0.5–1.5 μm thick can be deposited in pure silane.

Then the surface of the undoped layer must be oxidized. This can be done in two ways: (1) The sample can be exposed to air at room temperature for several days to permit the growth of a thin oxide layer and then the oxide thickness can be evaluated according to available data on the oxidation of a-Si:H as a function of time, already examined by Ponpon and Bourdon (1982). (2) The oxidation could be obtained by leaving the sample (Abeles *et al.*, 1981; Wronski *et al.*, 1981), at a given temperature (100–200°C) in a cleaned chamber, in the presence of pure O_2. Finally, a Pd film, whose thickness can range between 80 and 500 Å, can be thermally evaporated to complete the procedure. Photolithographic techniques and lift-off procedures can be used to define the catalytic area. Figure 6 shows a schematic of a typical MIS structure with the overall dimensions.

The $I-V$ characteristics of the MIS diode, determined by combining the diffusion (Wronski *et al.*, 1976) and tunneling through the thin oxide layer (Card and Rhoderick, 1971) can be described by the following relationship:

$$J = q\mu N_c E_s \exp(-\phi_B/kT)\tau[\exp(qV/nkT) - 1] = J_0[\exp(qV/nkT) - 1], \quad (1)$$

where μ is the electron mobility in the conduction band (CB), N_c the effective density of states in the CB, E_s the electric field at the surface

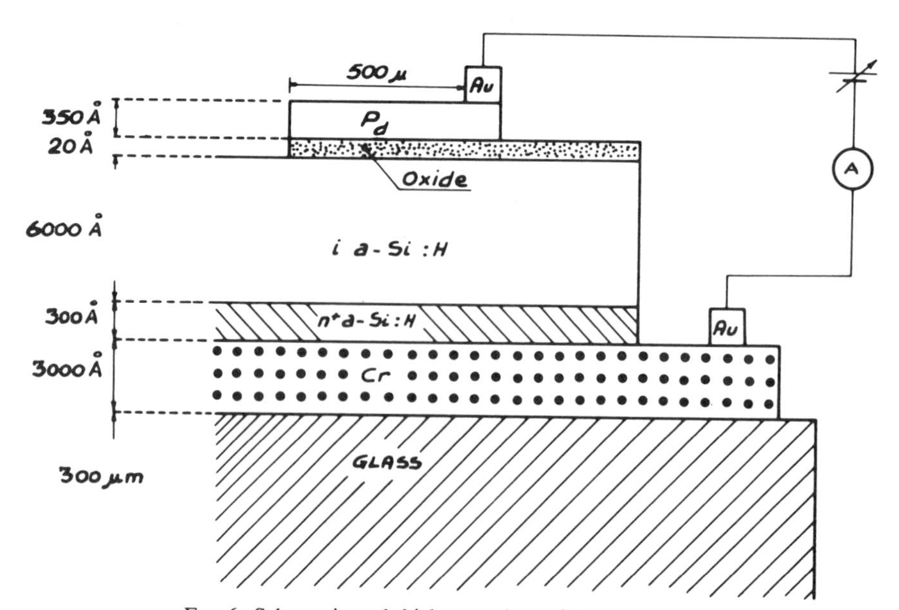

FIG. 6. Schematic and thickness values of the MIS diode.

of the semiconductor, ϕ_B the barrier height, τ the transmission coefficient for the oxide layer, and n the quality factor, which, in this structure, takes into account both the voltage drop across the insulating layer (Card and Rhoderick, 1971) and the generation–recombination processes in the depletion region.

Typical $I-V$ characteristics of an MIS diode with 20 Å of oxide as insulator are shown in Fig. 7 with reference to the device exposed to air at

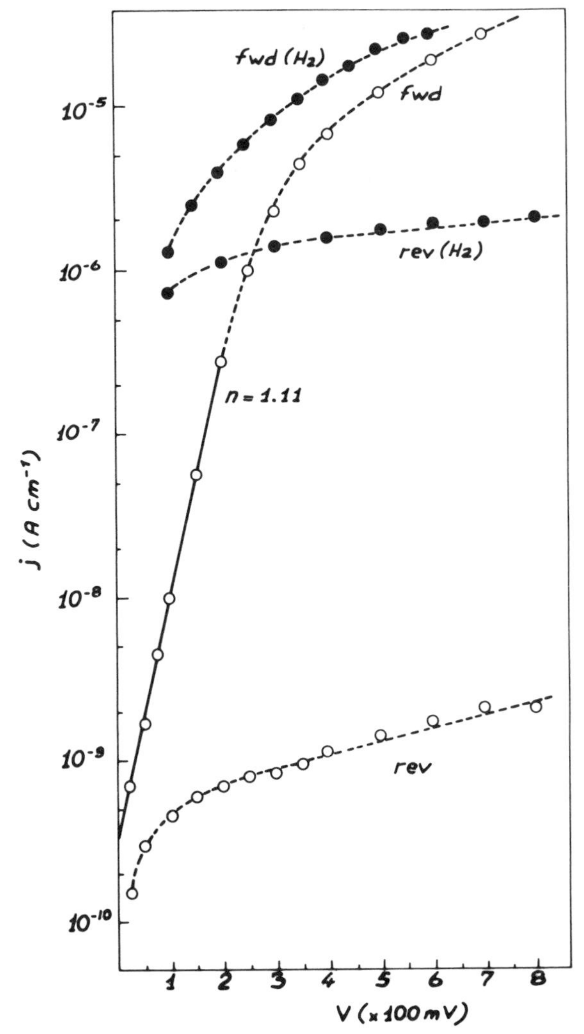

FIG. 7. $I-V$ curves of the MIS structure exposed to air (○) and to 280 ppm of H_2 in $N_2 + H_2$ mixture (●) at 295°K.

room temperature and to 280 ppm of H_2 in a $N_2 + H_2$ mixture. From the slope and the zero-bias intercept of the forward characteristic in air, it is possible to evaluate the n and J_0 values, which are equal to 1.11 and 3×10^{-10} A cm^{-2}, respectively. In the hydrogenated condition, the forward characteristic becomes limited by the series resistance and n_H, $J_{0,H}$ cannot be deduced with the aid of the preceding procedure, although a value of $J_{0,H}$ can be directly estimated from the reverse characteristics, bearing in mind that even if the reverse current does not completely saturate, the presence of a significant rectification allows a good estimate for the $J_{0,H}$ value (Goodman, 1964). The ratio $\Delta I/I$ is shown in Fig. 8 as measured at 0.4 V of reverse bias as a function of H_2 concentration in $N_2 + H_2$ fluxes for two types of devices, *A* and *B*, having oxide thickness values of 10 ad 20 Å, respectively. It is worthwhile to point out the high response of the B-type device, equivalent to the best observed in analogous single-crystal silicon-based structures (Rouths *et al.*, 1981). *A*-type devices show a relatively low sensitivity probably connected to the Pd silicide formation caused by the Pd migration through possible microchannels in ultrathin oxide regions. The sensitivity level saturates when the oxide layer approaches 20 Å, so this value should be considered a reliable reference thickness in designing MIS H_2-sensors. The changes in the contact potential, which is the main mechanism responsible

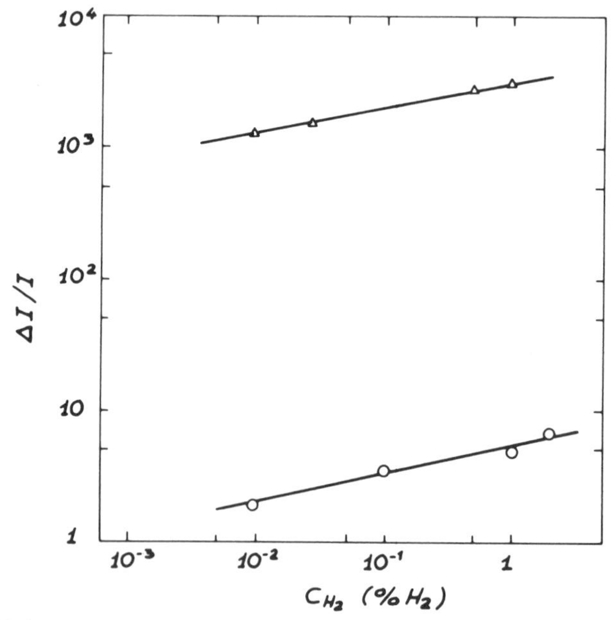

FIG. 8. Relative current responses of the type *A* (○) and type *B* (△) MIS structures versus hydrogen concentrations.

for the transport property variations (Fortunato *et al.*, 1984; D'Amico *et al.*, 1983b) in these devices can be evaluated by one of the following methods: (1) the temperature dependence of the saturation current, (2) $C-V$ measurements, (3) internal photoemission spectroscopy.

By method (1) it is possible to determine the barrier height ϕ_B as indicated in Eq. (1) and then to determine the ϕ_B change when the device is exposed to H_2.

In Fig. 9 the saturation-current values are reported as a function of $1000/T$ for the temperature range 25 – 120°C in air (J_0) and in the presence of 0.5% H_2 in a $H_2 + N_2$ mixture ($J_{0,H}$).

The slopes of the lines in Fig. 9 give ϕ_B (in air) equal to 1.04 eV and ϕ_{BH} in H_2 equal to 0.56 eV with a corresponding barrier height lowering by 0.48 eV.

By using $C-V$ measurements preformed at room temperature and at a frequency of 7 Hz, it is possible, plotting $1/C^2$ versus the applied voltage (for diodes with oxide thickness of about 20 Å and n close to unity) to estimate the built-in potential V_B (Card and Rhoderick, 1971; Wronski and Carlson, 1977), and the space-charge density (Wronski, 1977). The validity of this method is also supported by the absence of substantial disagreement between the V_B values obtained by $1/C^2$ versus V plots and those obtained from the extrapolated far-forward characteristic (Goodman, 1964). The value obtained for V_B (in air) is equal to 0.56 eV, while for 0.5% of H_2 in $H_2 + N_2$ mixture the V_{BH} value is 0.07 eV, with a lowering of 0.49 eV as determined by the data of Fig. 10.

From the slope of the lines in Fig. 10, the calculated space-charge density N is equal to 6×10^{15} cm^{-3} (in air) and seems to change only a little under H_2 exposure ($N_H = 5.3 \times 10^{15}$ cm^{-3}). These results, in conjunction with the observation that no feature appears in the $C-V$ curves when the device is exposed to H_2, confirm the change in the contact potential as the main mechanism for the H_2 sensitivity.

It is interesting to note that the values obtained for the reduction of the contact potential V_c, resulting from both the temperature dependence of the saturation current and $C-V$ measurements, are in good agreement with each other and are comparable to those obtained by other authors in single-crystal silicon MIS and MOSFET structures (Lundström, 1981). Furthermore, considering the values of ϕ_B and V_B or ϕ_{BH} and V_{BH}, it is possible to evaluate the position of the Fermi level in the a-Si:H used; it is about 0.5 eV below the conduction band, in agreement with the activation energy value $\Delta E \simeq 0.53$ eV obtained from the analysis of the series conductance of the diode. Figure 11 shows a simplified band diagram model of the MIS structure for the two conditions.

The ϕ_B variations can also be evidenced by phototransport measurements. In fact, by lowering ϕ_B the photoresponse of the MIS diode drops due

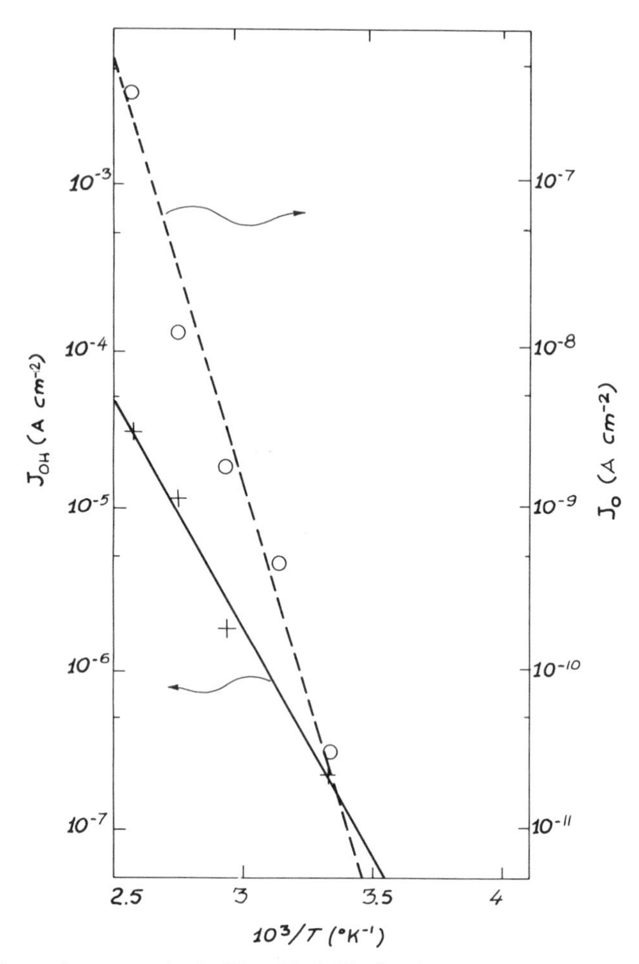

FIG. 9. Saturation current in air (O) and in 0.5% of H_2 in $N_2 + H_2$ mixture (+) as a function of $1000/T$. By the slope it is possible to determine the barrier height: 1.04 and 0.56 eV, respectively, for the two conditions.

to the reduction of the collection region. An estimate of the change in ϕ_B can be achieved by analyzing the internal photoelectric emission (IP) effect (Wronski *et al.*, 1980, 1981; Viktorovitch *et al.*, 1981). Under the hypothesis that the IP current dominates over the current due to photocarriers generated in the semiconductor, for photon energies $h\nu \leq 1.5$ eV, the photocurrent (i_{ph}) normalized to the number of incident photons (N_{ph}) onto the metal should follow the well-known relationship (Sze, 1969)

$$(i_{ph}/N_{ph})^{1/2} = A(h\nu - \phi_B),$$

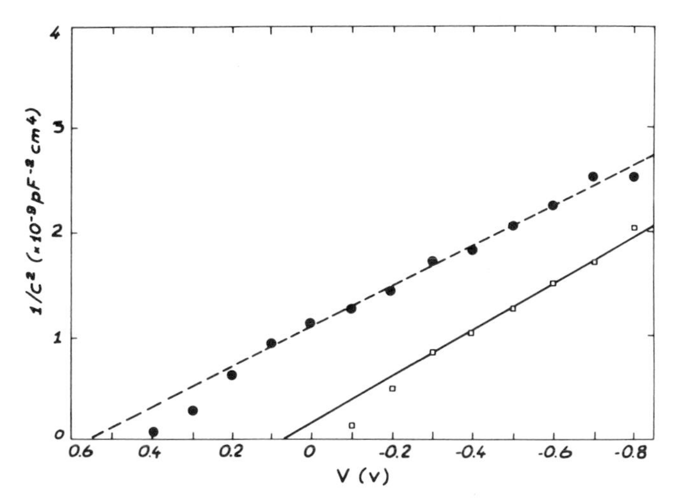

FIG. 10. $1/C^2$ versus V for the MIS diode in air (●) and in 0.5% H_2 in $N_2 + H_2$ mixture (□). The intercepts and the slopes are measures of the built-in potential and of the space-charge density, respectively. The values obtained for the two conditions are 0.56 eV for V_B and 0.07 eV for V_{BH}, while the slopes give for the space-charge density a value of 6.0×10^{15} cm^{-3} in air (N) and 5.3×10^{15} cm^{-3} in H_2 (N_H).

where A is a constant. The behavior of the MIS structure (in air) is a good fit to the above relationship, as is evident from curve (a) in Fig. 12, where the intercept gives a barrier height value of 0.98 eV.

During the admission of H_2, in order to restore the same photoelectron collection efficiency, a reverse bias (0.187 V in this case) must be applied. This is the bias needed to maintain constant the photocurrent in the region $h\nu > 1.5$ eV, where the photocurrent is determined by the photogeneration

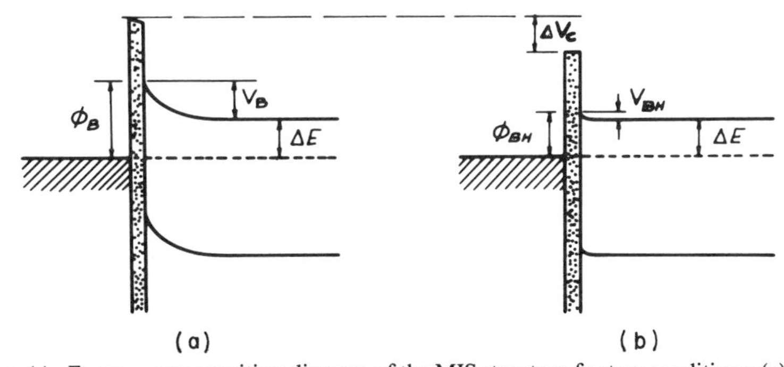

FIG. 11. Energy versus position diagram of the MIS structure for two conditions: (a) in air, (b) with absorbed hydrogen in Pd. The barrier heights (ϕ_B, ϕ_{BH}), the built-in potentials (V_B, V_{BH}), the depth of the Fermi level ΔE, and ΔV_c the change in the contact potential are indicated.

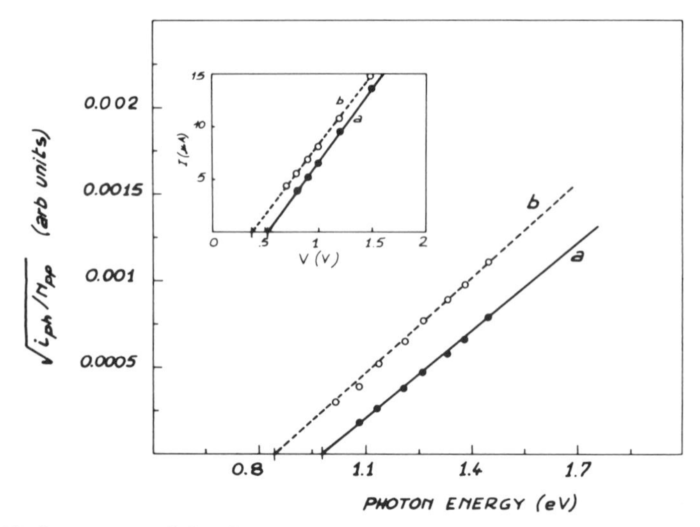

FIG. 12. Square root of the photocurrent normalized to the number of photons versus photon energy for two conditions: (a) in air, (b) in 1% H_2 in $H_2 + N_2$ mixture. The intercepts to zero photocurrent give $\phi_B = 0.98$ eV and $\phi_{BH} = 0.845$ eV. In the inset is shown the forward characteristic under the same conditions. The extrapolation gives for the built-in potential in air $V_B = 0.52$ eV and in H_2, $V_{BH} = 0.375$ eV.

inside the semiconductor (Fortunato *et al.*, 1983b). To eliminate induced frequency effects, dc measurements have been performed (Evangelisti *et al.*, 1983). To be able to measure the photocurrent down to 0.9 eV in the presence of the bias and of the H_2, a sample with an intermediate sensitivity ($\Delta I/I = 100$ at 0.5 V under 1% H_2 in $H_2 + N_2$ mixture) has been used. Curve (b) in Fig. 12 shows the internal photoemission current in the presence of 1% H_2 with 187 mV of reverse bias applied to the structure.

The ϕ_{BH} value obtained in this case is equal to 0.845 eV, resulting in a $\Delta\phi_B$ of about 135 meV. This value is in good agreement with that (145 meV) derived from the difference between the approximated value of the built-in potential obtained by the far-forward characteristic (Goodman, 1964) for the same sample under the same environmental conditions (see inset in Fig. 12). It is useful to note that determination of the change in contact potential by internal photoemission measurements is possible, bearing in mind that this process is driven by the photocurrent at least for photon energies less than 1.5 eV.

To end this section it is interesting to note that the variations observed in the saturation current, when the H_2 is admitted, are much lower than those expected following the thermionic theory with the known lowering in the barrier height. On the other hand, this fact confirms that the $I-V$ characteristic follows the diffusion theory; that is, the increase of the saturation current (resulting from the decrease in the barrier height) is damped by the decrease in the field near the interface.

As far as the response time is concerned, Fig. 13a shows typical output current response of the MIS diode, when a voltage polarization of −0.5V and an H_2 concentration of 0.5% are used as a function of time with the injection and removal of H_2, and subsequent injection of O_2 in the test chamber. The time required to reach 90% of the saturation value I_{max} is less than 1 min at 47°C. This time is normally controlled by the following parameters: hydrogen concentration, device temperature, volume of the test chamber, and gas flux.

During the desorption process in the presence of a flux of oxygen, the fall time is about a few seconds until $0.3I_{max}$; then there is an almost exponential decay, whose time constant is about 1.5 min, under the condition shown in Fig. 13a, that returns the device to the original equilibrium condition. Figure 13b shows another typical response when an H_2 concentration of 100 ppm is used. This time can be substantially lowered, for instance, by

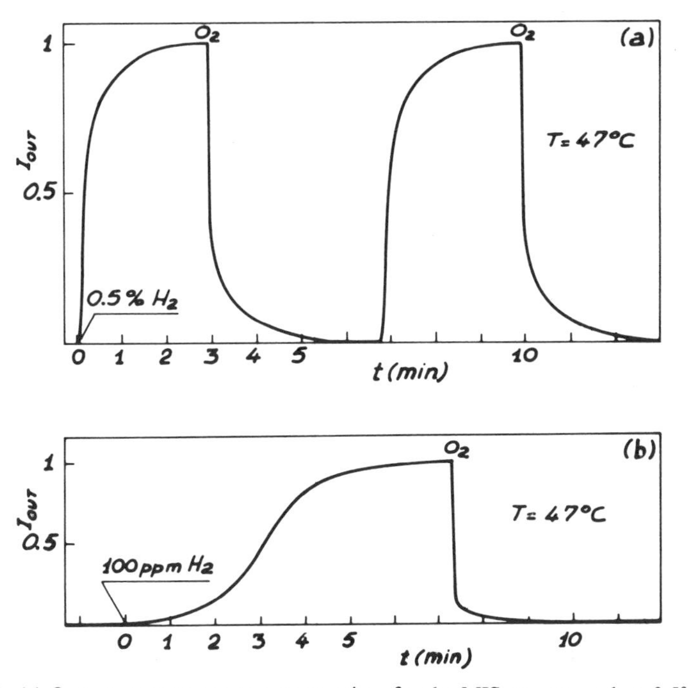

FIG. 13. (a) Output current response versus time for the MIS structure when 0.5% H_2 in $H_2 + N_2$ is used. (b) Output current response versus time for the MIS structure when 100 ppm H_2 in $H_2 + N_2$ is used.

reducing the chamber test volume, which is, in our experiment, about 10^3 cm^3.

The output current behavior observed by analyzing many absorption cycles shows that in the same day both the initial and the saturation levels are restored within an accuracy of less than 3% (short-time stability). A comparison of measurements made on different days separated by a week (long-time stability) shows that the accuracy decreases to about 8–12% when 0.5% H_2 is used and to about 5–9% for 100 ppm of H_2.

There is more than one reason for such instabilities. A first possibility can be the aging of the diode, and a second is Pd contamination and poisoning. Finally, another reason can be the diffusion of atomic hydrogen through the oxide and through the first few monolayers of a-Si:H with the formation of stable bonds, for instance, with dangling bonds present at the interface. To better understand this process, which can to a certain extent decrease the leakage current, further investigation is needed.

V. FET Structures for Ion and Gas Sensors

Two main types of devices that have been extensively investigated for chemical measurements are the single-crystal silicon field-effect transistors (FETs) (Bergueld, 1971, 1972; Matsuo and Wise, 1974; Bergueld and de Rooij, 1981; Ohta *et al.*, 1982; Akiyama *et al.*, 1982) and the ion-controlled diodes (ICD) (Wen *et al.*, 1978, 1979; Lauks 1981).

Because of its particular technological characteristics, the discussion of the ICD shall remain confined to those made of c-Si; therefore it will not be considered in this section. On the other hand, only those structures having a configuration adaptable to a-Si:H thin-film technology will be taken into consideration. Two kinds of FETs are referred to in the literature: the ion-selective FET (ISFET) and the gas-sensitive FET (CHEMFET).

In 1972 P. Bergueld first used the term ISFET to describe the unreferenced FET in which the gate was not present. Figure 14a shows the schematic of the prototype structure. After an appropriate hydration of the oxide layer, the response was nearly Nernstian to changes in the pH of the surrounding solution, and this was the first of a new generation of ion-sensitive solid-state devices. Later on, Matsuo and Wise (1974) introduced the concept of a reference electrode to provide a virtual gate in the solution (Fig. 14b), resulting in a substantial improvement in the performance of the ISFETs. As an example of CHEMFET, Fig. 14c shows the Pd-gate CHEMFET. For this structure, the basic response is controlled by the absorption of the H_2 by the Pd film (as already discussed in the preceding section). After dissociation at the Pd surface, the H diffuses to the interface

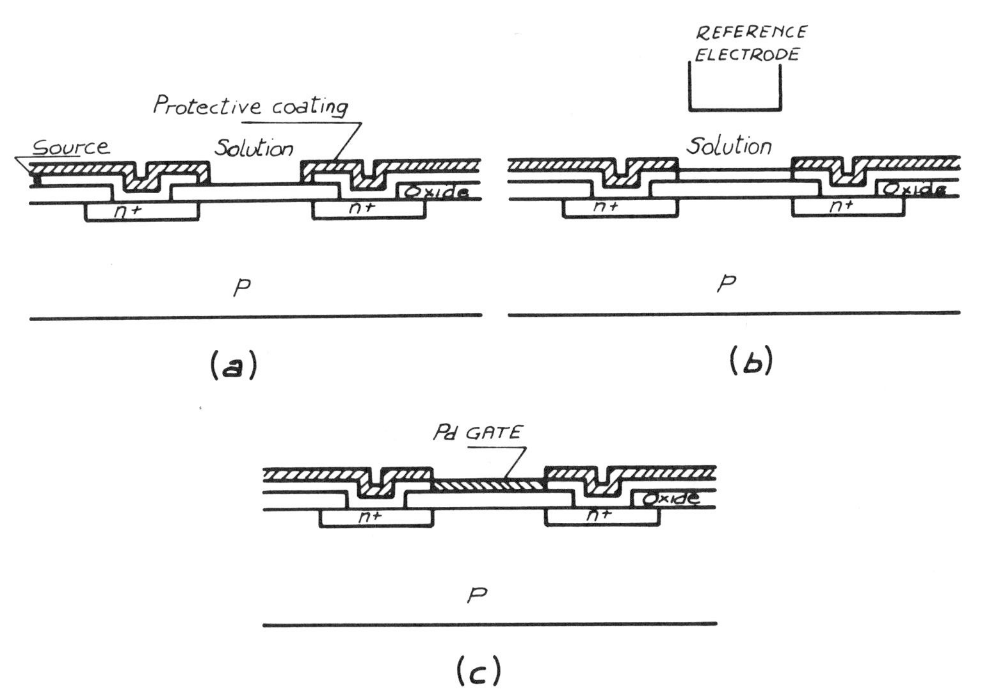

FIG. 14. (a) Schematic ISFET structure without reference. Source and drain are protected from solution; the oxide is hydrated. (b) Schematic of an ISFET structure with reference electrode. Source and drain are protected. (c) Schematic of a Pd gate CHEMFET. Source and drain are protected.

and through the change in the Pd work function, the operating characteristics of the CHEMFET are changed.

The ISFETs are important because they have significant advantages over conventional ion-selective electrodes in their faster response, small size, and low output impedance. Furthermore, the integrated circuit technology allows the fabrication of ISFETs for multi-ion detection, and this circumstance is extremely important for both biomedical and biological applications. The principles underlying the operation of these structures are similar to those of a MOSFET, in which the equation for the drain current I_d in the nonsaturated region ($V_d < V_g - V_t$) is given, according to first-order MOS transistor theory (Sze, 1969), by the equation

$$I_d = \mu(w/l)\, C_{ox}[V_d(V_g - V_t) - \tfrac{1}{2}V_d^2], \tag{2}$$

where μ is the mobility of the FET channel carriers, w and l are the gate dimensions, C_{ox} is the oxide capacity per unit area, V_d and V_g are the dc drain-to-source and gate-to-source voltages, respectively, and V_t is the threshold or turn-on voltage, given by

$$V_t = \phi_{MS} - \frac{Q_{ss} + Q_{oxf} + Q_{oxm}}{C_{ox}} + 2\phi_f - \frac{Q_b}{C_{ox}}$$

$$= V_f + 2\phi_f - \frac{Q_b}{C_{ox}}. \quad (3)$$

Here ϕ_{MS} is the work function difference between the metal and the semiconductor; $(Q_{ss} + Q_{oxf} + Q_{oxm})/C_{ox}$ is the voltage contribution due to the total charge of the interface states Q_{ss}, to the fixed charges in the oxide Q_{oxf}, and to the mobile charges Q_{oxm}; ϕ_f is the Fermi potential (dependent on the doping of the substrate); Q_b the charge in the depletion region; and V_f the flat-band voltage. In practice any change in the charge distribution at the gate caused by the presence of ions filtered by a selective membrane or any induced charge distribution due to a catalytic reaction between a metal gate and a certain gas will influence the V_t voltage. The V_t change will produce a variation in I_d, according to Eq. (2). In practice, the I_d is easily measured after the application of a constant drain-to-source voltage V_d.

In the case of an ISFET, in the presence of a reference electrode and a given solution, Eq. (2) becomes

$$I_d = \mu \frac{W}{L} C_{ox} \left\{ \left[V_g - (E_{ref} - \Delta\phi_j) - \left(\phi_0 + \frac{2.30RT}{ZF} \log(a_i + K_{ij}a_j) \right) - \phi_{Si} - \frac{Q_{ss} + Q_{ox(f+m)} + Q_b}{C_{ox}} + 2\phi_F \right] V_d - \frac{1}{2} V_d^2 \right\}, \quad (4)$$

where $E_{ref} - \Delta\phi_j$ is the voltage of the reference electrode including the liquid junction potential, V_g is the voltage applied between the reference electrode and the source, $\phi_0 + (2.30RT/ZF)\log(a_1 + K_{ij}a_j)$ is the Nernst potential (58 meV for each decade of ion activity at room temperature equal to 20°C) between the solution and the oxide interface, ϕ_0 is the standard potential, a_i is the activity of the principal ion (pi), a_j is the activity of the interfering ion (ii), K_{ij} is the selectivity coefficient of pi over ii, R is the gas constant, T is the temperature of the system at equilibrium, Z is the charge of the principal ion, F is Faradays constant, and ϕ_{Si} is the work function of silicon.

The advent of the a-Si:H as semiconductor material has opened new possibilities to produce chemical sensors in the form of FET structures suitable for both gas or ion detection. In fact, several kinds of a-Si:H

MOSFET devices have already been investigated in the literature and their overall characteristics seem to guarantee the possibility of producing either single or complex multifunction transducers.

Two of the most important FET configurations make use of a-Si:H as semiconductor material for the source–drain conduction channel. One uses a c-Si silicon substrate as gate material (Neudeck and Malhotra, 1975, 1976; Matsumura and Nara, 1980; Thompson *et al.*, 1982; Powell *et al.*, 1981; Hayama and Matsumura, 1980; Matsumura *et al.*, 1980; Abdulrida and Allison, 1983), the other uses glass only as a supporting substrate (Le-Comber *et al.*, 1979, 1981; Tuan *et al.*, 1982; Matsumura *et al.*, 1981; Ishibashi and Matsumura, 1982; Lloyd *et al.*, 1983; Nara and Matsumura, 1982; Matsumura and Hayama, 1980), and both employ the a-Si:H technology for the semiconductor channel. Merely to give an idea of the differences, schematics of the two types of devices are shown in Fig. 15a,b. The transfer characteristics are reported and discussed in the previously mentioned literature.

Typical ON–OFF drain current ratio in the best reported FETs can be as high as 10^5 or better. The remarkable rise in I_d is caused by an electron accumulation layer formed at the a-Si:H insulator interface, which creates an efficient path between the source and drain electrodes.

It is important to point out that the formation of the accumulation layer that determines the properties of the ON condition, is in a-Si:H strongly dependent on the following two properties, strictly related to the material quality: (1) a low density of localized gap states in the bulk and (2) a low density of interface states at the insulator–a-Si:H boundary.

For gas or ions, FET sensors response speed is not the main prerequisite because of the relatively long time constants associated with chemical measurements. However, the ON–OFF conductance ratio, as well as the transconductance I_d/V_g should be maximized. As far as noise is concerned, attention shold be paid to reducing both the noise-equivalent input current i_n and voltage v_n generators of the sensor. In fact, the total rms input noise for unity bandwidth of such a device can be expressed in a first approximation as (Motchenbacher and Fitchen, 1973)

$$v_{rms} = (v_n^2 + i_n^2 R_s^2 + 4kTR_s)^{1/2}$$

(if i_n and v_n are noncorrelated) and the reduction of v_{rms} can only be done by reducing the first two terms of the preceding equation. Because of the possible wide range of resistance values R_s of the virtual gate (reference electrode + solution + membrane), i_n reduction should be sought mainly in order to get a low $i_n R_s$ product.

To better characterize the FET sensor, the noise-equivalent temperature

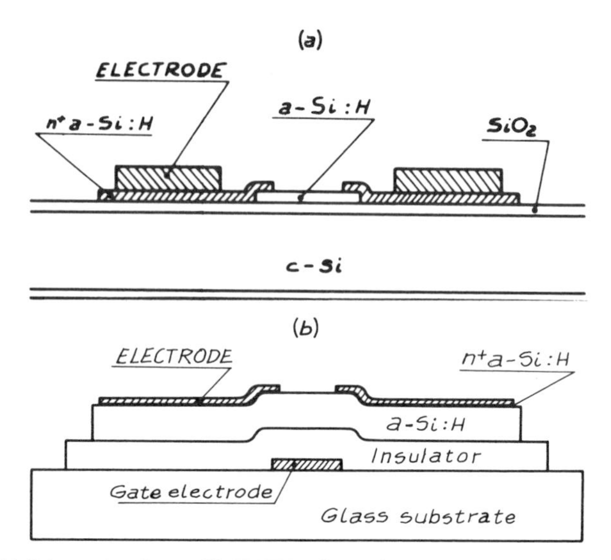

FIG. 15. (a) Schematic of an a-Si:H FET with c-Si as gate electrode. (b) Schematic of a a-Si:H FET on a glass substrate.

should be introduced; it is straightforwardly given by

$$T_{eq} = T(1 + v_n^2/4kR_sT + i_n^2R_s/4kT),$$

which becomes minimized when $R_s = v_n^2/i_n^2$, and in this particular case we have

$$T_{eq\,min} = T(1 + v_n i_n/2kT).$$

The best result can be obtained by designing a low-noise FET and a source resistance of the virtual gate matched with v_n and i_n. Following the theoretical analysis of the a-Si:H FET elaborated on by S. Kishida *et al.* (1983), it turns out that by the calculated characteristics the explicit dependence of I_d on the gate potential suggests a useful application of these devices as ISFET or CHEMFET sensors. We shall show that this theory leads to the conclusion that the I_d can be easily correlated to the virtual gate potential when the device is in a given solution.

In fact, in this theory the presence of a given localized state density in the gap and the Fermi–Dirac distribution of electrons trapped in the localized states contribute to give an I_d expression differing from that of c-Si FETs expressed by Eq. (2). In Kishida *et al.* (1983) two expressions of the drain current in a-Si:H FETs are given. One refers to the hypothesis of an exponential localized state density distribution, the other to the assumption of a uniform localized-state density distribution. As an example we shall mention only the consequence of the first of the two hypotheses with reference to the unsaturated region. In this case the drain current is given by

$$I_d = \tfrac{1}{2}(T/T_G)[(V_G - \Delta V_{GS})^{2T_G/T} - (V_G - V_D - \Delta V_{GD})^{2T_G/T}], \qquad (5)$$

where T_G is the temperature characteristic of the exponential distribution of localized states, V_G is the gate voltage supply, V_D is the drain voltage supply, and ΔV_{GS} and ΔV_{GD} are the V_G values determined at source and drain edges, respectively. For small values of V_D, i.e., $V_D \simeq 0$, Eq. (5) becomes

$$I_d = K(V_G - \Delta V_{GS})^2(T_G/T - 1)\, V_D.$$

In practice ΔV_{GS} is related to both the surface potential and to the flat-band voltage V_f. On the other hand, V_f is related to the virtual gate, which takes into account the presence of the solution as it has been shown in some details in Eqs. (3) and (4). A derivation of I_d that does not take into account one of the two hypotheses for the density-of-states distribution considered in Kishida *et al.* (1983) could be more difficult; nevertheless, a dependence on gate voltage will always be present and in any case could always be determined by experimental procedure. Also in this case, the I_d control will be exercised by the virtual gate, which contains, in ultimate analysis, the information on the ion concentration in solution.

Finally, a-Si : H CHEMFET sensors can also be made, and preliminary results in this direction are encouraging. Figure 16 shows the schematic of a Pd MOSFET used by the authors to measure H_2 concentrations, and Fig. 16 also shows the first results obtained by plotting I_d versus the gate voltage for constant drain-to-source voltage, in air and in the presence of 0.5% H_2 in $H_2 + N_2$ flux.

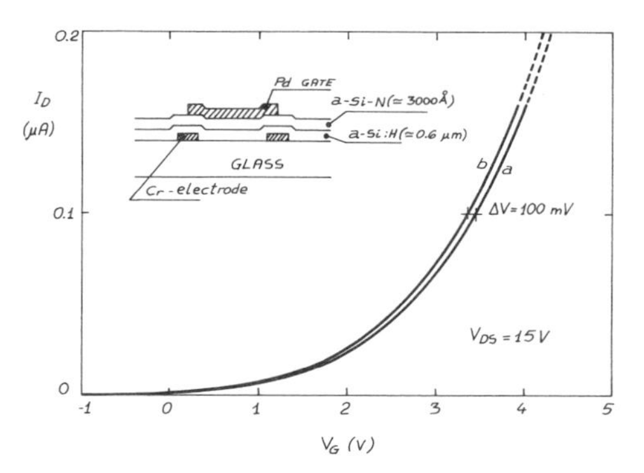

FIG. 16. Drain current versus gate voltage at fixed V_{DS} voltage for a Pd gate MOS a-Si : H structure in air [curve (a)] and in a flux of 0.5% H_2 in $H_2 + N_2$ [curve (b)]. The inset shows the schematic of the structure.

The data show that although the ΔI_d due to the admission of H_2 is not large, it is possible to measure the H_2 concentration. Improvement in the structure should be continued to enhance the device sensitivity.

VI. Conclusions

The importance of the characterization of the ambient, in terms of the measurements of those parameters capable of describing and/or controlling the ambient, is commonly recognized. These measurements can be done by special devices able to sense the physical or chemical quantities and possibly give an electric output signal.

It has been proved that these sensors can be made of a-Si : H, which can play a significant role because of its high performance-to-cost ratio. In this chapter some possible applications of a-Si : H have been presented.

Adsorbate effects on a-Si : H conductance have been shown to be of great interest because of the high sensitivity exhibited by the a-Si : H surface to particular gases under appropriate conditions, although both stability and selectivity problems are still far from being solved.

For applications in the thermistor area, one can point out the broad range of β values as well as the high values of α_T obtainable with a-Si : H. On the other hand, the many problems related to material instabilities, aging of the structure, and intrinsic and excess $1/f$ noise should not be underestimated. The removal of these drawbacks is of primary importance in determining the industrial interest of thermistors made of a-Si : H.

The studies made on the Pd MIS structure have demonstrated that H_2 sensors can be reliable devices, with high sensitivity and fast response. On the basis of the investigated structures and results, it has been established that the main mechanism responsible for the sensitivity is the change in contact potential. Once the a-Si : H film is in some way protected from contaminating gases or liquids, MIS H_2 sensor performances are mainly determined, controlled, and limited by the quality, poisoning, and stability of the Pd film or of any suitable catalytic material used.

Hydrogenated amorphous silicon FET structures are very important candidates for ACSs especially in the field of ISFETs. The long experience with c-Si FETs in the past years (involving membranes, electrode theory, and integrated readout systems) can be advantageously used.

The evolution in a-Si : H ISFETs is at the moment only at an early age; nevertheless it is felt that some good results could be achieved.

Much work must still be done, particularly in the fields of chemical-activity-to-electric-signal conversion, stability of the a-Si : H structures, and integrated preprocessing and simple processing (a-Si : H CCDs have already been fabricated but need improvement).

Also, the fundamental problem related to the synthesis of selective and technologically compatible membranes should be deeply considered. The contemporary solution of all these problems will probably determine success at the industrial level. Also of extreme importance is the study of linearization procedures and the development of algorithms for the deconvolution of multiple information at the outputs of multifunctional ACSs, whose future looks bright indeed.

Acknowledgments

The authors want to thank Prof. A. Paoletti, Director of the Istituto di Elettronica dello Stato Solido del Consiglio Nazionale della Ricerche, for continuous encouragement and Dr. C. Coluzza, Dr. C. Giovannella, and Dr. P. Fiorini of the Istituto di Fisica dell' Università di Roma for providing the a-Si:H materials and also for helpful discussions.

References

Abdulrida, M. C., and Allison, J. (1983). *Thin Solid Films* **102,** L43.
Abeles, B., Wronski, C. R., Goldstein, Y., Stasiervski, H. E., Gutkowicz-Krusin, D., Tiedje, T., and Cody, G. D. (1981). *AIP Conf. Proc.* **73,** 298.
Akiyama, T., Ujihira, Y., Okabe, Y., Sugano, T., and Niki, E. (1982). *IEEE Trans. Electron Devices* **ED-29,** 1936.
Bergueld, P. (1971). *IEEE Trans. Biomed. Eng.* **BME-17,** 70.
Bergueld, P. (1972). *IEEE Trans. Biomed. Eng.* **BME-19,** 342.
Bergueld, P., and de Rooij, N. F. (1981). *Sens. Actuators* **1,** 5.
Brattain, W. H., and Bardeen, J. (1953). *Bell Syst. Tech. J.* **32,** 1.
Card, H. C., and Rhoderick, E. H. (1971). *J. Phys. D* **4,** 1589.
D'Amico, A., Fortunato, G., Petrocco, G., and Coluzza, C. (1983a). *Appl. Phys. Lett.* **42,** 964.
D'Amico, A., Fortunato, G., Petrocco, G., and Coluzza, C. (1983b). *Sens. Actuators* **4,** 349.
Evangelisti, F., Fiorini, P., Fortunato, G., Frova, A., Giovannella, C., and Peruzzi, R. (1983). *J. Non-Cryst. Solids* **55,** 191.
Fonash, S. J., Huston, H., and Ashok, S. (1982). *Sens. Actuators* **2,** 363.
Fortunato, G., D'Amico, A., Petrocco, G., Salvati, M., and Cosmi, R. (1983a). *Proc. Ital. Vac. Soc. Conf. 8th, L'Aquila, 1983* p. 319.
Fortunato, G., D'Amico, A., Coluzza, C., and Fiorini, P. (1983b). *J. Non-Cryst. Solids* **59/60,** 1195.
Fortunato, G., D'Amico, A., Coluzza, C., Sette, F., Capasso, C., Patella, F., Queresima, C., Perfetti, P. (1984). *Appl. Phys. Lett.* **44,** 887.
Fritzsche, H., and Tanielian, M. (1981). *AIP Conf. Proc.* **73,** 318.
Fritzsche, H., and Tsai, C. C. (1979). *Sol. Energy Matter.* **1,** 471.
Goodman, A. (1964). *Surf. Sci.* **1,** 54.
Hayama, H., and Matsumura, M. (1980). *Appl. Phys. Lett.* **36,** 754.
Isibashi, K., and Matsumura, M. *Appl. Phys. Lett.* **41,** 454.
Keramati, B. (1980). Ph.D. Dissertation, University of Pennsylvania, Philadelphia.
Keramati, B., and Zemel, J. N. (1982). *J. Appl. Phys.* **53,** 1091.
Kishida, S., Nara, Y., Kobayashi, A., and Matsumura, M. (1982). *Appl. Phys. Lett.* **41,** 1154.
Kishida, S., Naruke, Y., Uchida, Y., Matsumura, M. (1983). *J. Non-Cryst. Solids* **59/60,** 1281.
Kishida, S., Naruke, Y., Uchida, Y., and Matsumura, M. (1983). *Jpn. J. Appl. Phys.* **22,** 511.

Lauks, I. (1981). *Sens. Actuators* **1,** 261.
LeComber, P. G., Spear, W. E., and Ghaith, A. (1979). *Electron. Lett.* **15,** 179.
LeComber, P. G., Snell, A. J., MacKenzie, K. D., and Spear, W. E. (1981). *J. Phys. Colloq. Orsay, Fr.* (1967) **42,** C4-423.
Lewis, F. A. "The Palladium–Hydrogen System," Chapters 1–4. Academic Press, New York.
Lloyd, J. P., Petty, M. C., Roberts, G. G., LeComber, P. G., and Spear, W. E. (1983). *Thin Solid Films* **99,** 297.
Lundström, I. (1981). *Sens. Actuators* **1,** 403.
Lundström, I., and Soderberg, D. (1981). *Sens. Actuators* **2,** 105.
Lundström, I., Shivaraman, M. S., Svensson, C., and Lundkvist, L. (1975). *Appl. Phys. Lett.* **26,** 55.
Matsumura, M., and Hayama, H. (1980). *Proc. IEEE* **68,** 1349.
Matsumura, M., and Nara, Y. (1980). *J. Appl. Phys.* **51,** 6443.
Matsumura, M., Hayama, H., Nara, Y., and Ishibashi, K. (1980). *IEEE Electron Devices Lett.* **EDL-1,** 182.
Matsumura, M., Kuno, S. I., and Uchida, Y. (1981). *J. Phys. Colloq. Orsay, Fr.* **42,** C4-519.
Matsuo, T., and Wise, D. K. (1974). *IEEE Trans Biomed. Eng.* **BME-21,** 485.
Motchenbacher, C. D., and Fitchen, F. C. (1973). "Low Noise Electronic Design." Wiley, New York.
Nara, Y., and Matsumura, M. (1982). *IEEE Trans. Electron Devices* **ED-29,** 1646.
Neudeck, G. W., and Malhotra, A. K. (1975). *J. Appl. Phys.* **46,** 2662.
Neudeck, G. W., and Malhotra, A. K. (1976). *Solid-State Electron.* **19,** 721.
Ohta, Y., Shoji, S., Esahi, M., and Matsuo, T. (1982). *Sens. Actuators* **2,** 387.
Ponpon, J. P., and Bourdon, B. (1982). *Solid-State Electron.* **25,** 875.
Powell, M. J., Easton, B. C., and Hill, O. F. (1981). *Appl. Phys. Lett. 38,* 794.
Rouths, F., Ashok, S., Fonash, J., and Rouths, J. M. (1981). *IEEE Trans. Electron Devices* **ED-28,** 1003.
Shivaraman, M. S., Lundström, I., Svensson, C., and Hammarsten, H. (1976). *Electron. Lett.* **12,** 483.
Steele, M. C., Hile, J. W., and McIver, B. A. (1976). *J. Appl. Phys.* **47,** 2537.
Suzuki, M., Maekawa, T., Nakao, A., Kumeda, M., and Shimizu, T. (1980). *Solid State Commun.* **36,** 393.
Sze, S. M. (1969). "Physics of Semiconductor Devices." Wiley (Interscience), New York.
Tanielian, M. (1982). *Philos. Mag. B* **45,** 435.
Tanielian, M., Fritzsche, H., Tsai, C. C., and Symbalisty, E. (1978). *Appl. Phys. Lett.* **33,** 353.
Tanielian, M., Chatani, M., Fritzche, H., Smid, V., and Persans, P. D. (1980). *J. Non-Cryst. Solids* **35/36,** 575.
Tanielian, M., Goodman, N. B., and Fritzsche, H. (1981). *J. Phys. Colloq. Orsay, Fr.* **42,** C4-375.
Thompson, M. J., Johnson, N. M., Nemanich, R. J., and Tsai, C. C. (1981). *Appl. Phys. Lett.* **39,** 274.
Thompson, M. J., Johnson, N. M., Moyer, M. D., and Lujan, R. (1982). *IEEE Trans. Electron Devices* **ED-29,** 1643.
Tuan, H. C., Thompson, M. S., and Johnson, N. M. (1982). *IEEE Electron Devices Lett.* **EDL-3,** 357.
Wen, C. C., Chen, T. C., and Zemel, J. N. (1978). *Proc. Int. Electron Device Meet., Washington, D. C., 1978* p. 108.
Wen, C. C., Chen, T. C., and Zemel, J. N. (1979). *IEEE Trans. Electron Devices* **ED-26,** 1945.
Viktorovitch, P., Moddel, G., and Paul, W. (1981). *AIP Conf. Proc.* **73,** 186.

Wronski, C. R. (1977). *IEEE Trans. Electron Devices* **ED-24,** 351.
Wronski, C. R., and Carlson, D. E. (1977). *Solid State Commun.* **23,** 421.
Wronski, C. R., Carlson, D. E., and Daniel, R. E. (1976). *Appl. Phys. Lett.* **29,** 602.
Wronski, C. R., Abeles, B., Cody, G. D., and Tiedje, T. (1980). *Appl. Phys. Lett.* **37,** 96.
Wronski, C. R., Goldstein, Y., Kelemen, S., Abeles, B., and Witzke, H. (1981). *J. Phys. Colloq. Orsay, Fr.* **42,** C4-475.
Zemel, J. N., Keramati, B., Spivak, C. W., and D'Amico, A. (1981). *Sens. Actuators* **1,** 427.

CHAPTER 12

Amorphous Light-Emitting Devices

Hiroshi Kukimoto

IMAGING SCIENCE AND ENGINEERING LABORATORY
TOKYO INSTITUTE OF TECHNOLOGY
YOKOHAMA, JAPAN

I. Introduction

Extensive studies of hydrogenated amorphous silicon (a-Si : H) in the past decade since conductivity control of this material was demonstrated by Spear and LeComber (1975) have been connected mainly with its applications to light-detecting devices, including thin-film solar cells, image pickup tubes, and electrophotographic receptors. One of the important properties of a-Si : H, which forms a background of the device physics, is its large optical absorption; its absorption coefficient is more than an order of magnitude larger than that of crystalline Si over the visible light range. This is due to the nature of the optical transition in a-Si, i.e., the momentum-nonconserved direct transition in contrast to the indirect transition in crystalline Si. The large optical transition probability in a-Si : H in turn is responsible for a more efficient luminescence than in crystalline Si, at least at low temperatures. Therefore photoluminescence has been widely measured in recent years to investigate electronic states, including tail states and midgap defect states in a-Si : H samples prepared by various methods.

The first attempt to utilize the luminescent property of a-Si : H for light-emitting devices can be traced back to the year 1976, when Pankove and Carlson (1976) demonstrated the electroluminescence in $p-i-n$ and Schottky junctions of a-Si : H at low temperatures. This was followed by

ISBN 0-12-752150-X

several studies (Street *et al.*, 1979; Nashashibi *et al.*, 1982; Konagai *et al.*, 1981). It was quite recently that room-temperature electroluminescence was observed in a-Si : H devices of a similar structure (Lim *et al.*, 1982).

All of the preceding observations, however, were of light emission in the infrared region, with the typical emission peak located around 1.3 eV. For practical applications, visible light emission from wider-band-gap materials is preferable. Candidates as wide-gap amorphous materials can be found in the paper by Anderson and Spear (1977), in which they described amorphous alloys such as a-Si_xC_{1-x} and a-Si_xN_{1-x}. Engemann and colleagues (1978) observed visible photoluminescence in a-Si_xC_{1-x}:H($x > 0.4$) prepared by glow-discharge (GD) decomposition of silane–ethylene mixtures; the main peak of the spectrum for a-$Si_{0.4}C_{0.6}$:H is located in the yellow (~2 eV). The author's group (Munekata *et al.*, 1980) has also demonstrated that a-Si_xC_{1-x}:H prepared by GD decomposition of tetramethylsilane shows white photoluminescence even at room temperature. Further studies of photoluminescence in a-Si_xC_{1-x}:H systems have been performed since then (Sussmann and Ogden, 1981; Munekata *et al.*, 1981; Watanabe *et al.*, 1982; Morimoto *et al.*, 1982; Nakazawa *et al.*, 1983; Masumoto *et al.*, 1983). Some typical preparation and properties of a-Si_xC_{1-x}:H will be dealt with in Part II.

Visible (white) electroluminescence at room temperature was observed for the first time by the author's group (Munekata and Kukimoto, 1983a) from an a-Si_xC_{1-x}:H ($x = 0.2-0.4$) layer sandwiched between two insulating layers of Y_2O_3 when driven by ac voltages higher than 100 V. Although such devices have yielded extremely low-efficiency luminescence, this observation is expected to encourage future development of higher-efficiency amorphous light-emitting devices. Part III deals with present and future aspects of a-Si_xC_{1-x}:H electroluminescent devices.

Another promising candidate to be noted as a room-temperature electroluminescent material is a-Si : H_x prepared by a unique method of homogeneous chemical vapor deposition(homoCVD), which has been demonstrated recently (Wolford *et al.*, 1983a,b; Meyerson *et al.*, 1983). Details of the method and the properties of the material appear in Volume 21A, Chapter 7, by Scott. The striking feature of the homoCVD a-Si : H_x films is room-temperature visible photoluminescence with high efficiency; the spectral peak is located at about 2 eV and the efficiency is comparable to that of conventional light-emitting diode materials such as GaAsP (Wolford *et al.*, 1983a,b). Furthermore, this unique material can be doped with *n*- and *p*-type impurities (Meyerson *et al.*, 1983). Part IV is concerned with future aspects of amorphous light-emitting devices utilizing such new materials as these, as well as with new device structures.

II. Preparation and Properties of a-Si_xC_{1-x}:H

1. Preparation and Hydrogenation

There are two typical deposition methods for preparing a-Si_xC_{1-x}:H films: glow discharge (GD) (e.g., Anderson and Spear, 1977) and rf reactive sputtering (RS) (e.g., Mogab and Kingery, 1968; Shimada *et al.*, 1979), both techniques being essentially the same as used for growing a-Si : H films. It is believed that the GD deposition tends to incorporate more hydrogen in a-Si:H films than does the RS deposition. This trend is also found for the case of a-Si_xC_{1-x}:H films (Katayama *et al.*, 1981). Therefore, a-Si_xC_{1-x}:H films are usually prepared by GD decomposition of various gas combinations: silane–ethylene (Anderson and Spear, 1977; Engemann *et al.*, 1978; Wieder *et al.*, 1979; Lee, 1980; Sussmann and Ogden, 1981), silane–methane (Tawada *et al.*, 1981, 1982; Watanabe *et al.*, 1982; Morimoto *et al.*, 1982), tetramethylsilane (Munekata *et al.*, 1980), and silane–tetramethylsilane (Munekata *et al.*, 1981).

The alloy composition can be changed over a wide range by controlling the mixing ratio of the starting gases, as exemplified in Fig. 1. The hydrogen content involved in GD samples depends mainly on the electric power fed to the glow discharge and on the substrate temperature, as is the case of a-Si : H. The hydrogen content depends also on the gases used. However, the general trend of hydrogen incorporation is more or less similar to the case of a-Si : H higher hydrogen content for higher powers and lower substrate temperatures.

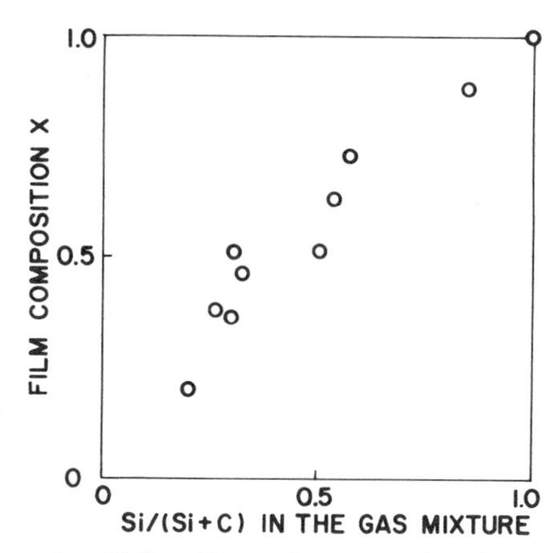

FIG. 1. Composition x in a-Si_xC_{1-x}:H samples prepared by GD decomposition of silane–tetramethylsilane (SiH_4–$Si(CH_3)_4$) mixtures at a substrate temperature of 200°C. [From H. Munekata and H. Kukimoto, unpublished.]

Hydrogen atoms are bonded with silicon and carbon as well in the films over a wide range of alloy composition, as is clearly seen in the infrared absorption spectra of Fig. 2. One thing to be noted is that hydrogen, once introduced into the film during the low-temperature deposition, tends to remain there even at high temperatures, probably owing to the more stable nature of the C–H bond compared to the Si–H bond (Munekata *et al.*, 1980).

2. Absorption Edge

Anderson and Spear (1977) have determined the change in the optical gap energy E_{opt} in a-Si_xC_{1-x}:H with composition. The value of E_{opt} increases monotonically with decreasing x, from ~ 1.6 eV at $x = 1.0$ to a maximum of ~ 2.8 eV at $x = 0.3$ and decreases down to ~ 2.4 eV at $x = 0$, for the films GD-deposited from silane–ethylene mixtures at about 200°C. Figure 3 shows the $(\alpha h\nu)^{1/2}$ versus $h\nu$ plot, where α is the absorption coefficient and $h\nu$ is photon energy, for several samples prepared by GD decomposition of silane–tetramethylsilane mixtures at the same substrate temperature. The

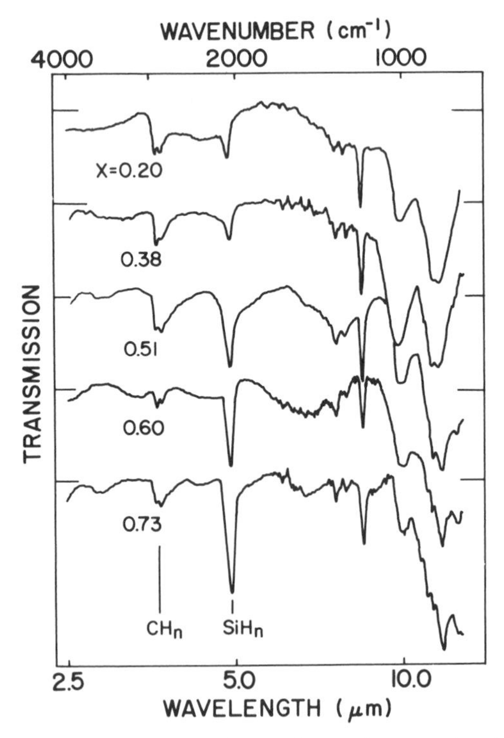

FIG. 2. Infrared absorption spectra of a-Si_xC_{1-x}:H samples grown by GD decomposition of SiH_4–$Si(CH_3)_4$ mixtures at 200°C. [From H. Munekata and H. Kukimoto, unpublished.]

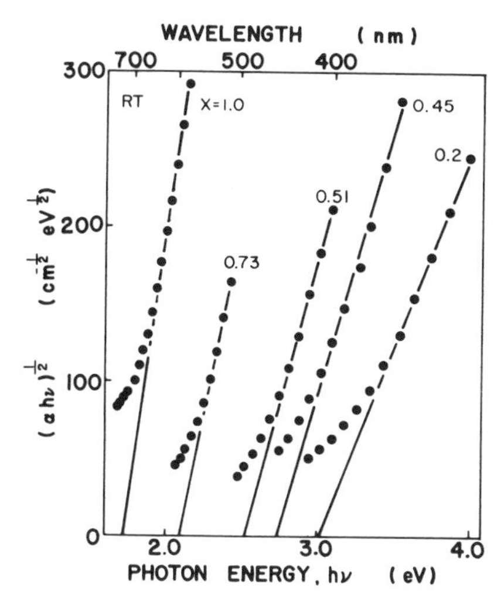

FIG. 3. $(\alpha h\nu)^{1/2}$ versus $h\nu$ plots for a-Si_xC_{1-x}:H. [From Munekata *et al.* (1981). Copyright North-Holland Publishing Company, Amsterdam, 1981.]

composition dependence of E_{opt}, which is determined by the intercept of the extrapolated straight line, is similar to the result of Anderson and Spear (1977). This indicates that E_{opt} is not strongly dependent on the starting materials if samples are prepared under similar conditions. The change of slope with composition may be related to the change in the density of states in the conduction and valence bands. One should also note that the samples with smaller x tend to exhibit longer tails, indicating an enhancement of the tailing of states near the band edges, probably because of the increased fluctuation of alloy composition.

3. PHOTOLUMINESCENCE

A striking contrast between a-Si_xC_{1-x}:H and a-Si:H can be found in their photoluminescent properties; e.g., emission spectra as well as decay time can be changed widely with composition, and efficient emission is obtained at room temperature for smaller x. Typical photoluminescence spectra at low temperature for several samples with different compositions are shown in Fig. 4 (Munekata *et al.*, 1981). The spectrum shifts toward higher energies with decreasing Si content x, reflecting the change in E_{opt}. At the same time, the spectrum tends to broaden, implying increased compositional fluctuations or an enhanced electron–phonon interaction. When the temperature is increased, the photoluminescence intensity for larger x samples decreases, as is known to be the case for a-Si:H. For smaller x samples, however, such temperature quenching is less pronounced. In particular, the large gap sample of a-$Si_{0.2}C_{0.8}$:H, prepared from tetramethylsilane only, exhibits

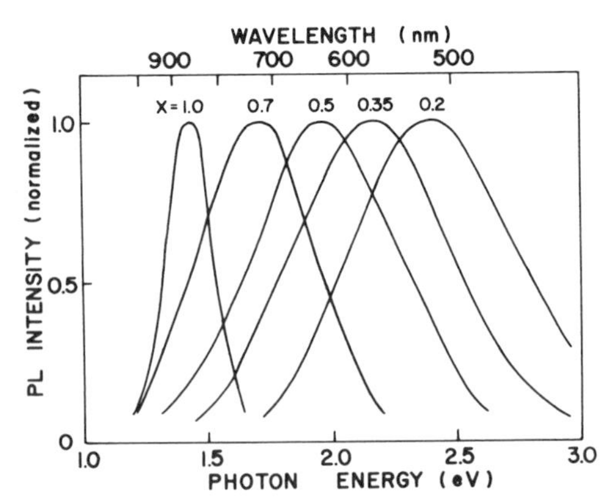

FIG. 4. Photoluminescence spectra of a-Si_xC_{1-x}:H at 77°K. [From Munekata *et al.* (1981). Copyright North-Holland Publishing Company, Amsterdam, 1981.]

reasonably efficient photoluminescence even at room temperature, with an intensity of approximately 70% of the 77°K intensity (Munekata *et al.*, 1980). This temperature-insensitive behavior of the photoluminescence is probably related to the deeper localized states responsible for the radiative transition. Furthermore, the emission intensity for larger gap materials, e.g., a-$Si_{0.2}C_{0.8}$:H, tends to increase monotonically with decreasing substrate temperature during the film preparation, which is similar to the case of homoCVD a-Si:H_x. These properties are particularly important from the viewpoint of device application.

On the other hand, the luminescence decay time for a-$Si_{0.2}C_{0.8}$:H has been found to be about 200 nsec (Nakazawa *et al.*, 1983), much shorter than that of a-Si:H, which ranges from 1 μsec to 1 msec (Kurita *et al.*, 1979). This observation is consistent with the temperature-insensitive and efficient radiative transition in C-rich a-Si_xC_{1-x}:H, in which excited electrons and holes are more tightly bound at the excited position and in the deeper band-tail states than in a-Si:H (Nakazawa *et al.*, 1983). Regardless of the reason, the fast decay time in a-Si_xC_{1-x}:H is also attractive for achieving fast-response light-emitting devices.

III. Electroluminescence of a-Si_xC_{1-x}:H

4. Structure and Preparation

Room-temperature electroluminescence (EL) in a-Si_xC_{1-x}:H was observed by Munekata and Kukimoto (1983a) for a thin-film EL device consisting of the amorphous film sandwiched between two insulating layers. The first insulating layer of Y_2O_3 3000 Å thick was deposited onto an indium-tin-oxide- (ITO) coated glass substrate held at 100°C. This was followed by a 2000-Å thick a-Si_xC_{1-x}:H($x = 0.2-0.4$) emitting layer, the second insulating layer of Y_2O_3, and finally 3-mm-diameter Al electrodes.

5. Characteristics

The device thus made shows white electroluminescence at room temperature when driven by a 1-kHz square-wave voltage greater than 100 V. The light output, together with the current, is observed only briefly at the leading and trailing edges of the square wave, where the polarity of the voltage changes as shown in Fig. 5. This behavior is typical of ac-electroluminescent devices like ZnS:Mn EL cells (Mach and Muller, 1982). It is noted, however, that the luminescence decay time is fast (~25 μsec) compared to that of ZnS:Mn cells (~1 msec), reflecting the short-lived photoluminescence of this amorphous material, as mentioned earlier. The electroluminescence (EL) spectrum is somewhat different from the photoluminescence (PL) spectrum for the same sample, as shown in Fig. 6. The energy of the EL peak is lower than that of PL, and the EL spectrum is broader than the PL spectrum. This has been explained by assuming different excited-carrier distributions in the band-tail states for these different excitation methods. This assumption is confirmed by the observation of the change in PL spectra with excitation photon energy (Munekata *et al.*, 1981).

The brightness of the devices of this kind is only about 1 fL at 180 V, and the estimated power conversion efficiency is extremely low ($10^{-6}-10^{-5}$) at present (Munekata and Kukimoto, 1983).

IV. Future Developments

The ac electroluminescence from the devices described in the preceding section is believed to result from impact-ionized carrier excitation. It has been shown that impact ionization can be enhanced in multilayered-hetero-

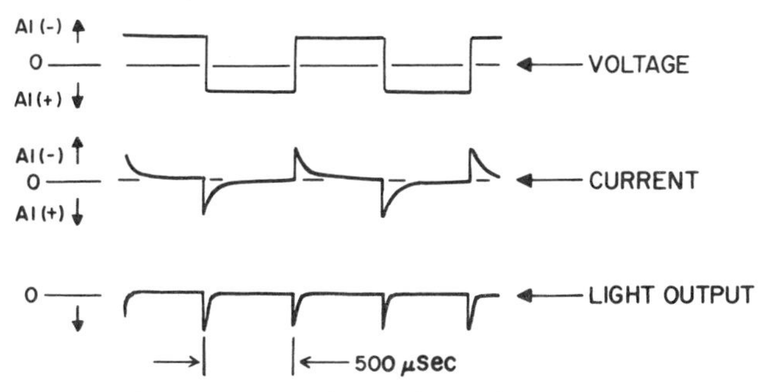

Fig. 5. Current and light output wave forms of an a-$Si_{0.3}C_{0.7}$:H electroluminescent device under the 1-kHz square wave exitation. [From Munekata and Kukimoto (1983a).]

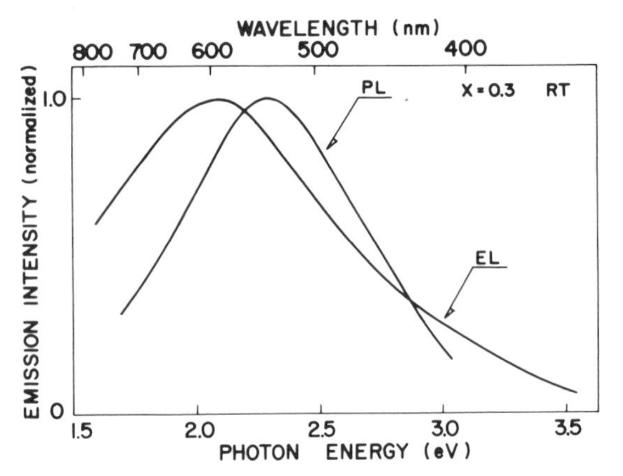

FIG. 6. Electroluminescence (EL) and photoluminescence (PL) spectra in a-$Si_{0.3}C_{0.7}$:H at room temperature. [From Munekata and Kukimoto (1983a).]

junction-semiconductor structures involving quantum wells, in which the carrier scattering is different from that in the bulk (Shichijo *et al.,* 1978; Holonyak *et al.,* 1980; Chin *et al.,* 1980). Therefore if one can obtain a similar structure for amorphous materials, an improvement in the impact-ionization efficiency and hence higher electroluminescence efficiency would be expected. In the case of crystalline semiconductors, such a structure must be made by choosing two materials having the same or similar lattice constants. Amorphous materials, on the other hand, are not restricted to this lattice matching requirement, and it therefore should be feasible to construct a variety of quantum-well structures. The author's group has studied optical properties of a-$Si_{0.2}C_{0.8}$:H/a-Si:H/a-$Si_{0.2}C_{0.8}$:H quantum-well structures prepared by GD deposition of silane (SiH_4) for a-Si:H and tetramethylsilane ($Si(CH_3)_4$) for a-$Si_{0.2}C_{0.8}$:H (Munekata and Kukimoto, 1983b). When the thickness of the a-Si:H well layer is reduced to about 25 Å, a notable change in optical gap as well as the shape of absorption spectra has been observed. The results have been explained on the basis of the quantization of the electronic states within the well layer. In view of this preliminary result, an ac electroluminescent cell consisting of multiple quantum-well layers of various amorphous materials such as GD a-Si:H, homoCVD a-Si:H_x, a-Si_xC_{1-x}:H, a-SiN_x, and a-SiO_x can be regarded as one of the future devices to be developed.

In order to realize junction light-emitting diodes (LEDs) having the same structure as that of practical III–V compound semiconductor LEDs, we must achieve *n*- and *p*-type conductivity control. The phosphorus and boron doping, which is known to be effective for obtaining high-conductivity *n*- and *p*-type a-Si:H, respectively, seems to be less effective for the wider-gap a-Si_xC_{1-x} : H. This is possibly related to the high density of native defects that form mid-gap states in these samples. Optimization of sample preparation conditions requires further investigation. As was pointed out in Part I, on the other hand, homoCVD has demonstrated the feasibility of wide-gap materials of a-Si : H_x that involve low defect density and can be effectively doped with P and B impurities. This unique deposition technique is expected to be applicable to other materials, including a-Si_xC_{1-x}, a-SiN_x, and a-SiO_x systems.

The final goal of amorphous light-emitting devices would be to couple them with their driving circuits, i.e., to make flat panel displays consisting of a light-emitting layer deposited onto a suitable Si integrated circuit or large-area array of a-Si thin-film transistors. Steady progress toward this goal can be expected with the introduction of new technologies developed in relation to other devices, along with the fundamental understanding of wide-gap amorphous materials.

REFERENCES

Anderson, A. D., and Spear, W. E. (1977). *Philos. Mag.* **35,** 1–16.

Chin, R., Holonyak, N., Jr., Stillman, G. E., Tang, J. Y., and Hess, K. (1980). *Electron. Lett.* **16,** 467–469.

Engemann, D., Fischer, R., and Knecht, J. (1978). *Appl. Phys. Lett.* **32,** 567–568.

Holonyak, N., Jr., Kolbas, R. M., Dupuis, R. D., and Dapkus, P. D. (1980). *IEEE J. Quantum Electron.* **QE-16,** 170–186.

Katayama, Y., Usami, K., and Shimada, T. (1981). *Philos. Mag. B* **43,** 283–294.

Konagai, M., Nishihata, K., Komori, K., and Takahashi, K. (1981). *Proc. IEEE Photovoltaic Specialists Conf., 15th, 1981* pp. 906–911.

Kurita, S., Czaja, W., and Kinmond, S. (1979). *Solid State Commun.* **32,** 879–883.

Lee, W. Y. (1980). *J. Appl. Phys.* **51,** 3365–3372.

Lim, K. S., Konagai, M., and Takahashi, K. (1982). *Jpn. J. Appl. Phys.* **21,** L437–L475.

Mach, R., and Muller, G. O. (1982). *Phys. Status Solidi A* **69,** 11–66.

Masumoto, Y., Shionoya, S., Munekata, H., and Kukimoto, H. (1983). *J. Phys. Soc. Jpn.* **52,** 3985–3991.

Meyerson, B. S., Scott, B. A., and Wolford, D. J. (1983). *J. Appl. Phys.* **54,** 1461–1465.

Mogab, C. J., and Kingery, W. D. (1968). *J. Appl. Phys.* **39,** 3640–3645.

Morimoto, A., Miura, T., Kumeda, M., and Shimizu, T. (1982). *Jpn. J. Appl. Phys.* **53,** 7299–7305.

Munekata, H., and Kukimoto, H. (1983a). *Appl. Phys. Lett.* **42,** 432–434.

Munekata, H., and Kukimoto, H. (1983b). *Jpn. J. Appl. Phys.* **22,** L544–L546.

Munekata, H., Murasato, S., and Kukimoto, H. (1980). *Appl. Phys. Lett.* **37,** 536–537. erratum (1981). *Appl. Phys. Lett.* **38,** 188.

Munekata, H., Shiozaki, A., and Kukimoto, H. (1981). *J. Lumin.* **24/25,** 43–46.

Nakazawa, E., Munekata, H., and Kukimoto, H. (1983). *Solid State Commun.* **45,** 925–927.

Nashashibi, T. S., Searle, T. M., Austin, I. G., Rhodes, A. J., Gibson, R. A., and LeComber, P. G. (1982). *Philos. Mag. B* **45,** 573–582.

Pankove, J. I., and Carlson, D. E. (1976). *Appl. Phys. Lett.* **29,** 620–622.

Shichijo, H., Kolbas, R. M., Holonyak, N., Jr., Dupuis, R. D., and Dapkus, P. D. (1978). *Solid State Commun.* **27,** 1029–1032.
Shimada, Y., Katayama, Y., and Komatsubara, K. F. (1979). *J. Appl. Phys.* **50,** 5530–5532.
Spear, W. E., and LeComber, P. G. (1975). *Solid State Commun.* **17,** 1193–1196.
Street, R. A., Tsang, C., and Knight, J. C. (1979). *Proc. Int. Conf. Phys. Semicond., 14th, Edingburgh, 1978* pp. 1139–1142.
Sussmann, R. S., and Ogden, R. (1981). *Philos. Mag. B* **44,** 137–158.
Tawada, Y., Okamoto, H., and Hamakawa, Y. (1981). *Appl. Phys. Lett.* **39,** 237–239.
Tawada, Y., Tsuge, K., Kondo, M., Okamoto, H., and Hamakawa, Y. (1982). *J. Appl. Phys.* **53,** 5273–5281.
Watanabe, I., Hata, Y., Morimoto, A., and Shimizu, T. (1982). *Jpn. J. Appl. Phys.* **21,** L613–L615.
Wieder, H., Cardona, M., and Guarnieri, C. R. (1979). *Phys. Status Solidi B* **92,** 99–112.
Wolford, D. J., Scott, B. A., Reimer, J. A., and Bradley, J. A. (1983a). *Proc. Int. Conf. Phys. Semicond., 16th, 1982 Montpellier,* pp. 920–922.
Wolford, D. J., Reimer, J. A., and Scott, B. A. (1983b). *Appl. Phys. Lett.* **42,** 369–371.

CHAPTER 13

Fast Detectors and Modulators

Robert J. Phelan, Jr.

NATIONAL BUREAU OF STANDARDS
BOULDER, COLORADO

I. Introduction

The use of crystalline silicon for $p-n$ junction optical detectors is well recognized. For the very fast devices and for other optoelectronic devices, there are reasons for not using silicon. The III–V compound semiconductors exhibit higher carrier mobilities, and higher mobilities are typically associated with faster devices. The mobilities reported for a-Si : H are particularly low, with values below 1 $cm^2\ V^{-1}\ sec^{-1}$. Silicon, being an indirect band-gap semiconductor, exhibits a relatively low absorption coefficient for photons with energies near the band-gap energy. Thus for a high quantum efficiency detector, silicon requires a greater optical path length than that required for a direct gap semiconductor. Depending on the structure, larger path lengths can lead to larger electron paths, longer transit times, and slower devices. Because of the crystalline symmetry of silicon, one does not expect a linear electrooptic effect (Garmire, 1975), and this excludes silicon from being considered as a material for an optical modulator.

In spite of all the reasons noted for not using silicon for optoelectronic devices, the potential economic advantages are very compelling. Silicon is a plentiful material and there is a large source of silicon device technology. If many optoelectronic devices could be made of silicon, we would be closer to making practical integrated optical circuits. To fuel this desire to use silicon technology where possible, we point out some unique properties exhibited by hydrogenated amorphous silicon. One can speculate that many of the arguments against using the common crystalline silicon may not be valid for

ISBN 0-12-752150-X

a-Si:H. The higher absorption coefficients and sharper absorption edge of a-Si:H, shown in Fig. 1, suggest the possibility that the structure of this material is sufficiently different from crystalline silicon that it might exhibit other properties similar to those of a direct gap semiconductor. Because a-Si:H can be deposited on a wide variety of substrates at relatively low temperatures, we can speculate on the possibilities of compensating mate-

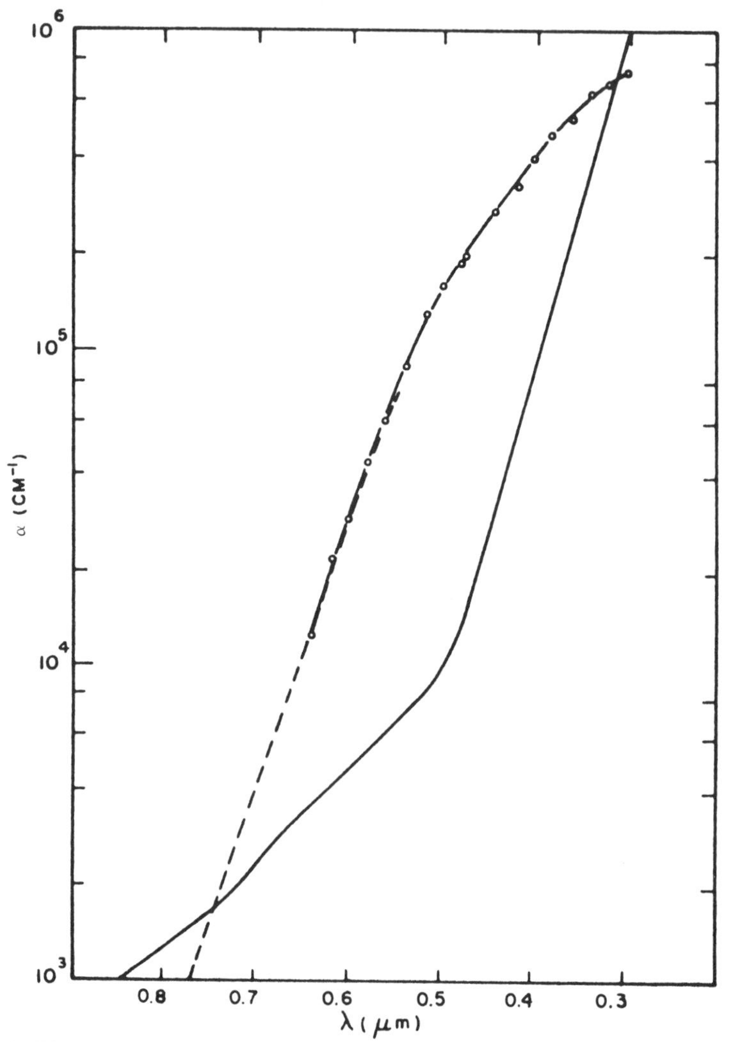

FIG. 1. The optical absorption coefficient as a function of wavelength for discharge-disposited a-Si (○) and single-crystal Si (solid curve). Dashed curve represents data of Loveland *et al.* (1973/1974). [From Carlson and Wronski (1976).]

rial properties with unique structures that allow for practical device performances. Although the crystalline symmetry of silicon excludes the linear electrooptic effect, a-Si : H on another substrate material is most likely to be under strain and might therefore exhibit this effect.

Sometimes, rather than attempting to justify a device theoretically beforehand, it is less expensive to make a device and evaluate it. The evaluation of the device may invalidate previous assumptions.

It is interesting to determine whether a-Si : H can be used to create useful optoelectronic devices with picosecond response times. Although one normally does not associate fast devices with low-mobility materials, subnanosecond optical detectors and modulators using hydrogenated amorphous silicon have been made. Fast speeds are achieved by using very short lifetime materials or by making the structure sufficiently small that transit time is the limiting factor. A major factor favoring a-Si:H is the fact that it can be deposited on a wide variety of substrates, and this allows for the fabrication of structures that would otherwise be very difficult to construct.

II. Devices

1. Detectors

Photovoltaic detectors made of hydrogenated amorphous silicon have been coupled to optical waveguides (Yumoto *et al.*, 1982). This effort took advantage of the fact that a-Si : H devices can be placed on a wide variety of substrates, and a reasonable amount of detail of the fabrication process is given. The detectors were deposited by using a plasma discharge onto glass and Ti-in-diffused $LiNbO_3$ optical waveguides. To obtain good adhesion, provide a transparent electrical contact, and act as the photovoltaic barrier, a 0.1-μm indium–tin–oxide (ITO) layer was sputtered onto the guides before the a-Si : H was deposited. A top 0.2-μm evaporated layer of aluminum was used as the other contact. The ITO films were sputtered at a power of 200 W in argon gas at a pressure of 2.7 Pa containing 9 mol % SnO_2. The 0.5-μm a-Si : H films were deposited by using a 3-W rf glow discharge with a pressure of 4 Pa of SiH_4 flowing at a rate of 5 cm^3 min^{-1} and a substrate temperature of 250°C yielding a growth rate of 0.3 μm $hr.^{-1}$.

From the fact that 633-nm light passing under the detector was not observed behind the detector, it was concluded that the beam was effectively absorbed. Operating in the photovoltaic mode, the detector voltage was a linear function of the guided wave power (Fig. 2). The rise time of these detectors was 350 nsec. From assumed values for the mobility, junction voltage, and depletion width, the dominant limiting factor for the speed of these detectors was the transit time.

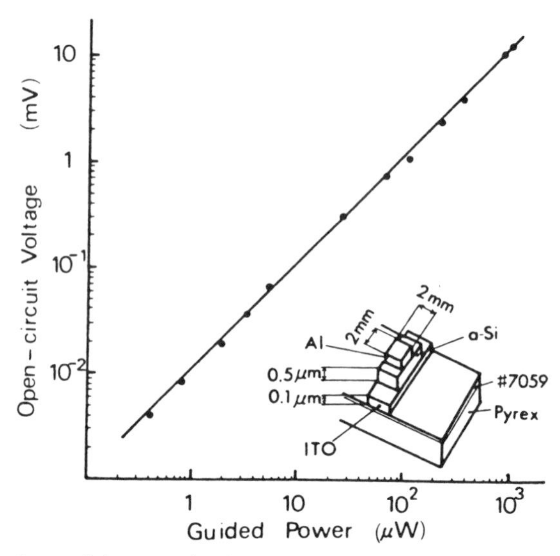

FIG. 2. Dependence of the open-circuit voltage of an a-Si photodetector on the power of a He–Ne laser light guided in a glass waveguide. Also shown is the device structure. [From Yumoto *et al.* (1982).]

The a-Si:H detectors described earlier were placed on waveguides of glass. Oliver and Peuzin (1978) showed that a-Si:H can be used as an optical guiding layer. The guiding layers of sputtered a-Si:H were deposited on glass. The waveguide measurements were made at a wavelength of 1.15 μm, and the a-Si:H exhibited an absorption coefficient of approximately 3 cm^{-1}. They measured an anomalously low value of refractive index of 3.04 and a birefringence corresponding to an index change of 0.02. This birefringence was attributed to a strong planar stress. It was also determined that the layer was not homogeneous throughout its thickness.

Again, Yumoto *et al.*, (1982) stated that the transit time was the major limiting factor to the speed of response of their detector. On the other hand, if the lifetime is shorter than the transit time, then the response time can be even shorter. Johnson *et al.* (1981) demonstrated photoconductive detectors using glow-discharge (GD) a-Si:H with relaxation times of 200 psec. These detectors had a 25-μm spacing between the electrodes. The 200 psec was observed at room temperature, but on cooling to 144°K the relaxation time decreased to 25 psec. In addition, other forms of amorphous silicon, vacuum evaporated (EV) and chemical vapor deposited (CVD), were evaluated for speeds of response. The a-Si:H exhibited the longest lifetime, and the shortest was measured with evaporated silicon, which exhibited a lifetime of 4 psec. These response times were measured using a correlation technique (Auston *et al.*, 1980) that has a resolution of approximately

1 psec. Results of the measurements are shown in Fig. 3. The GD sample of Fig. 3 was a 0.5-μm layer of a-Si : H deposited at a rate of about 1 μm hr^{-1}. The deposition was performed with a capacitively coupled glow-discharge system at a pressure of 48 Pa, an rf power density of 40 mW cm^{-2}, and a frequency of 13.56 MHz.

An alternative to decreasing the lifetime is decreasing the gap between electrodes, thus making the transit time smaller. Clearly, the capacitance must be kept small enough that the device is not *RC* limited. Glass *et al.* (1983) constructed a detector using a 30-nm-thick layer of sputtered a-Si : H sandwiched between two semitransparent Cr–Au electrodes. This device exhibited a rise time of approximately 25 psec (the response limit of the sampling oscilloscope) and material-limited decay time of 40 psec. The incident light passed through one of the contacts, and a fraction of the light was absorbed in the thin silicon layer. To enhance the optical absorption of this very thin layer, the structure had a glass substrate with a two-dimensional 250-nm periodic variation in the surface height. For a given wave-

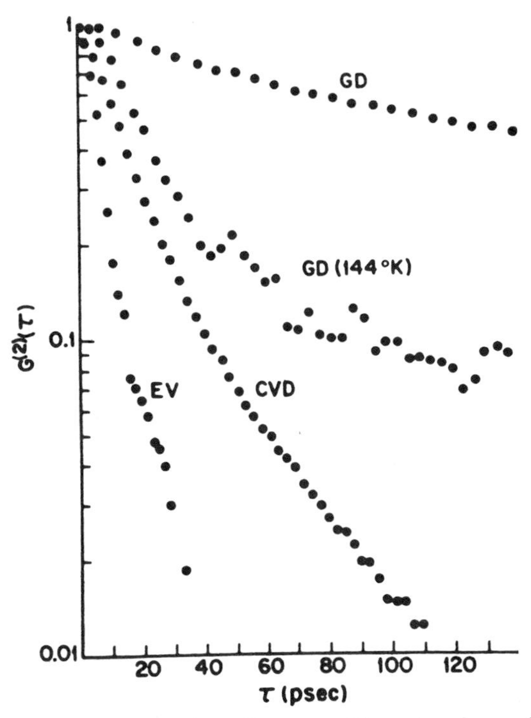

FIG. 3. Normalized electronic correlation functions of the various forms of a-Si as a function of the relative delay between the responses of two photoconductors. [From Johnson *et al.* (1981).]

length there is an angle of incidence at which the absorption is strongly enhanced. Up to 90% absorption was obtained for light of wavelength 0.7 μm.

Ideally, one would like a detector with a high quantum efficiency, a wide dynamic range, a fast response, a broad spectral response, and a uniform responsivity over a well-defined receiving aperture. A structure evaluated for these attributes used interdigitated metal contacts on a 1-μm layer of plasma-deposited a-Si : H with a sapphire substrate (Phelan *et al.,* 1983). To achieve short transit times, submicrometer spacings between the metal contacts were made using electron-beam lithography. Silicon that was not in the interdigitated area could absorb light and thus lead to longer transit times than determined by the gap spacing. This excess silicon was removed from the structure. The contacts were chromium Schottky barriers such that the device was equivalent to two back-to-back diodes. Dark currents were in the nanoampere range when the device was biased to 5 V. The current versus voltage curves of Fig. 4 show a horizontal line for no incident light and the saturated, reverse-biased diode characteristics with incident light. Saturation currents corresponding to quantum efficiencies exceeding 30% were reported for incident wavelengths shorter than that corresponding to the band gap. With light incident from the sapphire side, photoemission of carriers over the metal semiconductor extended the wavelength response

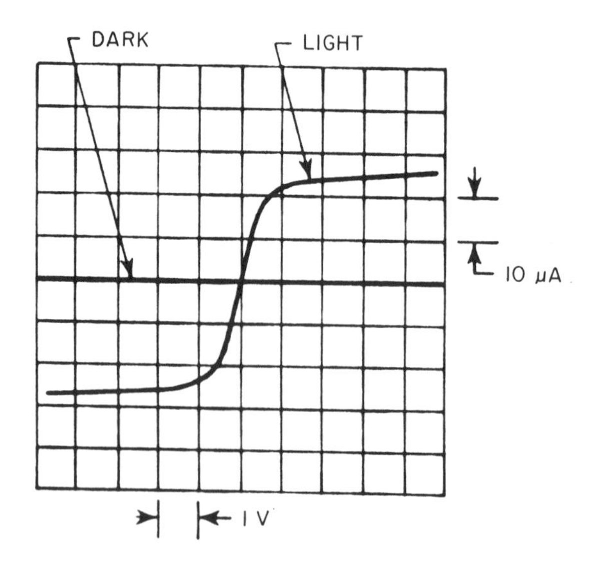

FIG. 4. The current–voltage curves for an interdigitated detector exhibiting the back-to-back diode characteristics.

beyond 2 μm into the infrared. Scanning the detector with a small light beam incident from the sapphire side yielded a uniform responsivity over the receiving area. The measured speed of response of these devices was limited by the resolution of the sampling oscilloscope. The approximately 1-μm a-Si : H was grown in a parallel plate capacitive system using a 0.02 W cm^{-3} plasma at partial pressures of silane and helium of 27 Pa with a substrate temperature of 250°C. The structure of these devices allows the use of other forms of silicon, but with the time resolution limited by the sampling oscilloscope, no clear rise time advantage was observed by using a-Si : H in place of CVD silicon on sapphire.

2. Modulators

Besides detecting and guiding light one can consider a-Si : H for modulating light. Many mechanisms can lead to the modulation of light (Willardson and Beer, 1972). There have been a number of reports using structures of the semiconductor gallium arsenide as optical modulators (Garmire, 1975). The modulation in these devices has been attributed to the linear electrooptic effect or to an electroabsorption. Typically, thermal effects are ruled out by the observed speeds of response of the devices. Because of the crystalline symmetry of silicon and the fact that it is not a direct band-gap semiconductor, the modulation effects in silicon are assumed to be much less significant than those for the III–V semiconductors. On the other hand, with a-Si : H one may fabricate some unique structures to enhance modulation effects.

Modulation depths of 50% have been demonstrated with light passing through a 1-μm layer of a-Si : H (Phelan *et al.*, 1981). This device used plasma-deposited a-Si : H sandwiched between two 25-nm gold contacts. Sapphire was used as the substrate (Fig. 5). The a-Si : H was deposited using

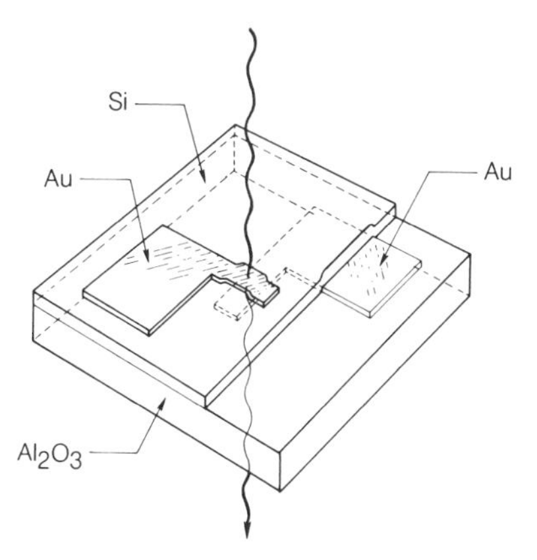

Fig. 5. Drawing of the modulator structure. [From Phelan *et al.* (1981).]

the same conditions as stated earlier for the interdigitated detectors (Phelan *et al.*, 1983). Again, electrically this device behaved like two back-to-back Schottky diodes. Optically, with the reflecting gold contacts, it was a thin interferometer. The transmission spectrum of this structure shows a periodic set of wavelengths of high transmittance (Fig. 6). An applied voltage changes the refractive index and thereby modulates the transmitted light. For wavelengths shorter than the band-gap wavelength, the change in the imaginary part of the index is dominant; this simply leads to changes in the peak transmittance. At longer wavelengths the change in the real part of the index is dominant; this leads to a shift in the wavelength of the transmission peaks.

The precise mechanism by which the a-Si:H modulates light has been questioned. Brodsky and Leary (1980), in discussing the change in reflectance of amorphous silicon with applied electric fields, gave transmittance versus temperature data supporting the view that the change could be a thermal effect. Freeman *et al.* (1980) stated that their applied powers were too low for a thermal effect and supported the view that the electroreflectance was of electronic origin. Phelan *et al.* (1981) support both views. They see changes that may be attributed to thermal effects and changes of electronic origin. A change in the band gap electronically or thermally can lead to changes in the refractive index.

Although speed of response arguments have been used to rule out thermal effects, with micrometer-thick layers thermal responses can occur in submicrosecond times or even subnanosecond times if sufficiently small filaments are involved. Thus it is difficult to rule out a thermal effect on the basis of speed only. To support the fact that thermal effects were seen under high

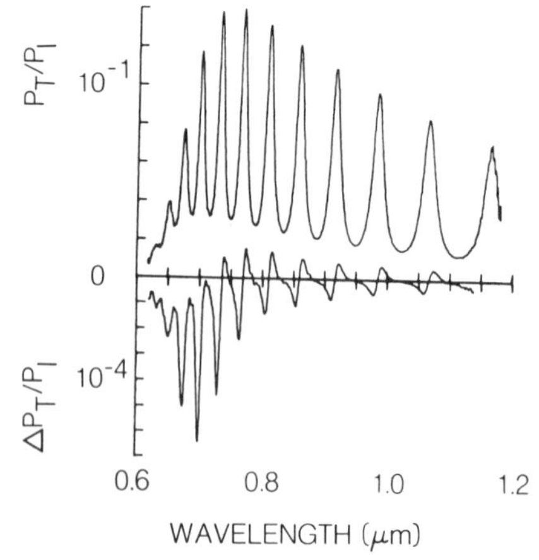

FIG. 6. Spectra of transmittance (upper curve) and change in transmittance with an applied field of 1.2×10^5 V cm^{-1} (lower curve). [From Phelan *et al.* (1981).]

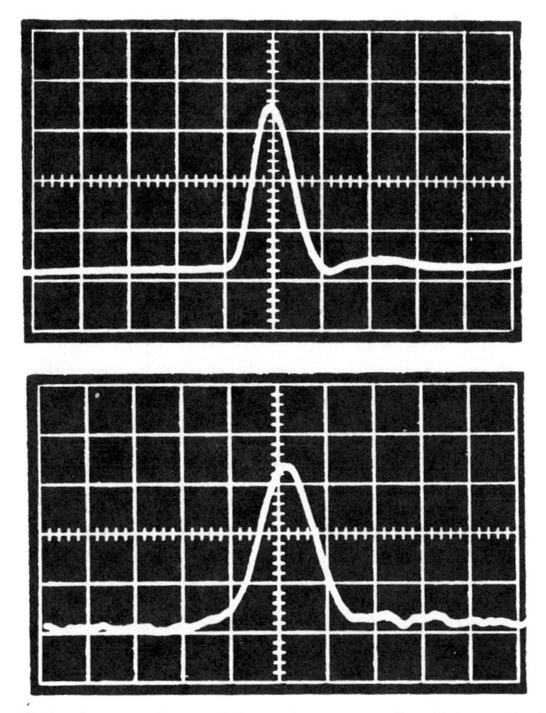

FIG. 7. The applied voltage to the modulator (upper trace) and the optical signal from the modulator (lower trace). The horizontal time scale is 0.5 nsec per major division. The peak applied voltage was 64 V, and the peak in the optical signal corresponds to a 1% depth of modulation. [From Phelan *et al.* (1981).]

applied power conditions, Phelan *et al.* (1981) studied the modulation versus the position where the light is passed through the structure. If the effect required an electric field or a change in the electronic charge, one would expect to see the modulation only if the light passed through the region where the electrical contacts overlap. With high applied powers, modulation was observed far from the overlap areas. In addition, one could detect hot filamentary regions within the overlapping areas. It was noted that in making such response maps one must vary both the optical wavelength and the modulation frequency. The wavelengths of the peaks in the transmittance for the contact area are different from the peak wavelengths for regions where there is only silicon on sapphire (adding the metal contacts not only enhances the peak-to-valley ratios of the transmittance spectrum but also shifts the wavelengths of the peaks). If one used the optimum wavelength for the contact region, he would expect to see very little modulation for another region. From the equations of thermal diffusion, as the incident spot moves away from the source of heat, the modulation frequency must be reduced to see a comparable amplitude change in the transmission. In addition, if a phase-sensitive detection system is used, there will be an electronic phase change with distance.

For low-power levels the modulation was uniform over the contact region and did not extend outside this area. When the optical wavelengths were sufficiently short to be absorbed and to generate electron–hole pairs, the decay time of the change in transmittance was dependent on whether or not there was a bias. Without a dc bias the decay time was a few nanoseconds in duration. With a bias the generated carriers were rapidly swept out of the device and subnanosecond decay times were obtained. With longer wavelength radiation the light was not absorbed and the modulation was attributed to the change in the real part of the index of refraction. At these wavelengths no bias was needed to obtain subnanosecond response times (Fig. 7). At these speeds the change in index was determined to be 4.7×10^{-4}, which was more than two orders of magnitude higher than previously reported for a-Si:H and not far from the highest values given for GaAs (Garmire, 1975). Although the Franz–Keldysh effect was suggested (Phelan *et al.*, 1981), other mechanisms may explain the observed optical modulation in a-Si:H.

III. Conclusion

The feasibility of making fast detectors and modulators using a-Si:H has been demonstrated. The most evident factor potentially affecting the device performance is the a-Si:H material itself. Materials prepared by different means exhibit different optical and electrical properties. Some devices are made by sputtering silicon in the presence of hydrogen. Others are made by using a glow discharge with silane gas. Many values of pressure and temperature are used during the deposition. The rate of deposition of the silicon, the partial pressures of added gases, and the specific geometry of the deposition system may or may not be important. Ideally, to prove the advantages and limitations of a particular material, one would make the same device out of other forms of silicon; but typically, one type of material, made in a specific manner, is reported for the device.

The particular metals and metallization processes may be very significant. For devices involving Schottky barriers the metal work functions may be important, but the process by which the metallization is created may be the dominant factor. Surface states at the silicon interface could make the particular metal insignificant. In some cases the metallization is governed by compatibility with available lithographic procedures.

The device structure is another factor affecting performance. Variations in the detailed geometry of the device can be very significant. For speed, the resistance, capacitance, and inductance of the device, in addition to its packaging, are important. If carrier transit times are a controlling factor, the ability to create close spacings between contacts can be significant. One advantage of using a-Si : H is that it can be deposited on a wide variety of substrates, thus allowing for the fabrication of unique structures. This may make it difficult to create the identical structure with another material if one desires to prove some other uniqueness property of a-Si : H.

Often reports leave out the detailed lithographic steps, the chemical etches, the plasma-etching conditions, cleaning solutions used, and temperature cycles involved. These factors, especially with regard to the reverse leakage current of $p-n$ junctions and Schottky barriers, may be very significant, but their significance is not known.

There are many experiments to be performed before we can clearly state the advantages and limitations of hydrogenated amorphous silicon for optoelectronic devices. Each laboratory has its unique means for creating the a-Si : H; this leaves some question about the equivalence of the materials. One test for equivalence of materials will be the duplication of device performance. As devices are reported that are sufficiently exciting, other groups will duplicate the results, and we shall develop a better understanding of the material and device fabrication parameters that are important.

REFERENCES

Auston, D. H., Johnson, A. M., Smith, P. R., and Bean, J. C. (1980). *Appl. Phys. Lett.* **25,** 509.

Brodsky, M. H., and Leary, P. A. (1980). *J. Non-Cryst. Solids* **35/36,** 487.

Carlson, D. E., and Wronski, C. R. (1976). *Appl. Phys. Lett.* **28,** 671.

Freeman, E. C., Anderson, D. A., and Paul, W. (1980). *Phys. Rev.* **21,** 472.

Garmire, E. (1975). *In* "Integrated Optics" (T. Tamir, ed.), p. 243. Springer-Verlag, Berlin and New York.

Glass, A. M., Liao, P. F., Johnson, A. M., Humphrey, L. M., Olson, D. H., Stern, M. B., and Lemons, R. A. (1983). *Conf. Lasers Electro-Opt., Baltimore, 1983.* Paper THL1.

Johnson, A. M., Auston, D. H., Smith, P. R., Bean, J. C., Harbison, J. P., and Adams, A. C. (1981). *Phys. Rev. B* **23,** 6816.

Loveland, R. J., Spear, W. E., and Al-Sharbaty, A. (1973/1974). *J. Non-Cryst. Solids* **13,** 55.

Olivier, M., and Peuzin, J.-C. (1978). *Appl. Phys. Lett.* **32,** 386.

Phelan, R. J., Jr., Larson, D. R., and Werner, P. E. (1981). *Appl. Phys. Lett.* **38,** 596.

Phelan, R. J., Jr., Larson, D. R., Frederick, N. V., and Franzen, D. L. (1983). *Proc. Soc. Photo-Opt. Instrum. Eng.* **439,** 207.

Willardson, R. K., and Beer, A. C., eds. 1972. "Semiconductors and Semimetals," Vol. 9. Academic Press, New York.

Yumoto, Y., Yajima, H., Seki, Y., Shimada, J., and Nakajima, M. (1982). *Appl. Phys. Lett.* **40,** 632.

CHAPTER 14

Hybrid Structures

Jacques I. Pankove

RCA/DAVID SARNOFF RESEARCH LABORATORIES
PRINCETON, NEW JERSEY

I. Introduction

There are applications that involve crystalline semiconductors in which the properties of a-Si:H can be used to great advantage in improving the performance of already well-known crystalline devices. In addition, new structures are possible that would not be readily achieved with crystalline silicon (c-Si).

The properties specific to a-Si:H that we propose to exploit in hybrid structures are the abundance of hydrogen, the wide energy band gap, and the high dielectric constant or refractive index.

II. Passivation of Crystalline Silicon

The surface of c-Si terminates in dangling bonds that absorb atoms from the ambient, usually oxygen, forming a thin oxide layer. However, it is well known that the natural oxide is not sufficient to passivate all the dangling bonds, since interfacial states are still present (Deal *et al.*, 1967). Dangling bonds at the surface of semiconductors are held responsible for generation–recombination (g–r) centers that have detrimental effects on device performance. Thus minority carriers disappear by surface recombination at such centers and become unavailable for signal processing. In the presence of an electric field, however, these centers generate unwanted carriers that contribute noise and leakage current. In fact, the reverse-bias characteristics of a $p-n$ junction in some cases are dominated by the leakage current occurring at the intersection of the $p-n$ junction with the crystal surface. Devices having $p-n$ junctions with large perimeter-to-area ratio are particu-

ISBN 0-12-752150-X

larly susceptible to surface contamination and to the presence of dangling bonds. Hence much effort has been devoted to the passivation of $p-n$ junctions. For silicon devices the most widely used passivation technique is the growth of a thermal oxide (SiO_2) over the $p-n$ junction.

Pankove *et al.* (1978) have demonstrated that it is possible to monitor the hydrogenation of dangling bonds by measuring the reverse-bias $I-V$ characteristics of a $p-n$ junction after various stages of hydrogenation and dehydrogenation. For this demonstration, $p-n$ junctions were fabricated by diffusing phosphorus in a p^+ substrate. Each junction had a cross section of 0.03 cm² and a surface periphery of 1.2 cm. The thermal oxide was etched off and the Si surface was cleaned by standard cleaning procedures (SC1, SC2)† and baked dry at 200°C in N_2. The $I-V$ reverse current characteristics were then measured (top curve in Fig. 1). Note that the diode was quite leaky, indicating the presence of a large number of dangling bonds.

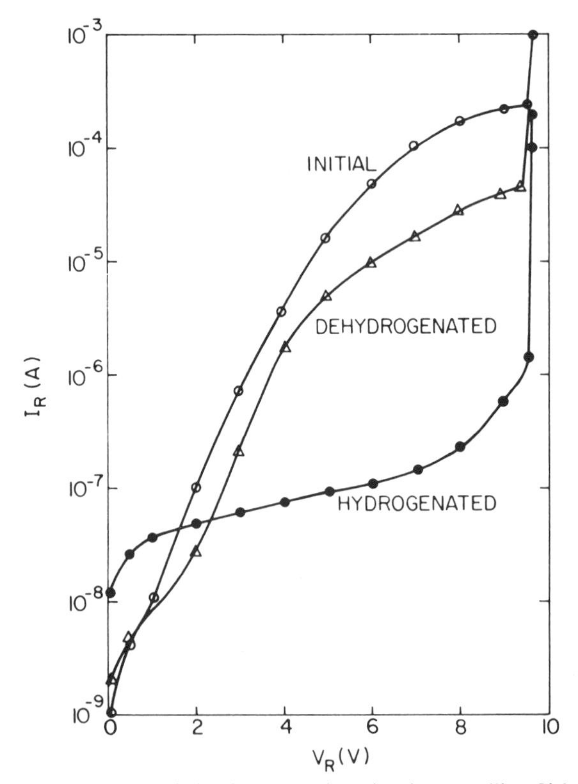

FIG. 1. Reverse $I-V$ characteristics for a $p-n$ junction in crystalline Si before and after hydrogenation and after dehydrogenation. [From Pankove *et al.* (1978).]

† SC1: $NH_4OH:H_2O_2:H_2O = 1:1:5$; SC2: $HCl:H_2O_2:H_2O = 1:1:5$.

Next, the $p-n$ junctions were hydrogenated in the system shown in Fig. 2. For this purpose the chamber is first evacuated and the sample is heated to 550°C, then hydrogen is admitted to a pressure of ~200 mTorr. A glow discharge is produced by a capacitively coupled high-frequency oscillator. The plasma generates atomic hydrogen. The sample is gradually cooled to room temperature with glow discharge on throughout the entire period of temperature drop. After this hydrogenation process, the reverse-bias $I-V$ characteristic of a typical hydrogenated $p-n$ junction (lower curve in Fig. 1) shows a dramatic drop in leakage current above 1.5 V and a sharpening of the breakdown. The $p-n$ junctions were then dehydrogenated by annealing at 550°C in vacuum and subjected to hydrogen floride etching and SC1 and SC2 cleaning. After this dehydrogenation, the $I-V$ characteristic (Fig. 1) reverts to a softer breakdown and high leakage. The slightly lower leakage is probably due to incomplete dehydrogenation.

Exposure of virgin diodes to molecular hydrogen (H_2) lowers the leakage current by a factor of five. Subsequent exposure to atomic hydrogen gives a further decrease in leakage current by two orders of magnitude.

Passivation by atomic hydrogen leaves the surface unprotected and therefore not acceptably passivated. Hence coating the devices with a-Si:H should provide not only the atomic hydrogen needed to passivate the dangling bonds, but also a layer of H-rich material that is highly insulating in the dark. This concept was successfully tested by Pankove and Tarng (1979). An array of $p-n$ junction diodes was made as described earlier. Part of the array was coated with a thermal oxide, while the rest of the array was coated with a layer of a-Si:H (0.36 μm thick) obtained by the glow-discharge decomposition of silane at 350°C. Standard photolithographic techniques were used to open portions of the layer for subsequent metallization. The inset of Fig. 3 shows a cross section of the structure. Several diodes in each

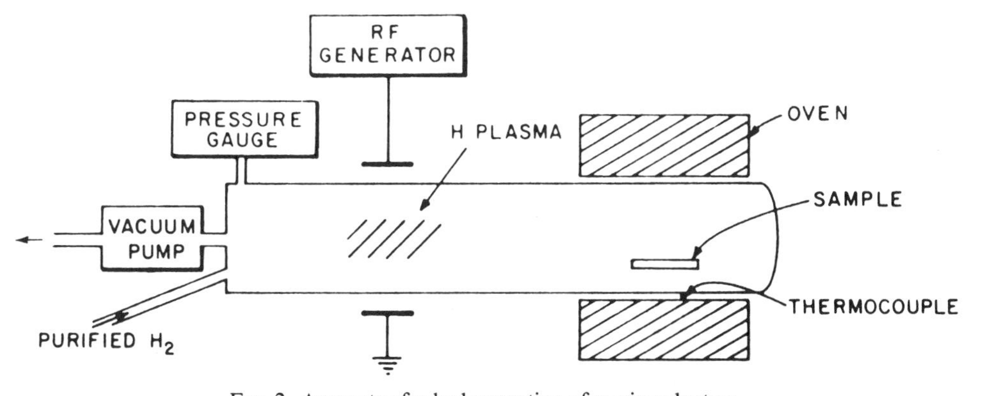

FIG. 2. Apparatus for hydrogenation of semiconductors.

group were tested, showing identical characteristics within each group.

The data of Fig. 3 show the reverse-bias $I-V$ characteristics for two representative $p-n$ junctions. The one labeled I_{ox} was passivated with thermal oxide, the one labeled $I_{a\text{-Si:H}}$ was passivated with a-Si:H. It is striking that the leakage current of the a-Si:H-passivated junction is two orders of magnitude lower than that of the SiO_2-passivated diode.

Note that the $I-V$ characteristics appear exponential over most of the bias

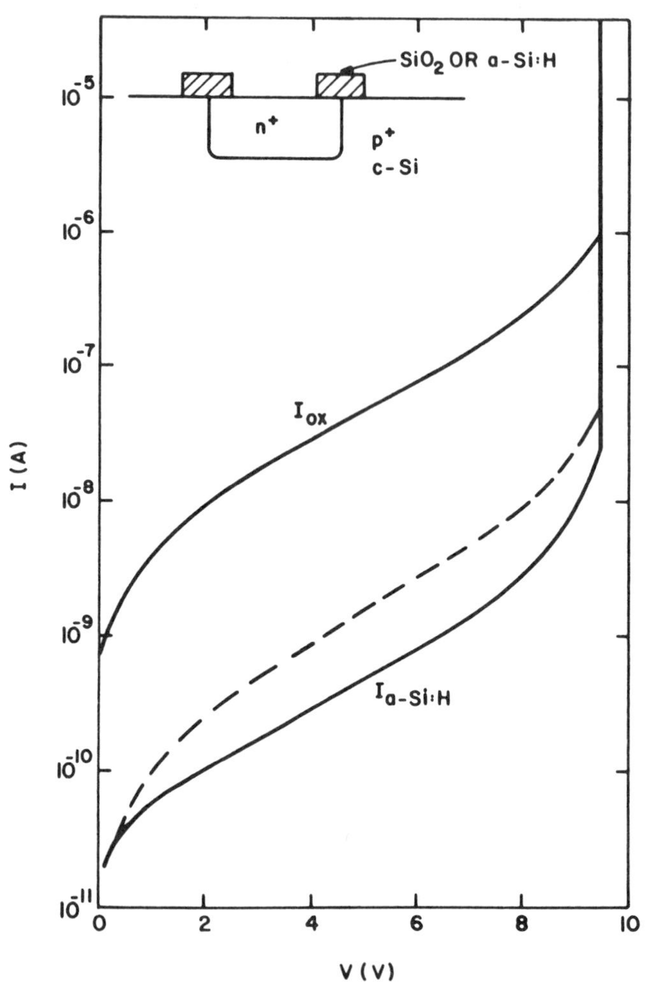

FIG. 3. $I-V$ characteristics of diodes under reverse bias: I_{ox}, SiO_2-passivated; $I_{a\text{-Si:H}}$, a-Si:H-passivated: The dashed line shows $I_{a\text{-Si:H}}$ after heating at 500°C for $\frac{1}{2}$ hr. The inset shows a cross-sectional diagram of the diode. [From Pankove and Tarng (1979).]

range. However, because of the many complex charge-transfer processes that can occur simultaneously when a reverse bias is applied, it is difficult to predict a functional dependence for the current. If we assume that the leakage is dominated by tunneling through generation–recombination centers, then Fig. 3 suggests that the number of such centers is much lower at the a-Si : H-passivated surface than at the SiO_2-passivated surface. It should be pointed out that the ohmic leakage through the two passivating layers, a junction-shunting current, is negligible, since the resistivities of the two layers are very high ($3 \times 10^{10}\ \Omega$ cm for a similar film of a-Si : H and $> 10^{14}\ \Omega$ cm for SiO_2). Furthermore, because the energy gap of a-Si : H (~ 1.7 eV) is much larger than that of c-Si (1.1 eV) (Zanzucchi *et al.* 1977, the thermal generation of carriers across the gap is much lower in a-Si : H. A further argument against the participation of a-Si : H in current transport is that, as shown in Fig. 4, the larger energy gap of a-Si : H induces a barrier between the crystalline and amorphous materials. This barrier blocks electrons from

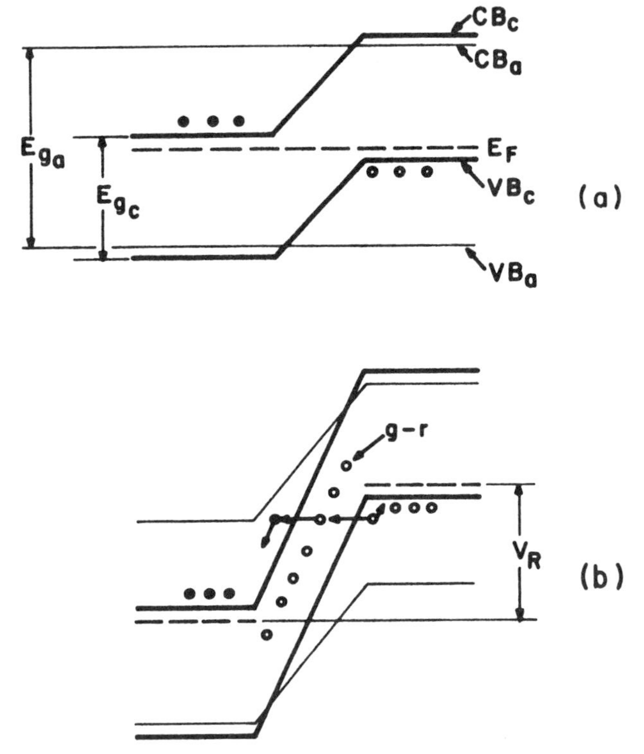

FIG. 4. Superposition of energy-band structures of a-Si : H and c-Si along the interface: (a) without bias and (b) with an applied reverse bias V_R.

the n-type region from reaching the conduction band of the a-Si : H [CB_a and a similar barrier blocks the holes from the p-type c-Si from reaching the valence band of a-Si : H (VB_a)].

At the higher-voltage range (above ~8 V), avalanche breakdown causes the current to increase more rapidly. At the lower-voltage range (below about 2 V), the main contribution to the current is the thermal generation of carriers through generation–recombination centers (two orders of magnitude less numerous on the a-Si : H-passivated surface than on the SiO_2-passivated surface). Note that diffusion of carriers generated thermally across the gap of c-Si, a process that should be independent of the passivant and independent of the applied bias above a few kT, is a negligible component of the reverse-bias current for these devices.

An important property of a passivating solid coating is that it should be stable in spite of subsequent thermal treatments, such as the sintering of metal connections. The problem of thermal stability is especially worrisome in the case of a-Si : H that is known to evolve hydrogen above about 350°C (Pankove and Carlson, 1977). Heating the sample for 30 min at 500°C caused the leakage current to increase by a factor of about three, as shown by the dashed line in Fig. 3. Note that the current for the heated sample is still about one order of magnitude lower than that of the oxide-passivated $p-n$ junction, which is well known for its thermal stability.

Another consideration is the bias–temperature (BT) stability i.e., the stability of the passivant when it is subjected simultaneously to an applied bias and to an elevated temperature. This question has been explored experimentally by Tarng and Pankove (1979). The diodes described earlier were subjected to BT stress in an SD-30 Ransco environmental chamber at $V_R = 5$ V and $T = 150$, 175, and 200°C for 126, 40, and 110 hr, respectively. Heat treatment was done in a closed-tube furnace in an N_2 ambient at 50°C increments between 300 and 500°C, for 30 min at each temperature. The (dark) $I_R - V_R$ characteristics were measured after each treatment using a Keithley 414A picoammeter.

A typical $I_R - V_R$ characteristic of a-Si:H passivated junction ($I_{a\text{-Si}}$) is shown in Fig. 5 together with that of a similar junction passivated with thermal oxide (I_{ox}). The effect of BT stress at 5 V and 200°C is shown by curve (1) in Fig. 5. Then the specimen was subjected to annealing treatments for 30 min each at successively higher temperatures up to 650°C as also shown in Fig. 5.

The leakage current I_R of a reverse-biased junction can be expressed as a sum of several components that have been discussed in detail by Tarng and Pankove (1979). Under the present experimental conditions, the major components are the surface generation–recombination current I_{gs} and the tunneling current I_t.

At low V_R, the tunneling current is negligible. The difference between

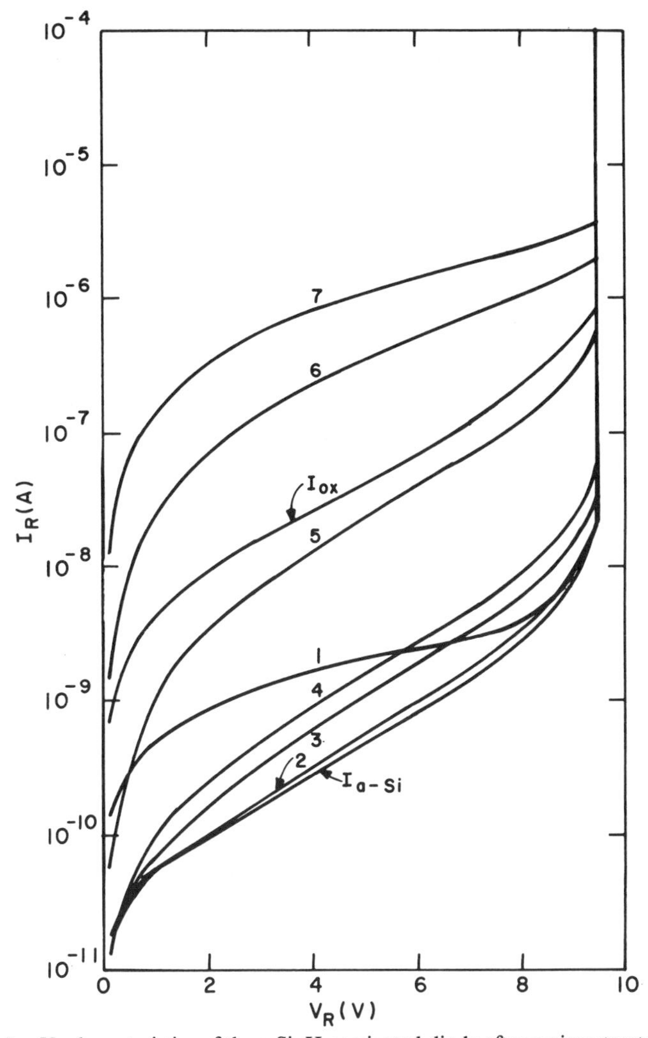

FIG. 5. $I_R - V_R$ characteristics of the a-Si : H-passivated diode after various treatments. $I_{a\text{-}Si}$: as deposited; (1) after BT at 200°C, 5 V; (2) after HT at 300°C; (3) after HT at 350°C; (4) after HT at 500°C; (5) after HT at 550°C; (6) after HT at 600°C; (7) after HT at 650°C; I_{ox}: thermal SiO_2 passivated diode. [From M. L. Tarng and J. I. Pankove, Passivation of *p–n* junction in crystalline silicon by amorphous silicon, *IEEE Trans. Electron Devices,* © 1979.]

$I_{a\text{-}Si}(0.1\text{ V}) = 2.0 \times 10^{-11}$ A and $I_{ox}(0.1\text{ V}) = 7.5 \times 10^{-10}$ A is due mainly to a difference in I_{gs}. The fortyfold (or more) reduction in I_{gs} must be largely due to a reduction in deep surface states at the c-Si and a-Si : H interface. At $V_R \simeq 1$ V, both $I_{a\text{-}Si}$ and I_{ox} increase exponentially with V_R, which indicates that significant tunneling is occurring. Figure 6 shows the corresponding I_R (5 V) versus $1/T$ plots. At room temperature, the apparent activation energy

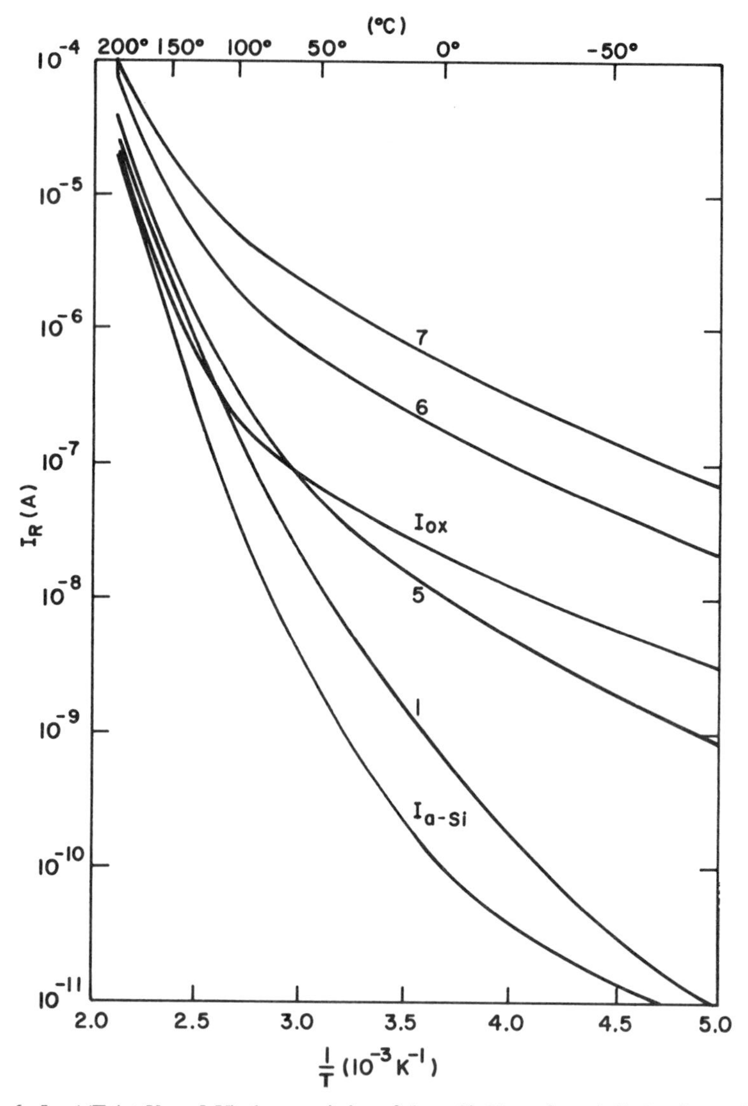

FIG. 6. $I_R - 1/T$ (at $V_R = 5$ V) characteristics of the a-Si:H-passivated diode after various treatments. Notations are similar to those of Fig. 5. [From M. L. Tarng and J. I. Pankove, Passivation of *p–n* junction in crystalline silicon by amorphous silicon, *IEEE Trans. Electron Devices*, © 1979.]

of I_{ox} is about 0.17 eV, which indicates that thermally assisted tunneling is dominated by transitions through shallow states. This is consistent with the fact that the two maxima in the surface-state density distribution of $Si-SiO_2$ interface occur at about 0.15 eV from the band edge (Goetzberger *et al.*,

1968). The activation energy for the $I_{a\text{-Si}}$ at room temperature is 0.41 eV. This suggests that the shallow surface states have been largely eliminated by hydrogenating the c-Si surface. The log $I_{a\text{-Si}}$ versus $1/T$ curve becomes linear at $T \gtrsim 150°C$; here $E_a = 1.05$ eV, which is close to the band-gap value of c-Si. For I_{ox}, however, no such linearity is reached up to 200°C, beyond which $E_a = 0.95$ eV. This is another indication that the I_{gs} and I_t components are much larger in I_{ox} than in $I_{a\text{-Si}}$. Therefore one can conclude that the a-Si : H passivation has reduced both the deep surface state and the shallow surface state densities.

Bias–temperature stress at 150°C and 5 V caused a sharp increase in leakage current for the first few hours, then saturation over 40 hr of stress. Bias–temperature stress at 200°C and 5 V caused a considerable reduction in I_R, whereas with thermal oxide passivation the leakage keeps increasing by several orders of magnitude, presumably due to an increased density of interfacial states. Here a-Si : H may act as a source of atomic hydrogen that diffuses toward the junction and helps maintain the low density of interfacial states. In fact, heat treatment at 300°C after the BT stress [curve (2) of Fig. 5] caused I_R to return almost completely to the prestress values. Subsequent higher temperature annealing, up to 500°C, increased the leakage by a factor of less than three, as shown by curves (3) and (4) of Fig. 5.

When the annealing temperature was raised to 550°C, where substantial dehydrogenation occurs, a much more drastic change in I_R was observed [curve (5) of Fig. 5]. As the annealing temperature was raised to 600 and 650°C, further large increases in I_R were observed [curves (6) and (7) in Fig. 5]. Most of this increase (in the form of tunneling current) was attributed to a higher density of surface states resulting from thermal dehydrogenation starting at 550°C.

Lithium-drifted silicon x-ray detectors designed to measure trace amounts of plutonium in soil samples have been fabricated by Walton *et al.* (1984). These diodes have been passivated with sputtered a-Si : H, with a resulting lowering of leakage current at the operating low temperatures.

To summarize the preceding section, compared with thermal oxide, a-Si : H is a superior passivant for c-Si junctions. The reverse leakage current is about two orders of magnitude lower than that of typical thermal–oxide-passivated junctions. The low leakage is achieved by producing a low interface-state density through hydrogenation of the c-Si surface during the a-Si : H deposition. Furthermore, the hydrogen present in the bulk of the a-Si : H appears to act as a hydrogen reservoir for rehydrogenation of the c-Si and a-Si : H interface. Such a rehydrogenation process limits the amount of leakage increase during BT stress and therefore provides the devices with an excellent operational life expectancy. Since thermal treatments up to 500°C (for 30 min) do not significantly affect the leakage, the a-Si : H passivation scheme is compatible with normal device fabrication processes, which may require metallization and alloying steps at ~450°C (~15 min) following passivation.

Let us point out that the atomic hydrogen generated during the glow-discharge decomposition of silane can passivate not only the dangling bonds at the surface of c-Si but also those lining grain boundaries (Seager *et al,*. 1980; Redfield, 1981; Robinson and D'Aiello, 1981) or slip planes such as those that occur in silicon on sapphire near the interface (Jastrzebski *et al.,* 1982). Atomic hydrogen can also affect point defects such as deep impurities in c-Si (Pearton and Tavendale, 1982) or even shallow acceptors—compensating the residual Si dangling bond adjacent to a substitutional B atom (Pankove *et al.,* 1983).

III. Heterojunctions

A heterojunction is a $p-n$ junction in which the two regions have different energy band gaps. Heterojunctions are of great practical interest, since they should allow efficient injection of minority carriers into the narrower gap material. Figure 7 illustrates a forward-biased ideal heterojunction between p-type c-Si and n-type a-Si:H. Electrons from a-Si:H are readily injected into c-Si, whereas the holes of p-type c-Si are blocked by a potential barrier.

Such efficient minority-carrier injectors have been proposed as emitters for high-power transistors (Kroemer, 1957) and demonstrated using an n-type oxygen-rich polycrystalline Si emitter on a p-type c-Si base (Oh-uchi *et al.,* 1979). The main problem expected to occur at heterojunctions between dissimilar materials is that associated with interfacial states that may either pin the Fermi level or act as generation-recombination centers. However, in the case of a-Si:H the abundance of atomic hydrogen should help eliminate the interfacial states.

Fuhs *et al.* (1974) studied a heterojunction between undoped a-Si:H and p-type c-Si. They measured the photovoltage generated by photons absorbed in the space-charge regions where the built-in fields separate the electron-hole pairs. By using light of different energies, and therefore different penetration depths, the locations of the space charges could be

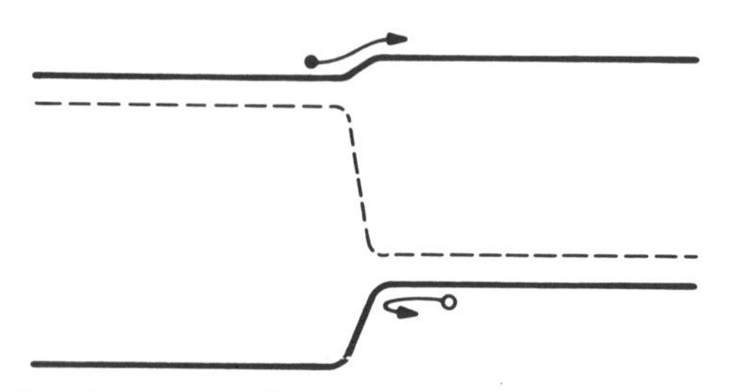

FIG. 7. Diagram of a forward-biased heterojunction.

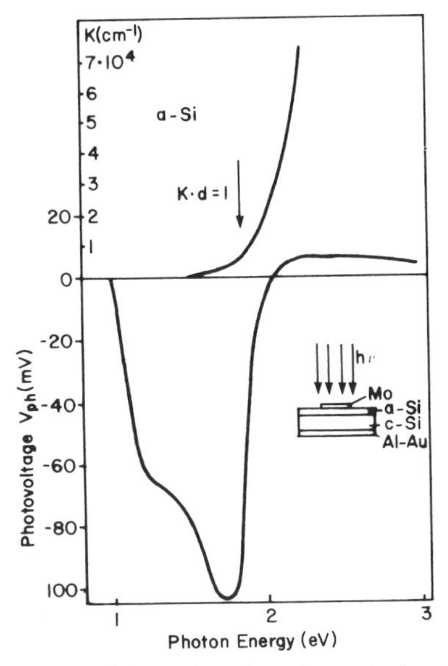

FIG. 8. Photovoltaic response of heterojunction shown in inset and absorption edge of a-Si:H. [After Fuhs *et al.* (1974).]

detected. The photovoltaic spectrum obtained by shining the light through the wider-gap a-Si:H is shown in Fig. 8. Below 2.0 eV, most of the light is absorbed in the c-Si. The negative polarity at the Mo contact indicates that there is a hole-depletion region near the interface in the c-Si (photoelectrons move toward the interface). Above 1.8 eV, the photovoltage drops rapidly because few photons reach the c-Si, since most of the absorption occurs in the a-Si:H. The photovoltage polarity reverses because of the opposite band bending near the Mo electrode. The saturation of the photovoltage at high levels of excitation defines the maximum band bending as 0.3 eV, leading to the conclusion that the Fermi level in a-Si:H is near midgap.

Fuhs *et al.* measured the $I-V$ characteristics at various temperatures finding some rectification. However, the reverse characteristic did not saturate, possibly indicating the presence of interfacial states.

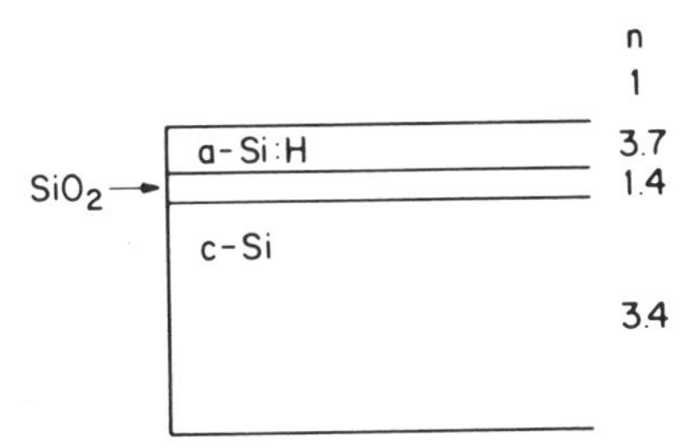

FIG. 9. Structure of optical waveguide.

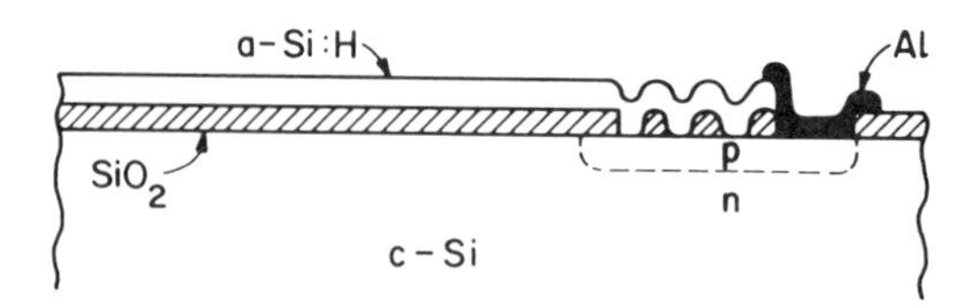

FIG. 10. Coupling of optical waveguide to a photodiode.

IV. Optical Waveguides

Because of its high refractive index $n \approx 3.7$ (Berger *et al.*, 1981), a-Si:H lends itself as a conveyor of optical information to light sensitive detectors incorporated in integrated circuits. In this application as an optical waveguide, an optical signal can be transmitted to various circuit components without cross-talk with signals in electrical interconnections.

A typical structure, shown in Fig. 9, would consist of a-Si:H deposited over SiO_2 that has a much lower optical refractive index, $n = 1.45$. The large steps in refractive index at the SiO_2 and the air interfaces confine the radiation to the a-Si:H layer that therefore acts as a waveguide. The coupling between the waveguide and a device (e.g., the detector) could be via a grating etched into the SiO_2, as shown in Fig. 10. The light-sensitive device may be a diode or a photoFET or the base of a phototransistor. The light source could be the end of an optical fiber or an LED. In the case of a GaAs integrated circuit, both the light source and the detector could be built into the same chip, while the a-Si:H overlay would provide the waveguiding. Although propagation losses of about 5 db cm^{-1} measured by J. Hammer (personal communications, 1979) are huge by optical fiber standards, they are no impediment to transmission over the dimensions of an IC chip.

REFERENCES

Berger, J. M., Ferraton, J. P., Yous, B., and Donnadieu, A. (1981). *Thin Solid Films* **86,** 337.
Deal, B. E., Sklar, M., Grove, A. S., and Snow, E. H. (1967). *J. Electrochem. Soc.* **114,** 266.
Fuhs, W., Niemann, K., and Stuke, J. (1974). *AIP Conf. Proc.* **20,** 345.
Goetzberger, A., Herne, V., and Nicollian, E. H. (1968). *Appl. Phys. Lett.* **12,** 95.
Jastrzebski, L., Lagowski, J., Cullen, G. W., and Pankove, J. I. (1982). *Appl. Phys. Lett.* **40,** 713.
Kroemer, H. (1957). *Proc. IRE* **45,** 1535.
Oh-uchi, N., Hayashi, H., Yamoto, H., and Matsushita, T. (1979). *IEDM IEEE* **79**(CH1504), 522.
Pankove, J. I., and Carlson, D. E. (1977). *Appl. Phys. Lett.* **31,** 450.
Pankove, J. I., and Tarng, M. L. (1979). *Appl. Phys. Lett.* **34,** 156.
Pankove, J. I., Lampert, M. A., and Tarng, M. L. (1978). *Appl. Phys. Lett.* **32,** 439.
Pankove, J. I., Carlson, D. E., Berkeyheiser, J. E., and Wance, R. O. (1983). *Phys. Rev. Lett.* **51,** 2224.

Pearton, S. J., and Tavendale, A. J. (1982). *Phys. Rev. B* **26,** 7105.
Redfield, D. (1981). *Proc. IEEE Photovoltaic Spec. Conf., San Diego, 15th, 1981* CH1644, 1179.
Robinson, P. H., and D'Aiello, R. V. (1981). *Appl. Phys. Lett.* **39,** 63.
Seager, C. H., Ginley, D. S., and Zook, J. D. (1980). *Appl. Phys. Lett.* **36,** 831.
Tarng, M. L., and Pankove, J. I. (1979). *IEEE Trans. Electron Devices* **ED-26,** 1728.
Walton, J. T., Pehl, R. H., Wong, Y. K., and Cork, C. P. (1984). *IEEE Trans. Nucl. Sci.* **NS-31,** 331.
Zanzucchi, P. J., Wronski, C. R., and Carlson, D. E. (1977). *J. Appl. Phys.* **48,** 5227.

CHAPTER 15

Electronic Switching in Amorphous Silicon Junction Devices

P. G. LeComber

CARNEGIE LABORATORY OF PHYSICS
UNIVERSITY OF DUNDEE
DUNDEE, SCOTLAND

A. E. Owen

DEPARTMENT OF ELECTRICAL ENGINEERING
UNIVERSITY OF EDINBURGH
EDINBURGH, SCOTLAND

W. E. Spear

CARNEGIE LABORATORY OF PHYSICS
UNIVERSITY OF DUNDEE
DUNDEE, SCOTLAND

J. Hajto

DEPARTMENT OF ELECTRICAL ENGINEERING
UNIVERSITY OF EDINBURGH
EDINBURGH, SCOTLAND

W. K. Choi

DEPARTMENT OF ELECTRICAL ENGINEERING
UNIVERSITY OF EDINBURGH
EDINBURGH, SCOTLAND

I. Introduction

Electronic switches are solid-state devices that can be changed from a nonconducting OFF state to a conducting ON state by an appropriate electrical signal. The importance of such devices in the development of solid-state digital electronics has been enormous and is likely to remain so in the foreseeable future, especially with the increasing demand for memory elements.

ISBN 0-12-752150-X

The present article is concerned with two-terminal switching devices. Generally, these have one or the other of the two types of current–voltage ($I-V$) characteristics shown schematically in Fig. 1a,b. In Fig. 1a the device switches from its OFF to its ON state at a critical threshold voltage V_{Th}, but if the ON state conditions fall below a critical holding point (I_h, V_h), the device reverts spontaneously to its OFF state. Devices of this kind are called *threshold* switches; they are nonpermanent, or "volatile," and they always revert to the OFF state in the absence of an appropriate bias. In Fig. 1b there is again a critical switching voltage for the OFF to ON transition, but both ON- and OFF-state characteristics extrapolate through the $I-V$ origin. Devices of this kind are therefore permanent, or nonvolatile, and they are called *memory* switches. Memory devices can remain in either the ON or OFF state more or less indefinitely, whether or not a bias is applied and the ON to OFF transition is usually triggered by a current pulse.

By the early 1970s many examples of threshold and memory switching had been reported in *homogeneous* thin films of a variety of amorphous materials, including simple oxides, transition-metal oxides, elemental selenium, and boron. By far the most important materials, however, were the chalcogenide glasses in which, depending on composition, reproducible characteristics of the kind illustrated in Fig. 1a (threshold switching) or Fig. 1b (memory switching) may be obtained (Ovshinsky, 1968; Owen and Robertson, 1973). A substantial specialist literature on electrical switching

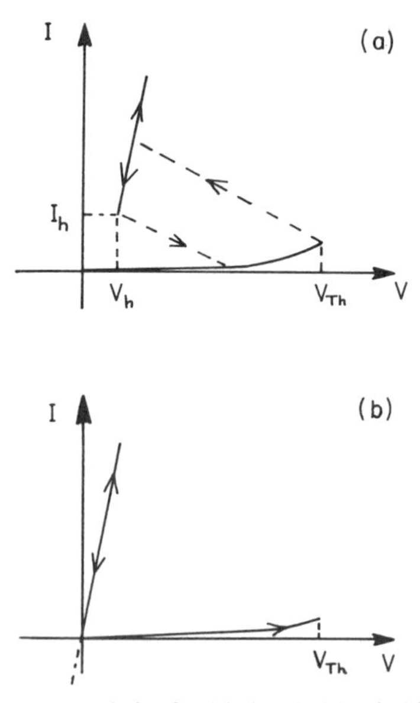

FIG. 1. Current–voltage characteristics for (a) threshold switching and (b) memory switching.

in the chalcogenide glasses has developed, and although there are still some controversial features, generally accepted models of at least a semiquantitative kind are now reasonably well established for both types of switching (Adler *et al.*, 1978; Owen *et al.*, 1979).

This article is concerned with recent studies of a rather different switching behavior in *heterogeneous* structures of amorphous hydrogenated silicon, which evolved from a collaborative project between the authors' groups at the Universities of Dundee and Edinburgh in the United Kingdom. A preliminary account of the work has already been published (Owen *et al.*, 1982), and in the following sections a more detailed description of the experimental observations will be presented.

Before proceeding, however, it is worth noting that, by contrast with the chalcogenide glasses and despite the almost unprecedented growth in research and development on amorphous silicon (a-Si) since the mid-1970s, very little has been reported on electrical switching in the latter material. To put the recent observations in context it is pertinent therefore to review briefly the relatively few previous reports of switching in a-Si.

II. Previous Work on Electrical Switching in Amorphous Silicon

Some cursory observations of threshold switching in homogeneous films of a-Si, with very tentative evidence for memory behavior, were described by Feldman and Moorjani (1970) and Moorjani and Feldman (1970) contemporaneously with some of the early literature switching in chalcogenide glasses. More detailed experiments on the same structures were reported later (Feldman and Charles, 1974; Charles and Feldman, 1975). The authors studied vacuum-evaporated films of a-Si in the range 0.3–2.0 μm thick, fitted with titanium electrodes. They also, incidentally, made similar observations on evaporated films of germanium, boron, and boron plus carbon. As threshold switches these a-Si structures had threshold voltages V_{Th} of 5–10 V, OFF resistances in the range 1–30 kΩ, and an ON resistance of about 100 Ω. In common with the chalcogenide glasses, there was a delay time before switching of 20–50 μsec or more (at room temperature) and the actual switching time was at least several microseconds. Feldman *et al.* did not, however, report any initial "forming" process, unlike the situation in chalcogenide glass switches (see also Part V). As already noted, there was some tentative indication of memory switching, but that feature was apparently not substantiated. Feldman and Charles (1974) interpreted the results in terms of a simple and qualitative electrothermal model involving the formation of a conducting filament; a similar, more quantitative model has been developed for switching in chalcogenide glasses (Owen *et al.*, 1979).

The work of Feldman and his colleagues, which originated in the early

1970s, seems to be the only investigation of switching in a-Si until the later studies of Dey and Fong (1977, 1979) and Dey (1980). These authors report results very similar to those of Feldman *et al.* They studied thin films of a-Si in the range 0.3–1.5 μm thick, deposited by electron-beam heating in a vacuum evaporator. Titanium contacts were again used, either in the form of evaporated films or as probes. Dey and Fong reported only threshold switching, with $I-V$ characteristics similar to those in Fig. 1a; they did not mention any evidence for memory behavior. In contrast to Feldman *et al.*, however, Dey and Fong did observe forming effects; that is, the initial threshold voltage was relatively large but it decreased to a more or less constant value after a number of switching cycles. In Dey and Fong's devices the threshold voltage varied systematically from about 6 V for the thinner films ($\simeq$0.3 μm) to about 9 V for the thicker films ($\simeq$1.2 μm). The delay time before switching was in the range 2–60 μsec, varying in a systematic way with pulse height, pulse duration, and repetition rate, again in a manner very similar to switching in calcogenide glasses (Adler *et al.*, 1978; Owen *et al.*, 1979). Dey and Fong also interpreted their results in terms of a simple one-dimensional electrothermal model, but one developed a little more quantitatively than that by Feldman *et al.*

It should be noted that both Feldman *et al.* and Dey and Fong used a-Si films deposited by vacuum evaporation. This probably accounts for the relatively low OFF-state resistances which they both found (100–100 kΩ). It is now well established that vacuum-evaporated a-Si is a very different material from the hydrogenated form of a-Si obtained, for example, by the carefully controlled glow-discharge decomposition of silane (e.g., Spear, 1977).

Three papers concerned specifically with switching in hydrogenated amorphous silicon by Gabriel and Adler (1982), den Boer (1982), and Owen *et al.* (1982) appeared almost concurrently early in 1982, each reporting very different effects observed in different a-Si structures. Our own work (Owen *et al.*, 1982), including recent results, is described in detail in the following sections.

Den Boer studied n^+-i-n^+ structures of a-Si prepared by the glow-discharge decomposition of SiH_4 ("i" stands for "intrinsic" or undoped material). The n^+ layers were 50 nm thick and prepared by adding 1% PH_3 to the SiH_4 gas flow; the i layer in different devices ranged in thickness from 2.5 to 5 μm. Den Boer found that the n^+-i-n^+ devices functioned as threshold switching devices with nonpolar characteristics similar to those sketched in Fig. 1a. For the *first* switching cycle the threshold voltage was in the range 40–100 V, but for *all* subsequent operations it was only 10–35 V, depending on the i layer thickness (as the i layer thickness increased the threshold voltage also increased). The OFF-state resistance of the n^+-i-n^+ switches was about 1 MΩ, and the ON-state resistance about 1 kΩ. There was again an observable delay time before switching, ranging from a few microseconds when the applied voltage was about 8 V greater than V_{Th} to about a millisecond for voltages within 1 V of V_{Th}. The n^+-i-n^+ threshold switches could be cycled through at least 10^9 stable switching operations. Den Boer also compared structures with chromium or a combination of chromium and n^+ contacts (i.e., $Cr-n^+-i-Cr$ and $Cr-i-Cr$). The former had rectifying characteristics while the latter switched but were very unstable.

Gabriel and Adler (1982) prepared their a-Si films by sputtering from a polycrystalline silicon target in an argon–hydrogen plasma. In all cases their devices were notionally homogeneous thin films of intrinsic a-Si:H with molybdenum contacts. The samples were fabricated under a wide range of deposition conditions in two sputtering systems, and although results from some of the devices were rendered rather doubtful because of contamination problems, in no case did Gabriel and Adler observe any evidence of reversible switching. They concluded that, in contrast to the chalcogenide glasses, a-Si does not have the electronic and structural properties required for reversible switching.

III. Device Structure and Fabrication

We turn now to the work on electronic switching carried out in the authors' laboratories. Although a number of different a-Si multilayer structures have been investigated, all the results discussed in the following refer to p^+-n-i devices deposited in this sequence by the glow-discharge technique, with gas-phase doping. Stainless steel substrates were generally used and the total thickness of the deposited a-Si layers was between 0.5 and 1.0 μm. After completion of the a-Si deposition a series of gold (Au), aluminum (Al), or nichrome (NiCr) dots (up to approximately 1 mm in diameter) was evaporated onto the surface of the samples, and the top contact was completed either by a probe or by a thin wire attached to the metal dots with conducting silver paste.

IV. Static Current–Voltage Characteristics of Virgin Devices

Typical current–voltage ($I-V$) characteristics for a freshly prepared (unswitched) device are illustrated in Fig. 2a in both the forward and reverse directions. (The forward direction is defined such that the substrate, and hence the p^+ region, is positively biased.) It must be emphasized here that these measurements were taken "by hand," point by point, in a manner that required a few seconds for each point to be measured. (The significance of

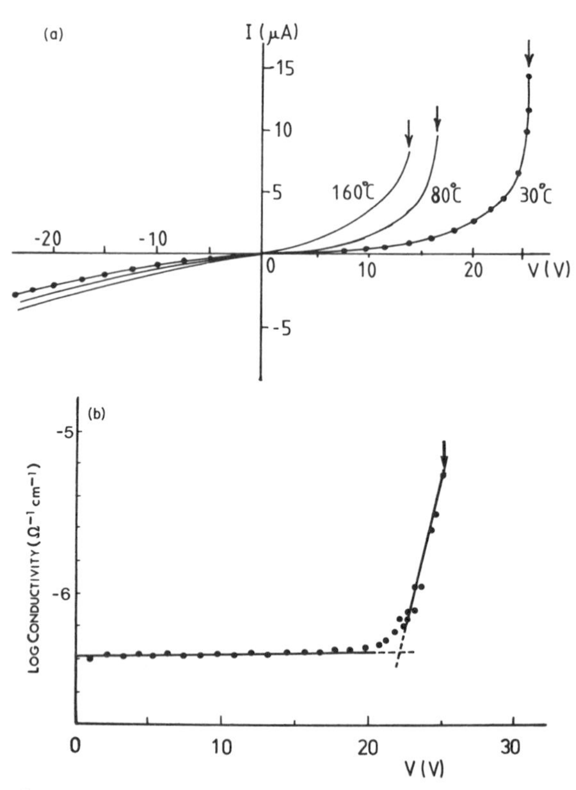

FIG. 2. (a) Static current–voltage characteristics of a virgin (unformed) p^+-n-i device. The positive quadrant corresponds to a positive bias applied to the p^+ layer. (b) The data for 30°C in (a) plotted in the form of log conductivity versus voltage and illustrating the change from ohmic to nonohmic behavior.

this remark will become apparent in the next sections.) In the forward direction there is a region of ohmic behavior over a limited voltage range followed by an abrupt change to a markedly nonohmic region until, at the point indicated by the arrow, the device is unstable and it is impossible to continue with point-by-point measurements. The change from ohmic to nonohmic behavior is more clearly apparent in the effective conductivity versus applied voltage plot of Fig. 2b. As the ambient temperature increases, the onset of nonohmic behavior moves to lower voltages. In the reverse direction, corresponding to a negative potential applied to the p^+ side, there is an initial ohmic region which is symmetrical for positive and negative voltages. However, the change to nonohmic behavior is much more gradual in the reverse direction, leading to the eventual breakdown of the device.

V. Forming: Static Characteristics

As noted in the previous section, during point-by-point measurements under forward bias, the a-Si p^+-n-i device tends to become unstable when the applied bias is about 24 V at room temperature. At higher temperatures the instability occurs at lower voltages, as indicated by the arrows in Fig. 2a. On attempting to increase the voltage still further, the device switches into a low-resistance ON state. Typical $I-V$ characteristics for both polarities in the ON state are shown in Fig. 3; the $I-V$ curve is ohmic, it extrapolates through the origin (i.e., the ON state is permanent), and it is slightly asymmetrical. Note that the current is now measured in milliamperes and that the voltages across the device are small. On increasing the voltage in the forward direction the ON state current continues to increase apparently indefinitely, subject only to any current-limiting resistor, and the device is eventually destroyed, presumably by Joule heating. In the reverse direction, however, another instability is observed, and at about -1 V (typically) the

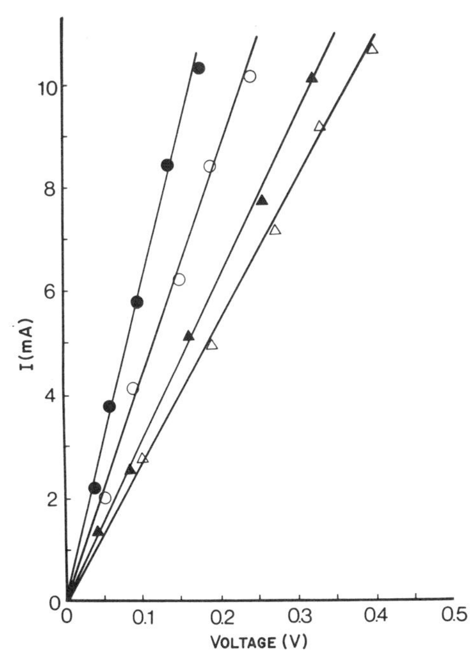

FIG. 3. Current–voltage characteristics of a device in the ON state, at different temperatures, in the forward and reverse directions. The forward direction corresponds to p^+ layer positively biased. (●) forward bias, 80°C; (▲) forward bias, 30°C; (○) reverse bias, 80°C; (△) reverse bias, 30°C.

device switches back into a high-resistance OFF state. The OFF–ON transition may now be repeated by biasing in the forward direction, but on the second and all subsequent switching operations the forward threshold voltage V_{ThF} occurs at a much lower voltage than the first operation, e.g., at ≈ 5 V compared with the 25 V *observed under the conditions obtaining for the measurements shown in Fig. 2.* The first OFF–ON transition, occurring at a relatively high voltage, seems therefore to be unique and by analogy with the usage of switching in chalcogenide glasses, it is referred to as "forming."

The formed a-Si $p^{+}-n-i$ device may be cycled through ON and OFF states by a sequence of biasing in forward and reverse directions with critical points at V_{ThF} and, in the reverse direction, V_{ThR}. A complete and typical characteristic obtained on a curve tracer is illustrated in Fig. 4. On occasions the device appears to go through a number of intermediate ON states during the OFF–ON transition, and this is indicated in the figure. In addition, there is often an observable and appreciable region of negative resistance in the reverse-biased OFF state characteristic of a formed device, denoted by N in Fig. 4.

VI. Forming: Dynamic Characteristics

The forming process (i.e., the first OFF–ON transition) does not occur instantaneously when a voltage step or pulse is applied to the a-Si $p^{+}-n-i$ device. Initially there is a delay time t_{D} during which the device current *remains essentially constant at the OFF-state value appropriate to the voltage across the device.* Only after this delay does the current begin to increase, and it then rises almost instantaneously to its ON-state value. The forming delay time is an extremely sensitive function of the applied forming

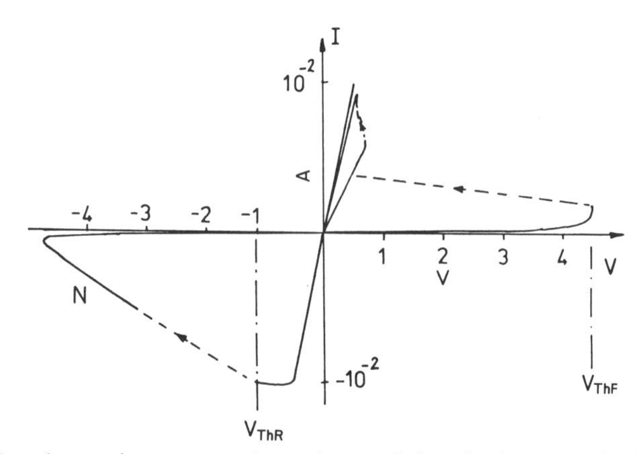

FIG. 4. Complete static current–voltage characteristics of a formed a-Si $p^{+}-n-i$ device, showing the forward and reverse threshold voltages, V_{ThF} and V_{ThR}, respectively.

voltage V_F, and typical data, obtained at three temperatures, are given in Fig. 5. The forming delay time varies over nearly 10 orders of magnitude, from a few hundred seconds at low forming voltages to about 10 nsec at high V_F. In particular, at a temperature-dependent critical forming voltage V_{CR} there occurs virtually a discontinuous change in t_D. The forming voltages V_{CR} indicated in Fig. 5 are approximately the same as the voltages at the points of instability marked by the arrows in Fig. 2a; V_{CR} also corresponds to the forming voltage obtained in experiment with a curve tracer, operated in ac mode at a frequency of 50 Hz. It can also be seen in Fig. 5 that above and below V_{CR} the delay time tends to a value that seems to be approximately independent of both voltage and temperature; for the particular results illustrated, t_D is in the range 10^2–10^3 sec for $V < V_{CR}$ and lies between 10 and 100 nsec for $V > V_{CR}$.

The results plotted in Fig. 5 for $V < V_{CR}$ correspond of course to voltages less than the point of instability indicated in Fig. 2a. There does appear to be a lower limit to the forming voltage, however, and present results indicate that the limiting voltage coincides with the bias at which the $I-V$ characteristics change from their ohmic to nonohmic behavior (see Fig. 2b). Several

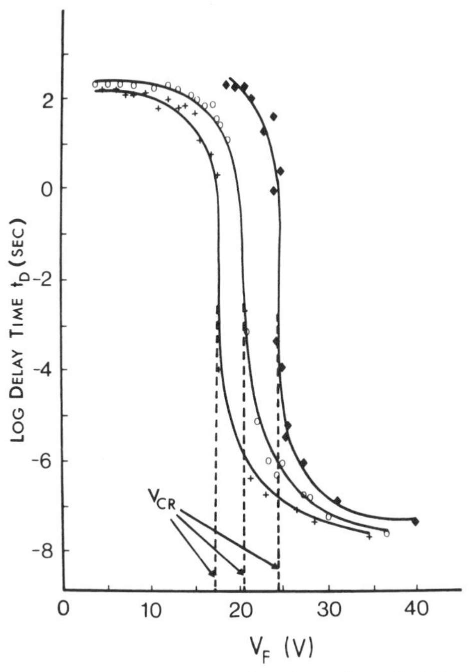

FIG. 5. The delay time for forming t_D as a function of applied forming voltage at three different temperatures: (◆) 30°C, (O) 80°C, (+) 160°C.

experiments have shown that virgin devices fail to switch (form) even if held for many hours at a forward bias only just below the nonohmic region. In other words, forming occurs at any forward bias within the nonohmic region of the $I-V$ characteristics, but at voltages below the point of instability t_D is comparatively long. It must also be emphasized again that the device current remains constant at its OFF-state magnitude during the delay time, even when t_D is 100 sec or more.

Preliminary experiments have been carried out to determine the effects of device geometry on the forming voltage V_F. It was found that V_F increases linearly with the thickness d_n of the n layer. Data for typical p^+-n-i devices are plotted in Fig. 6. In this case V_F was measured by applying a voltage ramp and its value coincides with the voltage V_{CR} indicated in Fig. 5. It can be seen that the results do not extrapolate to zero voltage for zero n layer thickness.

The charge $Q = \int_0^{t_D} I\,dt$, which flows through or into the device during the forming delay time, has been determined for $V_F > V_{CR}$. Figure 7 shows that in this range of V_F the ratio Q/d_n is approximately independent of V_F and d_n for n layer thicknesses between 215 and 780 nm. This could mean that forming occurs when a critical volume charge has accumulated in the n region.

VII. Dynamic Switching of Formed Devices

The principal experimental features of the pulsed operation of formed a-Si p^+-n-i devices have already been described by Owen *et al.* (1982). A representative diagram drawn from an oscilloscope trace of the OFF → ON

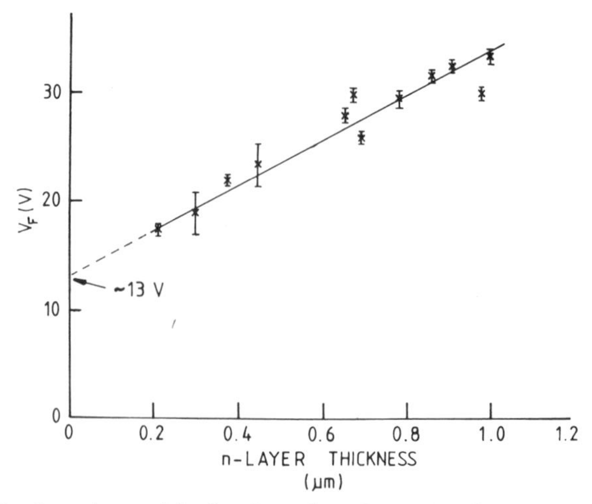

FIG. 6. The dependence of the forming voltage V_F on the thickness of the n layer.

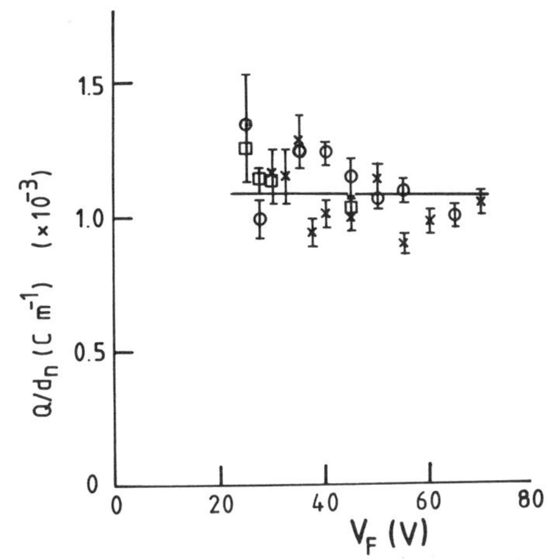

FIG. 7. The ratio of charge to n layer thickness (Q/d_n) as a function of forming potential. $d_n = 215$ nm (x), 445 nm (○), 780 nm (□).

transition on applying a fast voltage ramp is shown in Fig. 8. The main points to note are as follows:

(i) When biased with a pulse in the forward direction, the device switches ON, provided the pulse height exceeds the static threshold voltage V_{ThF} as defined in Fig. 4.

(ii) On the time scale of ~ 1 nsec or less, there is no observable delay time in the response of a formed device. The device current follows the applied voltage instantaneously on this time scale.

(iii) Provided the pulse is long enough, the ON state is permanent and the pulse duration required for switching to a memory state increases as the

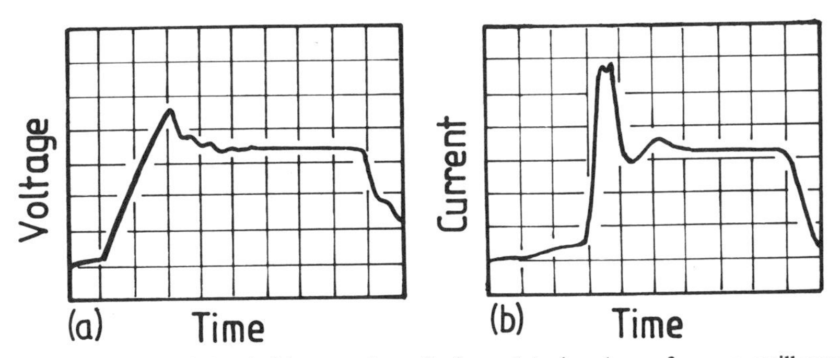

FIG. 8. The OFF–ON switching transient of a formed device, drawn from an oscilloscope trace: (a) applied bias, vertical scale: 2 V per division, horizontal scale: 20 nsec per division; (b) device current, vertical scale: approximately 6 mA per division, horizontal scale: 20 nsec per division.

pulse height decreases toward V_{ThF}. In typical cases a permanent ON state is obtained with pulse durations of a few tens of nanoseconds and magnitude ~5 V in excess of V_{ThF}.

(iv) Similarly, on biasing in the reverse direction with a pulse of height $> V_{ThR}$ the device switches from ON → OFF and again there is no observable delay in response.

(v) The ON-state appears to be truly permanent. No detectable changes have been observed in devices stored in their ON state for a year or more.

VIII. Discussion of Possible Switching Mechanisms

The mechanisms underlying the switching phenomena in the a-Si devices are not understood at present; clearly, more data will be required to explain these exceptional properties. In the following we therefore draw some comparisons with other related switching devices and only briefly speculate on possible mechanisms.

Although there is no observation of memory switching in analogous crystalline Si (c-Si) devices, *threshold* switching is well known in c-Si p^+-n-i, n^+-p-i, and related structures (Yamamoto and Morimoto 1972, Yamamoto *et al.*, 1976; Buxo *et al.*, 1978; Sarrabayrouse *et al.*, 1980; Simmons and El-Badry, 1978; Kroger and Wegener, 1973, 1975). The *i*-layer in these devices is usually a SiO_2 film (≤ 40 Å), thin enough to pass appreciable tunneling currents, but it may be significant that these metal–insulator–semiconductor–semiconductor (MISS) devices can also be fabricated in an "all-Si" form using polycrystalline Si as the *i*-layer (Kroger and Wegener, 1975). These devices switch to a nonpermanent ON state when the p^+-n (or n^+-p) junction is forward biased, which is the same polarity producing the memory ON state in the a-Si structures. In the MISS device the switching action is associated with minority-carrier injection from the $p-n$ junction and accumulation at the interface of the *i* layer, normally leading to punchthrough to the injecting contact, which causes the device to switch ON. The values of the threshold voltage are similar to those observed in the a-Si forming process, and it is conceivable that for this operation the processes are similar. The experimental results described in Part VI, which suggest that the forming process is likely to be charge controlled, would not be inconsistent with this model.

For the c-Si MISS p^+-n-i threshold device the values of threshold voltage are predicted (Simmons and El-Badry, 1978) to depend on $(d_n - W)^2$, where d_n is the thickness of the n layer and W is the width of the depletion region, and for low donor concentrations this dependence is supported experimentally. In contrast, the results in Fig. 6 show that in the a-Si devices the forming voltage varies linearly with d_n. It is difficult to decide at present whether this disagreement suggests a different mechanism or whether it arises from our attempt to extrapolate from a model developed for a crystalline threshold device to an amorphous memory junction.

An alternative model, based on a regenerative process, has also been suggested for the c-Si MISS devices (Sarrabayrouse *et al.*, 1980), taking into account a carrier multiplication mechanism at the "inverted" $Si-SiO_2$ interface. The model correctly accounts for a number of MISS properties; for example, it predicts a threshold independent of d_n, which agrees with the observations on p^+-n-i MISS structures when the n layer is more heavily doped. But this model is also in disagreement with the data for the a-Si devices shown in Fig. 6, and just as for the "punchthrough" model, it is therefore difficult at the present stage to decide whether the regenerative model could be relevant to the understanding of a-Si memory switching.

In the field of amorphous semiconductors much attention has been given over the past 10–15 years to memory switching devices fabricated from multicomponent chalcogenide glasses in which the reversible memory action is associated with the growth and destruction of a crystalline filament (Ovshinsky, 1968; Cohen *et al.*, 1972; Steventon, 1974; Owen and Robertson, 1973; Owen *et al.*, 1979). Although it is very likely that in the a-Si devices some form of filament formation (not necessarily crystalline) is taking place in the OFF–ON transition, the switching phenomena are clearly very different from those in the chalcogenides, at least operationally. The most obvious difference is the completely nonpolar character of switching in chalcogenide glass devices, in contrast to the marked polarity dependence of the a-Si memory switches. More important, perhaps, the switching and setting times for the a-Si device are much faster (~ 10 nsec for either the OFF–ON or ON–OFF transition, compared with at least several milliseconds in chalcogenide devices) and the energy involved in the switching process is considerably lower (1 μJ or less, compared with 1 mJ or more). Also chalcogenide glass devices require voltage pulses of magnitude 25–30 V (for a device $\simeq 1$ μm thick) to establish the ON state, and very often they need 100 or more "forming" cycles before reasonably stable operation is achieved. This again contrasts with the operation of the a-Si memory switch, in which (for a total device thickness of $\simeq 1$ μm) there is a single forming step with a threshold voltage of about 30 V, and for all subsequent operations the forward threshold (V_{ThF}) is 4–6 V.

The closest parallel to the a-Si devices described in this paper seems to be the observation of memory switching in heterojunctions of n-type ZnSe grown epitaxially on p-type (single-crystal) Ge substrates, reported by Hovel (1970) and by Hovel and Urgell (1971). The ZnSe–Ge heterojunction devices are polar and the transition times for the OFF–ON and ON–OFF operation are both in the region of 100 nsec or less. Similar, but not so well-substantiated, memory switching characteristics have also been briefly reported in devices fabricated by forming Schottky contacts on *n*-type GaAs and Si (single crystal) (Moser, 1972). Hovel and Urgell (1971), have tentatively and qualitatively explained switching in the heterojunction by a model involving the filling and emptying of traps in the ZnSe, with the formation of a current filament in the ON state. However, even in this case of superficially similar characteristics, there are notable differences. Most significantly, the polarity required for switching in the ZnSe–Ge heterojunction is the opposite to that found in the a-Si devices, and the OFF–ON threshold voltage for the ZnSe–Ge switch decreases substantially with temperature (from about 1 V at 200°K to less than 0.1 V at 400°K), whereas V_{ThF} for the a-Si devices investigated so far is at the most only weakly temperature dependent. In addition, the memory state of the ZnSe–Ge heterojunctions is generally lost within a few weeks, whereas no change in the characteristics of the a-Si memory devices has been observed after storage over a period of 18 months.

To conclude this chapter it is perhaps worthwhile to note that the switching phenomena observed in the a-Si memories are not the only nanosecond processes known for this material. Drift mobility studies, which show that electron transit times across about 1-μm-thick films are of the order of 10 nsec or so, have been known for over a decade (LeComber and Spear, 1970; Spear, 1983). More recently, it has been demonstrated that hydrogenated amorphous silicon can be used to modulate light at subnanosecond speeds (Phelan *et al.,* 1981; see also Chapter 13 by Phelan of this volume). In pulsed laser annealing of a-Si it has been suggested that the electron–hole plasma generated by the laser could produce rapid second-order phase transitions (van Vechten *et al.,* 1979). The challenge and excitement in understanding the a-Si memories lies in discovering whether the origins of the fast switching processes are electronic, structural, or both.

Acknowledgment

The authors are grateful to the Venture Research Unit of British Petroleum International PLC for a grant supporting the research described in this paper.

References

Adler, D., Henisch, K. H., and Mott, N. F. (1978). *Rev. Mod. Phys.* **50,** 209.
Buxo, J., Owen, A. E., Sarrabayrouse, G., and Sebaa, J. P. (1978). *Rev. Phys. Appl.* **17,** 767.
Charles, H. K., and Feldman, C. (1975). *J. Appl. Phys.* **46,** 819.
den Boer, W. (1982). *Appl. Phys. Lett.* **40,** 812.
den Boer, W. (1983). "Interference Effects and Space-Charge-Limited Conduction in Amorphous Silicon Devices." D. Tech. Thesis, Technische Hogeschool, Delft, The Netherlands, September 1983.

Cohen, M. H., Neale, R. G., and Paskin, A. (1972). *J. Non-Cryst. Solids* **8/10,** 885.
Dey, S. K. (1980). *J. Vac. Sci. Technol.* **17,** 445.
Dey, S. K., and Fong, W. T. J. (1977). *Proc. Int. Vacuum Congress, 7th, and Int. Conf. Solid Surfaces, 3rd, Vienna, 1977.* Vol. 3.
Dey, S. K., and Fong, W. T. J. (1979). *J. Vac. Sci. Technol.* **16,** 240.
Feldman, C., and Charles, H. K. (1974). *Solid State Common.* **15,** 551.
Feldman, C., and Moorjani, K. (1970). *J. Non-Cryst. Solids* **2,** 82.
Gabriel, M. C., and Adler, D. (1982). *J. Non-Cryst. Solids* **48,** 297.
Hovel, H. J. (1970). *Appl. Phys. Lett.* **17,** 141.
Hovel, H. J., and Urgell, J. J. (1971). *J. Appl. Phys.* **42,** 5076.
Kroger, H., and Wegener, H. A. R. (1973). *Appl. Phys. Lett.* **23,** 397.
Kroger, H., and Wegener, H. A. R. (1975). *Appl. Phys. Lett.* **27,** 303.
LeComber, P. G., and Spear, W. E. (1970). *Phys. Rev. Lett.* **25,** 509.
Moorjani, K., and Feldman, C. (1970). *J. Non-Cryst. Solids* **4,** 248.
Moser, A. (1972). *Appl. Phys. Lett.* **20,** 244.
Ovshinsky, S. R. (1968). *Phys. Rev. Lett.* **21,** 1450.
Owen, A. E., LeComber P. G., Sarrabayrouse, G., and Spear, W. E. (1982). *IEE Proc. (Part I Solid State Electron Devices)* **129,** 51.
Owen, A. E., and Robertson, J. M. (1973). *IEEE Trans. Electron Devices,* **ED-20,** 105.
Owen, A. E., Robertson, J. M., and Main, C. (1979). *J. Non-Cryst. Solids* **32,** 29.
Phelan, R. J. Jr., Larson, D. R., and Werner, P. E. (1981). *Appl. Phys. Lett.* **38,** 596.
Sarrabayrouse, G., Buxo, J., Owen, A. E., Munoz-Yague, A., and Seboa, J.-P. (1980). *IEE Proc. (Part I, Solid State & Electron Devices)* **127,** (3), 119.
Simmons, J. G., and El-Badry, A. A. (1978). *Radio and Electron. Eng.* **48,** 215.
Spear, W. E. (1977), *Adv. Phys.* **26,** 811C.
Spear, W. E. (1983). *J. Non-Cryst. Solids* **59/60,** 1.
Steventon, A. G. (1974) in Stuke, J., and Brenig, W. eds. "Amorphous and Liquid Semiconductors," Vol. 1, p. 675. Taylor and Francis, London.
Van Vechten, J. A., Tsu, R., and Saris, F. W. (1979). *Phys. Lett.* **74A,** 422.
Yamomoto, T., Kawamura, K., and Shimizu, H. (1976). *Solid State Electron.* **19,** 701.
Yamomoto, T., and Morimoto, H. (1972). *Appl. Phys. Lett.* **20,** 269.

Index

Contents of Previous Volumes

Volume 1 **Physics of III – V Compounds**

Volume 2 **Physics of III – V Compounds**

Volume 3 **Optical of Properties III – V Compounds**

Edward D. Palik and George B. Wright, Free-Carrier Magnetooptical Effects
Richard H. Bube, Photoelectronic Analysis
B. O. Seraphin and H. E. Bennett, Optical Constants

Volume 4 **Physics of III–V Compounds**

N. A. Goryunova, A. S. Borschevskii, and D. N. Tretiakov, Hardness
N. N. Sirota, Heats of Formation and Temperatures and Heats of Fusion of Compounds $A^{III}B^{V}$
Don L. Kendall, Diffusion
A. G. Chynoweth, Charge Multiplication Phenomena
Robert W. Keyes, The Effects of Hydrostatic Pressure on the Properties of III–V Semiconductors
L. W. Aukerman, Radiation Effects
N. A. Goryunova, F. P. Kesamanly, and D. N. Nasledov, Phenomena in Solid Solutions
R. T. Bate, Electrical Properties of Nonuniform Crystals

Volume 5 **Infrared Detectors**

Henry Levinstein, Characterization of Infrared Detectors
Paul W. Kruse, Indium Antimonide Photoconductive and Photoelectromagnetic Detectors
M. B. Prince, Narrowband Self-Filtering Detectors
Ivars Melngailis and T. C. Harman, Single-Crystal Lead–Tin Chalcogenides
Donald Long and Joseph L. Schmit, Mercury–Cadmium Telluride and Closely Related Alloys
E. H. Putley, The Pyroelectric Detector
Norman B. Stevens, Radiation Thermopiles
R. J. Keyes and T. M. Quist, Low Level Coherent and Incoherent Detection in the Infrared
M. C. Teich, Coherent Detection in the Infrared
F. R. Arams, E. W. Sard, B. J. Peyton, and F. P. Pace, Infrared Heterodyne Detection with Gigahertz IF Response
H. S. Sommers, Jr., Microwave-Based Photoconductive Detector
Robert Sehr and Rainer Zuleeg, Imaging and Display

Volume 6 **Injection Phenomena**

Murray A. Lampert and Ronald B. Schilling, Current Injection in Solids: The Regional Approximation Method
Richard Williams, Injection by Internal Photoemission
Allen M. Barnett, Current Filament Formation
R. Baron and J. W. Mayer, Double Injection in Semiconductors
W. Ruppel, The Photoconductor–Metal Contact

Volume 7 **Application and Devices: Part A**

John A. Copeland and Stephen Knight, Applications Utilizing Bulk Negative Resistance
F. A. Padovani, The Voltage–Current Characteristics of Metal–Semiconductor Contacts
P. L. Hower, W. W. Hooper, B. R. Cairns, R. D. Fairman, and D. A. Tremere, The GaAs Field-Effect Transistor
Marvin H. White, MOS Transistors
G. R. Antell, Gallium Arsenide Transistors
T. L. Tansley, Heterojunction Properties

Volume 7 **Application and Devices: Part B**

Volume 8 **Transport and Optical Phenomena**

Volume 9 **Modulation Techniques**

Volume 10 **Transport Phenomena**

Volume 11 **Solar Cells**

Volume 12 **Infrared Detectors (II)**

Volume 21 **Hydrogenated Amorphous Silicon: Part A**

Volume 21 **Hydrogenated Amorphous Silicon: Part B**

Volume 21 **Hydrogenated Amorphous Silicon: Part C**